Three-Dimensional Magnonics

Three-Dimensional Magnonics

LAYERED, MICRO- AND NANOSTRUCTURES

edited by

Gianluca Gubbiotti

JENNY STANFORD
PUBLISHING

Published by

Jenny Stanford Publishing Pte. Ltd.
Level 34, Centennial Tower
3 Temasek Avenue
Singapore 039190

Email: editorial@jennystanford.com
Web: www.jennystanford.com

British Library Cataloguing-in-Publication Data
A catalogue record for this book is available from the British Library.

Three-Dimensional Magnonics: Layered, Micro- and Nanostructures

ISBN 978-981-4800-73-0 (Hardcover)
ISBN 978-0-429-29915-5 (eBook)

To Roberta, Matteo, and Andrea with love.

Contents

Preface

Magnonics, a research field that uses spin waves, the collective excitations of ordered magnetic materials, or magnons (their quanta), as a tool for signal processing, communication, and computation, has rapidly grown over the past decade because of its low-energy-consumption property and potential compatibility with next-generation circuits beyond CMOS electronics [1].

Magnonics systems investigated up to now are mainly planar systems based either on insulating or on metallic materials. In the former case, yttrium iron garnet (YIG) films with an artificial micron-size periodicity have been mainly used as model systems to demonstrate new magnonic functionalities due to their low damping and long spin-wave decay length [2, 3]. Metallic materials are polycrystalline magnetic alloys (NiFe and CoFeB) that offer a better perspective to meet the industrial demand for miniaturization and integration. Their drawback is the relatively high spin-wave damping, which prevents propagation above a few microns [4, 5], but this can be compensated by downscaling the lattice periodicity to the nanometric size, thanks to the new possibility of launching spin waves with submicrometric wavelength by ultrashort laser pulse [6], spin-transfer torque [7, 8], and the spin Hall effect [9]. Several other strategies have been proposed to reduce the spin-wave wavelength, such as the grating-coupler effect [10, 11]; magnetic textures, including vortices [12]; domain walls [13]; and the spin-wave refraction at the thickness step between films of different thickness [14–16].

This book provides an overview of recent developments in the exploitation of the third dimension in magnonics, with special attention to the propagation of spin waves in layered nanostructures, spin textures, curved surfaces, 3D nano-objects, and cavity magnonics.

The interest in 3D magnonic nanostructures follows the latest trend in CMOS electronics based on the expansion from 2D planar to 3D vertically integrated structures. To remain at the same technological level, a similar expansion should be realized in magnonics.

Following this trend, researchers are proposing different strategies to build the next generation of magnonic systems. The first of them is an extension of planar patterned nanostructures, where arrays of patterned magnetic dots or antidots have a layered structure. In this case, the vertical stacking of ferromagnetic materials, placed in direct contact or separated by a nonmagnetic spacer, adds new functionality and degree of freedom for controlling the spin-wave band structure [17–19].

A very promising alternative to physical patterning is based on the use of hybrid heterostructures in the form of bilayer systems. These include metal–insulator [10], metal–heavy metal [20], metal–antiferromagnet [21, 22], and metal–ferroelectric [23] systems, where new properties of spin waves, such as confining and filtering, guiding and steering, nonreciprocity and reconfigurability, and direct and indirect gap in the magnonic band structure, emerge from interaction between the continuous (flat) magnetic film, in which spin waves propagate, and the nonmagnetic component, which induces periodic modulation of either the static or the dynamic internal magnetic field on the magnetic film itself. Interestingly, the efficient excitation of exchange-dominated short-wavelength (down to about 50 nm) spin waves has already been demonstrated for metal–insulator hybrid nanostructures [24] and layered nanostructures [25].

Very recently, curved surfaces and 3D micro- and nano-objects, such as nanotubes, nanospheres, and nanocaps, have been proposed as model systems to control and manipulate the spin configuration [26, 27] and spin-wave dynamics [28]. The ground states of these 3D magnetic structures exhibit peculiar and unexpected spin textures, which are normally not observed in planar nanostructures and can dramatically affect the dynamic properties of the systems.

Finally, recent frontiers concerning cavity magnomechanics and optomagnonics are presented and discussed.

This book contains 11 chapters written by outstanding international research groups in the field of magnonics, and it can be considered as the continuation and extension of the book entitled *Magnonics: From Fundamentals to Application*, edited by S. O. Demokritov and A. N. Slavin, published in 2013 by Springer.

The book is structured as follows:

In Chapter 1, Cottam et al. carry out a predominantly theoretical study of the spin-wave properties in ferromagnetic nanostructures consisting of single- and multilayer (or composite) ferromagnetic nanowires, along with the 1D magnonic crystal arrays formed from them. The results of the theory are compared with the experimental findings obtained by Brillouin light scattering spectroscopy and ferromagnetic resonance. The described dipole exchange theory is based on a microscopic (or Hamiltonian) approach with a discrete lattice of interacting effective spins. The long-range dipole–dipole interactions are evaluated as direct lattice sums and can be combined with the short-range intralayer exchange and interlayer exchange (due to RKKY and biquadratic interactions).

Chapter 2 by Graczyk et al. presents the results of spin-wave band structure calculations of planar magnonic crystals, which are also inhomogeneous along their thickness. Cases of bicomponent antidot lattices and arrays of elliptical Permalloy nanodots deposited on top of a continuous Permalloy (Py) film and of a bilayer system composed of a dipolarly coupled CoFeB and Py film with the periodically corrugated surface, in the presence of a Cu spacer, are presented. In these latter structures, the spin-wave propagation in a uniform ferromagnetic film is modified by the vertical dynamic coupling with the patterned element placed above. The chapter ends with an example of band structure calculations of an fcc 3D magnonic crystal obtained for nanometric magnetite/maghemite spheres embedded in the Co matrix.

Beginin et al. in Chapter 3 explore both experimentally and theoretically single-layer and vertically coupled magnonic crystals with the meander configuration based on YIG films grown onto a structured nonmagnetic gadolinium–gallium garnet substrate. Propagation of different waves such as backward-volume, forward-volume, and surface magnetostatic waves has been considered. These vertically integrated 3D magnonic structures open the perspective to increase the density of magnonic and spintronic elements for the fabrication of scalable and configurable magnonic networks. The authors propose a new strategy of how magnonics can go in three dimensions, thus staying on the same technological platform as conventional electronics.

In Chapter 4, Liu, Chen, and Yu address the recent experimental results based on the hybrid magnonic crystals composited of a YIG thin film with an array of ferromagnetic Co nanowire arrays deposited on top. Short-wavelength spin waves in the exchange-dominated range of the spectrum are observed. Results show that the interlayer magnon–magnon coupling strength can be tuned by different magnetic configurations of the Co/YIG bilayer, that is, parallel and antiparallel orientations of the magnetization in the two layers.

In Chapter 5, Gallardo et al. study the dynamics of spin waves under the influence of bulk and interfacial Dzyaloshinskii–Moriya interaction (DMI), either in thin films or in magnonic crystals. Calculations are performed within the micromagnetic theory and using the plane-wave method. Results are substantiated by micromagnetic simulations. Nonreciprocal properties induced by different classes of DMI in thin-film magnets are presented and discussed, with special emphasis on the band structure of a chiral magnonic crystal created with a periodic DMI.

Hämäläinen, Qin, and van Dijken review in Chapter 6 their recent work on strain-coupled multiferroic heterostructures and discuss their potential as programmable hybrids in magnonics research. They show that multiferroic heterostructures permit the imprinting of magnetic domain patterns with a regular modulation of magnetic anisotropy, strong pinning of straight magnetic domain walls, and electric field control of magnetic anisotropy and magnetic domain wall motion. All these features represent an attractive platform for the emission of short-wavelength spin waves, the confinement of spin-wave modes in structurally uniform films, and the manipulation of spin-wave transmission.

Albisetti et al. present in Chapter 7 their recent work on spin-wave propagation control using nanopatterned spin textures in continuous magnetic films. First, they discuss thermally assisted magnetic scanning probe lithography and its use for stabilizing 2D magnetic domains, 1D domain walls, and 0D topological spin textures with highly controlled properties. A new method for realizing reconfigurable magnonic crystals, based on periodic magnetization patterns, is then discussed. Finally, they shift their focus from micron-size magnetic domains to nanoscopic spin textures and discuss the use of tailored domain walls for confining at the nanoscale and steering propagating spin waves. The properties of confined spin-wave modes have been studied experimentally and via micromagnetic simulations, and a strategy for realizing tunable spin-wave circuits based on domain walls for finely controlling the interference of confined spin-wave modes is presented.

In Chapter 8, Sluka and Wintz give an overview of contemporary methods for spin-wave excitation and propagation. They focus on the exploitation of spin textures for the generation and propagation of spin waves with very short wavelengths as well as on the 3D nature of such excited waves. In particular, a layered vortex pair system is discussed as a generator for directional spin waves with nanometric wavelengths, while domain walls are presented as 1D spin-wave propagation channels. The spatial profiles of the emitted spin waves were imaged by means of time-resolved scanning transmission X-ray microscopy.

In Chapter 9, Mondal et al. present the recent development and precessional magnetization dynamics and spin waves in different classes of 3D micro- and nano-objects. This is followed by a discussion of the advantages of topological magnetic objects with 3D spin texturing, patterned magnetic multilayers, spin configuration in a 3D magnetic curved surface, and spin ice over the 2D planar arrays.

Otálora et al. present an analytical description of the magnetostatic-induced chiral effects on the spin-wave transport in magnetic nanotubes and compare it to their thin-film counterparts in Chapter 10. The nanotubes are in a vortex-like configuration, and spin-wave propagation is considered to be along the tube length for different nanotube radius. The curvature-induced nonreciprocities in the spin-wave dispersion relation and in frequency linewidth and decay length have been investigated.

In Chapter 11, Leo et al. give an overview of the emerging field of cavity magnonics, analyzing the concepts, key results and experiments, and perspectives. They first present the basics ideas about magnon–photon coupling and then discuss the fingerprints of a magnon–photon strongly coupled system. Key results are discussed concerning respectively coupling to high-order magnetostatic modes, in the case of multiple YIG spheres, and to trasmon qubits. Finally, recent frontiers concerning cavity magnomechanics and optomagnonics are presented and discussed.

I would like to thank all the participating authors who have enthusiastically accepted and contributed to this book. Without their efforts and dedicated time, this project could not have been successful. I gratefully acknowledge Helmut Schulteiss for preparing the cover book figure.

I sincerely thank my parents, Angelo and Maria, who supported me throughout my life and career as a physicist.

Gianluca Gubbiotti
Perugia
2019

References

1. Chumak, A., et al. (2015). Magnon spintronics, *Nat. Phys.*, **11**, 453.
2. Serga, A. A., et al. (2010). YIG magnonics, *J. Phys. D: Appl. Phys.*, **43**, 264002.
3. Soumah, L., et al. (2018). Ultra-low damping insulating magnetic thin films get perpendicular, *Nat. Commun.*, **9**, 3355.
4. Krawczyk, M., et al. (2014). Review and prospects of magnonic crystals and devices with reprogrammable band structure, *J. Phys.: Condens. Matter*, **26**, 123202.
5. Tacchi, S., et al. (2017). Brillouin light scattering studies of 2D magnonic crystals, *J. Phys.: Condens. Matter*, **29**, 073001.
6. Lenk, B., et al. (2011). The building blocks of magnonics, *Phys. Rep.*, **507**, 107.
7. Demidov, V. E., et al. (2010). Direct observation and mapping of spin waves emitted by spin-torque oscillators, *Nat. Mater.*, **9**, 984.
8. Madami, M., et al. (2011). Direct observation of a propagating spin wave induced by spin-transfer torque. *Nat. Nanotechnol.*, **6**, 635.
9. Demidov, V. E., et al. (2012) Magnetic nano-oscillator driven by pure spin current, *Nat. Mater.*, **11**, 1028.

10. Yu, H., et al. (2013). Omnidirectional spin-wave nanograting coupler, *Nat. Commun.*, **4**, 2702.
11. Liu, C., et al. (2018). Long-distance propagation of short-wavelength spin waves, *Nat. Commun.*, **9**, 738.
12. Wintz, S., et al. (2016), Magnetic vortex cores as tunable spin-wave emitters, *Nat. Nanotechnol.*, **11**, 948.
13. Van de Wiele, B., et al. (2016). Tunable short-wavelength spin wave excitation from pinned magnetic domain walls, *Sci. Rep.*, **6**, 21330.
14. Stigloher, J., et al. (2016). Snell's law for spin waves, *Phys. Rev. Lett.*, **117**, 037204.
15. Toedt, J., et al. (2016). Interface modes at step edges of media with anisotropic dispersion, *Phys. Rev.B*, **93**, 184416.
16. Stigloher, J., et al. (2016). Spin-wave wavelength down-conversion at thickness steps, *Appl. Phys. Express*, **11**, 053002.
17. Gubbiotti, G., et al. (2018). Reprogrammable magnonic band structure of layered permalloy/Cu/permalloy nanowires, *Phys. Rev.B*, **97**, 134428.
18. Graczyk, P., et al. (2018). Magnonic band gap and mode hybridization in continuous permalloy films induced by vertical dynamic coupling with an array of permalloy ellipses, *Phys. Rev.B*, **98**, 174420.
19. Popov, P. A., et al. (2019). Spin wave propagation in three-dimensional magnonic crystals and coupled structures, *J. Magn. Magn. Mater.*, **476**, 423.
20. Gallardo, R., et al. (2019). Flat bands, indirect gaps, and unconventional spin-wave behavior induced by a periodic Dzyaloshinskii-Moriya interaction, *Phys. Rev. Lett.*, **122**, 067204.
21. Albisetti, E., et al. (2016). Nanopatterning reconfigurable magnetic landscapes via thermally assisted scanning probe lithography, *Nat. Nanotechnol.*, **11**(6), 545.
22. Albisetti, E., et al. (2018). Nanoscale spin-wave circuits based on engineered reconfigurable spin-textures, *Commun. Phys.*, **1**, 56.
23. Hämäläinen, S. J., et al. (2018). Control of spin-wave transmission by a programmable domain wall, *Nat. Commun.*, **9**(1), 4853.
24. Liu, C., et al. (2018). Long-distance propagation of short-wavelength spin waves, *Nat. Commun.*, **9**, 738.
25. Wintz, S., et al. (2016). Magnetic vortex cores as tunable spin-wave emitters, *Nat. Nanotechnol.*, **11**, 948.
26. Fernández-Pacheco, A., et al. (2017). Three-dimensional nanomagnetism, *Nat. Commun.*, **8**, 15756.

27. Donnelly, C., et al. (2017). Three-dimensional magnetization structures revealed with X-ray vector nanotomography. *Nature*, **547**, 328.
28. Otálora, J. A., et al. (2016). Curvature-induces asymmetric spin-wave dispersion, *Phys. Rev. Lett.*, **117**, 227203.

Chapter 1

Dipole Exchange Theory of Magnons in Structured Composite Nanowires and Magnonic Crystal Arrays

M. G. Cottam,[a] Z. Haghshenasfard,[a] A. O. Adeyeye,[b] and G. Gubbiotti[c]

[a] *Department of Physics and Astronomy, University of Western Ontario, London, Ontario, N6A 3K7, Canada*
[b] *Information Storage Materials Laboratory, Department of Electrical and Computer Engineering, National University of Singapore, 117576, Singapore*
[c] *Istituto Officina dei Materiali del CNR (CNR-IOM), Sede Secondaria di Perugia, c/o Dipartimento di Fisica e Geologia, Università di Perugia, I-06123 Perugia, Italy*
cottam@uwo.ca

Motivated by recent experimental developments for fabricating and characterizing nanometer-scale magnetic elements, either singly or in arrays, we present a mainly theoretical study of some spin-wave properties in ferromagnetic nanostructures. In doing so, we make comparisons with related experiments that employ Brillouin light scattering spectroscopy and ferromagnetic resonance. The nanostructures considered here are single- and multilayer (or composite) ferromagnetic nanowires, along with the 1D magnonic crystal arrays formed from them. For all these structures we use a discrete-lattice approach to establish a microscopic (or

Three-Dimensional Magnonics: Layered, Micro- and Nanostructures
Edited by Gianluca Gubbiotti

ISBN 978-981-4800-73-0 (Hardcover), 978-0-429-29915-5 (eBook)
www.jennystanford.com

Hamiltonian-based) theory for describing the dependence of the spin-wave frequencies on the applied magnetic field and on the wave vector in the dipole exchange regime. In all cases the effective lattice parameter a, which describes the cell size of the effective spins used for the calculations, is chosen to be smaller than the exchange correlation length. The Hamiltonian includes both the short-range exchange and the long-range magnetic dipole–dipole interactions, as well as the effects of an external magnetic field, single-ion anisotropy, biquadratic exchange, and the so-called Ruderman–Kittel–Kasuya–Yosida (RKKY) exchange interactions, as appropriate. In some structures, where there may be competing interactions in a multicomponent nanowire and/or in an array, it becomes necessary to solve first for the distribution of the static magnetization before considering the spin-wave dynamics.

One aspect of particular note that we emphasize through examples later in this chapter relates to the increased interest in nanowires (and their magnonic crystal arrays) that are fabricated to have a multilayered structure with different thicknesses of the layers and thus multiple reprogrammable magnetization configurations and dynamic responses. This makes layered structures interesting as prototypes of multiple magnetic storage layers and as nonvolatile vertical magnetic logic gates.

The exploration of the third dimension in magnonics also follows trends in CMOS electronics.

1.1 Introduction

Ferromagnetic nanowires (NWs), fabricated either individually or in interacting arrays, have recently generated much attention for their spin-wave (SW) properties due to both their technical applications and fundamental scientific interest. This is a result of advances in the techniques for nanoscale materials growth (see, for example, Ref. [1]). A topic of particular interest regarding applications is that of reprogrammable magnonic devices [2–6] in which several different magnetic states might be realized in a NW array with switching between states on demand (by varying an applied field, for example), as we discuss later.

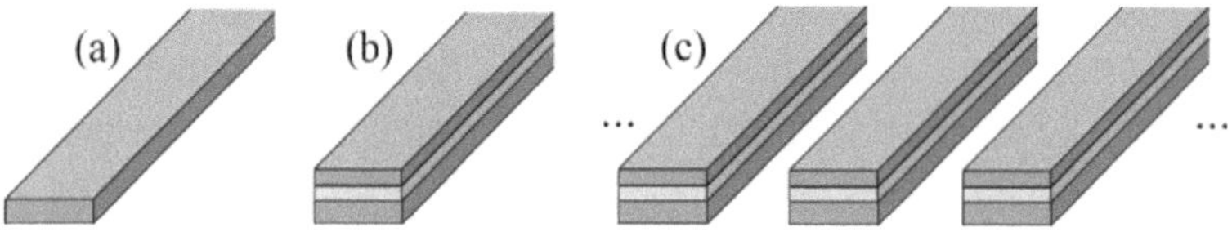

Figure 1.1 Schematic of some magnetic NW structures: (a) a simple rectangular NW stripe, (b) a multilayer NW stripe, and (c) an MC array formed as a lateral (side-by-side) arrangement of multilayer NW stripes with spacers.

Some of the magnetic NW structures relevant to this chapter are depicted schematically in Fig. 1.1. First in Fig. 1.1a we show a simple NW or stripe (with a rectangular cross section) composed of a single uniform magnetic material. Figure 1.1b shows another NW that is composite in structure, being formed as a multilayer. In this case there are just three layers, allowing for various possibilities. For example, the top and bottom layers could be magnetic, while the middle layer could be composed of another magnetic material or it could act as a nonmagnetic spacer or it could be a nonmagnetic metal that is sufficiently thin for long-range exchange coupling of the Ruderman–Kittel–Kasuya–Yosida (RKKY) type (to be discussed later) to occur across it [7]. Figure 1.1c shows part of a lateral array of NWs (that may be of uniform or multilayer structure) separated horizontally by nonmagnetic spacers, across which there may exist dipolar fields. In some of the other magnonic crystals (MCs) considered later, the NWs may be grown directly in contact with another NW of a different material to create so-called bicomponent MCs. To help characterize the SW modes, comparisons may usefully be made in some cases with complete films or multilayers: these are just the limit where the lateral widths of the stripes become large such that edge effects are negligible.

The SWs of a bulk ferromagnet are characterized by a 3D wave vector $\mathbf{q} = (q_x, q_y, q_z)$. By contrast, in nanostructures the SWs are spatially confined and are required to satisfy boundary conditions at the surfaces or interfaces, which leads to qualitative changes in the SW spectrum. The SWs in complete thin films and multilayers are characterized by a 2D wave vector $\mathbf{q} = (q_x, q_z)$ corresponding to the two directions of translational symmetry. In the direction perpendicular to the film surfaces, the SWs may show a standing-mode behavior or be localized with an amplitude decaying away from one or both surfaces. This is a consequence of the lack of symmetry in the direction perpendicular to the film. Likewise, in NWs, the modes are characterized by a 1D wave vector $\mathbf{q} = (q_z)$ along the longitudinal symmetry axis (taken as z) and the SWs may either be localized or take a standing-wave form in the other two directions.

Consequently, the behavior of the SWs in nanostructures is quite different from that in bulk ferromagnets. The lack of translational symmetry perpendicular to surfaces or interfaces leads to boundary conditions and modification of the usual Bloch's theorem [8]. Therefore, the SW spectrum becomes discrete, consisting of separate dispersion branches.

The individual nanostructures can be fabricated in periodic arrays. Specifically, we may consider MCs that consist of lateral arrays of NWs arranged side-by-side with an intervening nonmagnetic spacer, as in Fig. 1.1c, or with another magnetic material. These artificial crystals exhibit novel features such as discreteness of the frequencies and the formation of magnonic band gaps that do not exist in natural structures. MCs offer the possibility for control of the collective SWs (through their frequencies and propagation characteristics), by analogy with the control of light in photonic crystals. Hence, they are the subject of extensive investigation for their potential in a wide range of magnetic device applications [4, 9–15].

Various experimental techniques such as Brillouin light scattering (BLS) spectroscopy and ferromagnetic resonance (FMR) have been used to study these SW bands and gaps [16–25]. In most cases only a few of the lowest modes show a dispersive behavior and can be used for information processing in magnonic devices [26]. Many experiments have been made on NWs in arrays, including NWs with different cross sections (such as L-shape, U-shape, saw-tooth-shape, etc.) [27–31], composite NWs (made of more than one material) [32–37], NWs with curved sections [38, 39], magnetized modulated NWs [40], width-modulated NWs [41, 42], and multilayer NWs with inclusion of exchange [29, 30, 32] and dipole–dipole [6, 33–35] interactions. In addition, several strategies have been developed to create artificial ground states where the magnetization configuration in adjacent NWs are either parallel (P) or antiparallel (AP) aligned and one can switch from one configuration to the other by the application of an external magnetic field [43–45].

On the theoretical side, the SW properties of nanostructures generally have involved using a macroscopic (or continuum) approach by solving the linearized torque equation of motion for the magnetization supplemented by Maxwell's equations of electromagnetism (see, for example, Refs. [46–51]). Another approach, which we employ here, is based on a microscopic theory where the dipolar and exchange interactions (as well as other interactions) between spins are considered within a Hamiltonian approach with a diagonalization transformation. Instead of using Maxwell's equations, the dipolar fields are calculated using discrete lattice sums [52, 53]. This method is well suited to systems with inhomogeneous magnetization [54–57] as well as in NWs [58, 59] and films [60–63] with lateral dimensions or thicknesses of only a few atomic spacings.

In this chapter we present a theoretical study of dipole exchange SWs in ferromagnetic nanostructures. The above-mentioned microscopic method is followed in which we include Hamiltonian terms for short-range bilinear exchange, long-range dipole–dipole interactions, and the Zeeman energy of the applied magnetic field, as well as (optionally) the single-ion anisotropy and long-range interlayer exchange interactions (biquadratic and RKKY). We investigate first the SW dispersion relations for ultrathin films and multilayers in Section 1.2. Next in Section 1.3 we generalize the methodology to apply to NWs, to include consideration of inhomogeneously magnetized NWs in the presence of a transverse applied field. In Section 1.4 we study the magnonic band structure in MCs where the NWs have a multilayer structure, taking Py/Cu/Py as an example (where Py is shorthand for Permalloy, $Ni_{80}Fe_{20}$). Finally, an application is made to the magnetization dynamics in layered Py/Ru/Py NWs in Section 1.5, where we explore the role of interlayer exchange interactions across a thin Ru layer. Throughout we make comparisons with experimental SW data, where available.

1.2 Hamiltonian Theory for Films

For comparison purposes, and for simplicity, we start by considering ultrathin films and multilayers. We model an ultrathin film with (010) surfaces as having N atomic layers of spins arranged on a simple cubic lattice structure with a lattice parameter a. The overall structures may consist of more than one film together with intervening spacers. The simplest multilayer is a double film in which one film consists of N_1 layers (with thickness $d_1 = N_1a$) and the other one has N_2 layers (with thickness $d_2 = N_2a$) and the two films are separated by a nonmagnetic spacer layer s. We denote $N = N_1 + N_2$ as the total number of spin layers. The cross sections of these structures are shown in Fig. 1.2, along with the choice of coordinate axes. Typically, the static field H_0 is applied parallel to the film surfaces (taken as the z direction, coinciding with direction of easy magnetization). We assume that the film surfaces are in the xz plane and the y axis is perpendicular to the surfaces.

The system will be described by a Hamiltonian that includes both the short-range exchange and the long-range dipole–dipole interactions, together with the Zeeman energy due to the applied field H_0. The total Hamiltonian is $H = H_{\text{ex}} + H_{\text{dip}} + H_z$, where the first term H_{ex} describes the Heisenberg exchange interaction:

$$H_{\text{ex}} = -\frac{1}{2}\sum_{in,jm} J_{in,jm}\mathbf{S}_{in}\cdot\mathbf{S}_{jm}. \tag{1.1}$$

The indices i and j label the spin sites within any one atomic layer, while n and m (= 1, 2, 3, ..., N) label the different layers of the film in the y direction. Thus $\mathbf{S}_{in}$ is the spin operator at site i in layer n. Typically, the exchange interaction $J_{in,jm}$ is taken to be J for all nearest neighbors and zero otherwise.

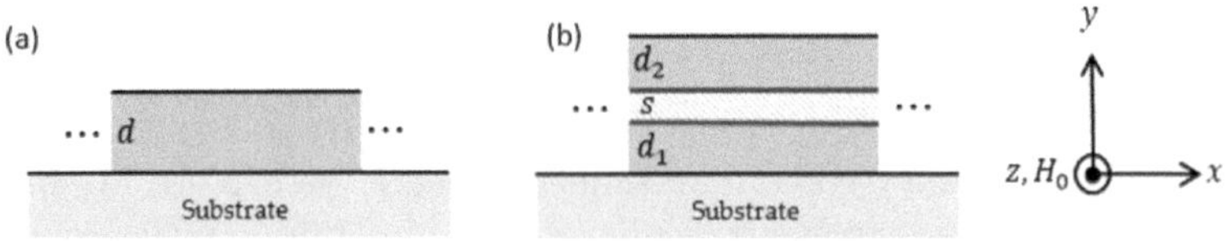

Figure 1.2 Schematic cross section of (a) a single ferromagnetic film of thickness d and (b) two ferromagnetic films with thicknesses d_1 and d_2 coupled via dipolar fields across a nonmagnetic spacer of thickness s.

The second term in the Hamiltonian describes the long-range dipolar interactions between a pair of spins $\mathbf{S}_{in}$ and $\mathbf{S}_{jm}$ at a distance apart specified by the separation vector $\mathbf{r}_{in,jm} = \mathbf{r}_{in} - \mathbf{r}_{jm}$:

$$H_{\text{dip}} = \frac{1}{2}(g\mu_{\text{B}})^2 \sum_{in,jm} \sum_{\alpha,\beta} D^{\alpha,\beta}_{in,jm} S^{\alpha}_{in} S^{\beta}_{jm}. \tag{1.2}$$

Here g is the Landé factor, μ_{B} denotes the Bohr magneton, and $D^{\alpha,\beta}_{in,jm}$ (with α, β = x, y, z) represents the components of the coupling tensor:

$$D^{\alpha,\beta}_{in,jm} = \frac{\left|\mathbf{r}_{in,jm}\right|^2 \delta_{\alpha,\beta} - 3r^{\alpha}_{in,jm} r^{\beta}_{in,jm}}{\left|\mathbf{r}_{in,jm}\right|^5}, \tag{1.3}$$

which has a long-range character. The last term in the Hamiltonian is the Zeeman energy due to the z-directed static field H_0:

$$H_z = -g\mu_{\text{B}} H_0 \sum_{in} S^z_{in}. \tag{1.4}$$

Other Hamiltonian terms will be considered later.

1.2.1 SW Frequencies

To study the SW frequencies, we first apply the Holstein–Primakoff representation [64] to rewrite the spin Hamiltonian in terms of boson creation and annihilation operators $a^{\dagger}_{in}$ and a_{in} at the sites. The transformed Hamiltonian at low temperatures ($T \ll T_{\text{C}}$, with T_{C} being the Curie temperature) can be expanded as $H = H^{(2)} + H^{(3)} + H^{(4)} + \ldots$, apart from a constant, with $H^{(m)}$ denoting the term with a product of m boson operators. The linear SW spectrum is described by the second-order Hamiltonian $H^{(2)}$ which takes the form:

$$H^{(2)} = \sum_{n,m,\mathbf{q}} \Big[A^{(2)}_{n,m}(\mathbf{q}) a^{\dagger}_{\mathbf{q},n} a_{\mathbf{q},m} + B^{(2)}_{n,m}(\mathbf{q}) a_{\mathbf{q},n} a_{-\mathbf{q},m} + B^{*(2)}_{n,m}(\mathbf{q}) a^{\dagger}_{\mathbf{q},n} a^{\dagger}_{-\mathbf{q},m} \Big], \tag{1.5}$$

where $\mathbf{q} = (q_x, q_z)$ is the 2D in-phase wave vector and we have defined a 2D Fourier transformation of the boson operators from a position-dependent form to $a_{i,n} = 1/\sqrt{N_0} \sum_{q} a_{\mathbf{q},n} e^{-i\mathbf{q}\cdot\mathbf{r}_i}$ (with N_0 being the total number of magnetic sites in any layer). The new boson operators are in a "mixed" representation specified by $\mathbf{q}$ and the layer number l. The coefficients $A^{(2)}_{n,m}(\mathbf{q})$ and $B^{(2)}_{n,m}(\mathbf{q})$ are given by [60]

$$A^{(2)}_{n,m}(\mathbf{q}) = g\mu_{\mathrm{B}} H_0 + S\sum_{l=1}^{N}\left[J_{n,l}(0) - (g\mu_{\mathrm{B}})^2 D^{z,z}_{n,l}(0)\right]\delta_{nm} - S\,J_{n,m}(\mathbf{q}) + \frac{1}{2}(g\mu_{\mathrm{B}})^2 S D^{z,z}_{n,m}(\mathbf{q}), \tag{1.6}$$

and

$$B^{(2)}_{n,m}(\mathbf{q}) = \frac{1}{4}(g\mu_{\mathrm{B}})^2 S\left[D^{x,x}_{n,m}(\mathbf{q}) - 2i\,D^{x,y}_{n,m}(\mathbf{q}) - D^{y,y}_{n,m}(\mathbf{q})\right]. \tag{1.7}$$

Here we have introduced $J_{n,m}(\mathbf{q})$ and $D^{\alpha,\beta}_{n,m}(\mathbf{q})$ as the Fourier transforms of the exchange and dipole–dipole interactions, respectively.

The final step to determine the SW frequencies involves diagonalizing the second-order Hamiltonian $H^{(2)}$. To achieve this, we make a canonical (generalized Bogoliubov) transformation from operators $\{a^{\dagger}_{\mathbf{q},n}, a_{\mathbf{q},n}\}$ to a new set of boson operators $\{\alpha^{\dagger}_{\mathbf{q},n}, \alpha_{\mathbf{q},n}\}$ using

$$a_{\mathbf{q},n} = \sum_{l=1}^{N}\{\sigma_{n,l}(\mathbf{q})\alpha_{\mathbf{q},l} + \sigma^{*}_{n+N,l}(\mathbf{q})\alpha^{\dagger}_{-\mathbf{q},l}\}, \tag{1.8}$$

and the Hermitian conjugate expression for $a^{\dagger}_{\mathbf{q},n}$. Here $\sigma_{n,l}(\mathbf{q})$ is an element of the $2N \times 2N$ transformation matrix $\sigma(\mathbf{q})$ calculated in Ref. [51]. The non-interacting SW frequencies $\omega_l(\mathbf{q})$ at wave vector $\mathbf{q}$ and mode branch $l\,(= 1, 2, \ldots, N)$ can be found by solving the following matrix equation:

$$\begin{pmatrix} A^{(2)}(\mathbf{q}) & 2B^{(2)}(\mathbf{q}) \\ -2B^{*(2)}(-\mathbf{q}) & -\tilde{A}^{(2)}(-\mathbf{q}) \end{pmatrix}\sigma_l(\mathbf{q}) = \pm\omega_l(\mathbf{q})\sigma_l(\mathbf{q}), \tag{1.9}$$

where the $+$ $(-)$ sign is taken for $l = 1, 2, \ldots, N$ $(l = N+1, N+2, \ldots, 2N)$, and $\sigma_l(\mathbf{q})$ is the lth column of the matrix $\sigma(\mathbf{q})$. The diagonalized form of $H^{(2)}$ can then be expressed as

$$H^{(2)} = \sum_{\mathbf{q},l}\omega_l(\mathbf{q})\alpha^{\dagger}_{\mathbf{q},l}\alpha_{\mathbf{q},l}. \tag{1.10}$$

The dispersion relations $\omega_l(\mathbf{q})$ for the SWs in the magnetostatic regime (ignoring exchange) are well known [65] for single magnetic films from the macroscopic (continuum) theory. In particular, there is a surface mode (the Damon–Eshbach, or DE, mode) propagating perpendicularly to the in-plane magnetization direction and with angular frequency given by

$$\omega_{\rm DE}(\mathbf{q}) = \left[(\omega_0 + \frac{1}{2}\omega_M)^2 - \frac{1}{4}\omega_M^2 \exp(-2qd)\right]^2, \qquad (1.11)$$

when $q_z = 0$, $\omega_M = 4\pi g\mu_{\rm B} M_{\rm S}$ and $\omega_0 = g\mu_{\rm B} H_0$. Two limiting cases of Eq. 1.11 are $\omega_{\rm DE}(\mathbf{q}) = [\omega_0(\omega_0 + \omega_M)]^{1/2}$ when $qd \ll 1$ and $\omega_{\rm DE}(\mathbf{q}) = (\omega_0 + \omega_M/2)$ when $qd \gg 1$. The magnetostatic theory for a double film was developed in Ref. [66].

1.2.2 Applications

Now we present some numerical examples as applications of the theory. The material parameters are the exchange stiffness D, the saturation magnetization $M_{\rm S}$, the spin quantum number S, and $g\mu_{\rm B} = (\gamma/2\pi)$ The expressions $D = SJa^2/g\mu_{\rm B}$ and $M_{\rm S} = g\mu_{\rm S}/a^3$ provide the correspondence between these parameters and the terms in the spin Hamiltonian, where a is a lattice constant. In some work the micromagnetic stiffness $\mathcal{A}$ is used, which is related to the parameter D by $D = 2\mathcal{A}/M_{\rm S}$. Additionally, we need to choose the effective lattice constant a to be less that the exchange correlation length $a_{\rm ex} = \sqrt{D/4\pi M_{\rm S}}$. As mentioned, we employ Permalloy in many examples, for which $a_{\rm ex}$ is about 5 or 6 nm.

First, we show the dispersion relation for a 40-nm-thick film of Permalloy with in-plane direction of magnetization (taking $a = 5$ nm and $N = 8$ layers of spins). It is well known that the SWs depend on the magnitude $q = |\mathbf{q}|$ as well as on the in-plane propagation direction $\theta = \tan^{-1}(q_x/q_z)$, and hence they appear in bands corresponding to different values of $l (= 1, 2, \ldots, 8)$. In Fig. 1.3 we show the lowest two SW bands with the frequencies plotted as a function of q for different angles. We see that at $q \approx 0$ the SW frequencies (for any value of the propagation angle θ) of the first branch ($l = 1$) have a common value of 13.3 GHz, which is very close to the value 13.6 GHz predicted by the magnetostatic Eq. 1.11. Also, when the in-plane wave vector and film thickness satisfy $qd \gg 1$, this equation gives $(\omega_0 + \omega_M/2) \approx 17.8$, which is roughly where the top of the first band, representing the DE mode, levels off at $q \approx 0.038\,\text{nm}^{-1}$.

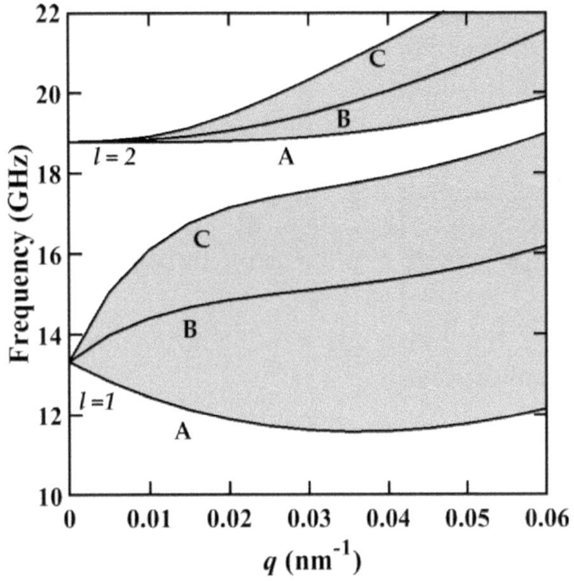

Figure 1.3 Examples of dispersion relation showing the SW frequencies $\omega_l(\mathbf{q})/2\pi$ as a function of $|\mathbf{q}| = q$ for Permalloy single film magnetized parallel to the surfaces with $N = 8$ ($d = 40$ nm with $a = 5$ nm) layers at several values of the in-plane direction θ We take $\mu_0 H_0 = 0.2$ T, $D = 24$ T nm^2, $M_s = 0.0645$ T, and $g\mu_B = 29.5$ GHz/T. The bands are shaded and only two lowest branches for $l = 1$ and 2 are shown. The lines correspond to A, $\theta = 0$; B, $\theta = \pi/4$; and C, $\theta = \pi/2$.

The dispersion relation results for two Permalloy films that are coupled across a nonmagnetic spacer is presented in Fig. 1.4a, where we have considered continuous Permalloy (10 nm)/Cu (10 nm)/Permalloy (30 nm) films. At $q = 0$, the lowest two modes shown are degenerate at 6.3 GHz. We also compare our numerical calculations with experimental data from BLS measurements [67], and it is seen that very good agreement is obtained. Another example for a trilayer magnetic film with the composition Py/Cu/Py/Cu/Py is presented in Fig. 1.4b, again giving good agreement with experiment [37]. The weakly dispersive mode at higher frequency (near 23 GHz) in panel (b) is a perpendicular standing SW resonating across the thickness of the 30 nm Py film.

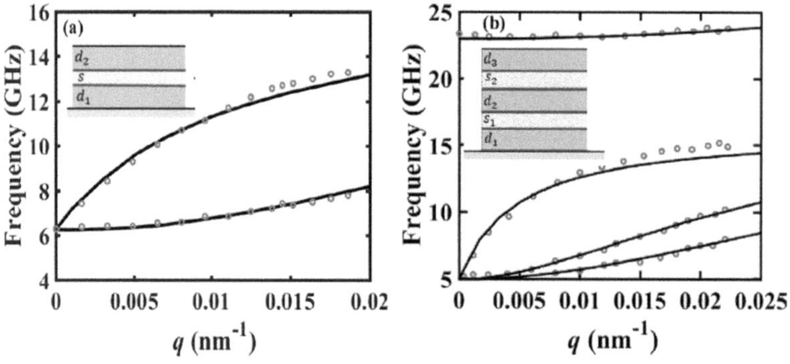

Figure 1.4 Theoretical (solid lines) and experimental (circles) results for the SWs in (a) two Permalloy films coupled across a Cu spacer. We take $a = 5$ nm, $d_1 = 10$ nm, $d_2 = 30$ nm, $s = 10$ nm, $\mu_0 H_0 = 0.05$ T, $D = 30$ T nm^2, $M_s = 0.069$ T, and $g\mu_B = 29.5$ GHz/T. (b) The trilayer film Py(30 nm)/Cu(10 nm)/Py(15 nm)/Cu(10 nm)/Py(30 nm) as a function of the transferred in-plane wave vector $\mathbf{q}$ in the x direction. We take $a = 1.25$ nm, $\mu_0 H_0 = 0.03$ T, $D = 35$ T nm^2, $M_s = 0.072$ T, and $g\mu_B = 29.5$ GHz/T. The insets show a schematic of the geometry.

1.3 Hamiltonian Theory for Nanowires

Next, we generalize the preceding theory for a complete film to apply to a NW (or stripe), where we will have the additional effects of confinement and spatial quantization of the SWs with respect to the width of the NW. We model a NW in the simplest case as having a rectangular cross section (in the xy plane) with width W and thickness d, extending along the z direction. The cross section consists of N atomic spins, where $N = Wd/a^2$. For a two magnetically coupled NWs, separated by a spacer s the number of atomic spins in a cross section is $N = (N_1 + N_2)$, where now N_1 and N_2 denote the number of *spin sites* in the individual NWs. The cross-sectional representation of a multilayer NW with spacers is depicted in Fig. 1.5.

1.3.1 SW Frequencies

The total spin Hamiltonian for the system, whether it is a single- or a multilayer NW, can be written down as before with exchange,

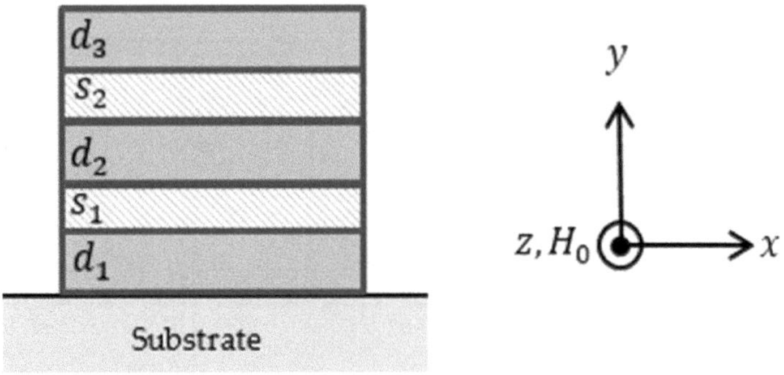

Figure 1.5 Schematic cross section of a multilayer NW composed of three magnetic stripes with thicknesses d_1, d_2, and d_3 and two intervening spacer layers with thicknesses s_1 and s_2 The NWs are magnetized along their length (z axis), which corresponds to their easy direction of magnetization.

dipole–dipole terms, and Zeeman energy terms, as given formally by Eqs. 1.1–1.4. We note for NWs that i and j now label the spin sites along the z-directed lines of spins, while n and m label the individual spin sites in any one cross section of the NW.

The theory now proceeds by analogy with the film case described in Section 1.2.1. The essential difference here is that the NWs have only one direction of translational symmetry (chosen to be along the longitudinal z axis of the NWs). Therefore $\mathbf{q} = (q_z)$ is the 1D wave vector characterizing the SW excitations. Equations of the same form as Eq. 1.5 and Eqs. 1.8–1.10 still apply, but with some redefinitions and change of notation. Thus n and m in Eq. 1.5 now label each of the N sites in any cross section of the NW, and the definitions of the coefficients $A^{(2)}_{n,m}$ and $B^{(2)}_{n,m}$ in Eqs. 1.6 and 1.7 become generalized (see, for example, Refs. [54, 55, 57, 58]).

The formalism described above allows for the evaluation of other SW properties, apart from the mode frequencies. In particular, it is of interest in NWs to study the localization or standing-mode character of the modes associated with the position dependence of the SW amplitudes. The square of these amplitudes for the transverse S^x and S^y spin operators are related to the BLS intensities for scattering from those parts of the NW where the light penetrates. The connection is made through the transformation eigenfunctions $\sigma_l(\mathbf{q})$ in Eqs. 1.8 and 1.9, as explained in Refs. [37, 54, 55, 58].

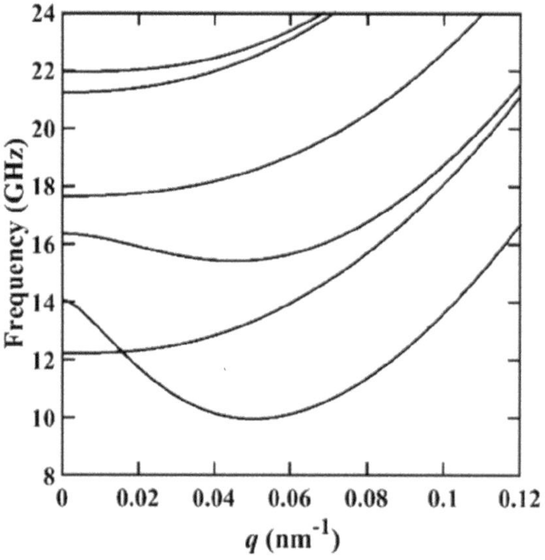

Figure 1.6 SW frequencies versus longitudinal wave vector **q** for a Permalloy NW with square cross section ($W = d = 60$ nm). The assumed parameters are $\mu_0 H_0 = 0.05$ T, $D = 30$ T nm^2, $M_s = 0.069$ T, and $g\mu_B = 29.5$ GHz/T.

1.3.2 Applications

Some numerical calculations are presented next for single- and multilayer NWs. First, in Fig. 1.6 we show the dispersion relation for the lowest SW frequencies as a function of the longitudinal wave vector $q = q_z$ for a single Permalloy NW with a square cross section for dimensions $W = d = 60$ nm. The sample is magnetized along its length (z axis of easy magnetization). We see that the spectrum consists of discrete branches instead of bands, as was the case for complete films. Some of the branches show a minimum at nonzero q, which is a consequence of the interplay between the exchange and dipole–dipole interactions. We also see some effects of hybridization and mode crossing or repulsion for the first and the second branches.

Another example is depicted in Fig. 1.7a, where we show the calculated and measured SW frequencies for a single Permalloy NW (magnetized along z) with cross section 175 nm × 20 nm.

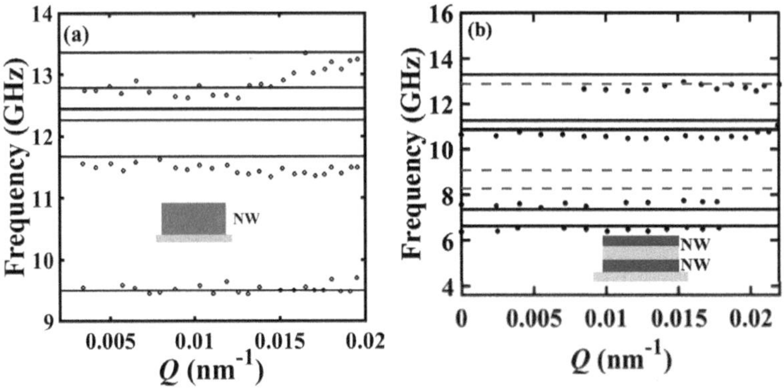

Figure 1.7 Theoretical (solid lines) and experimental (points) SW frequencies versus wave vector **Q** for (a) a Permalloy NW with width $W = 175$ nm and thickness $d = 20$ nm (magnetic applied field is $\mu_0 H_0 = 0.05$ T) and for (b) a 150-nm-wide NW with the composition Py(30 nm)/Cu(10 nm)/Py(60 nm)/Cu(10 nm)/Py(30 nm) as a function of the transferred in-plane wave vector with **Q** (magnetic applied field is $\mu_0 H_0 = 0.03$ T).

In this case the effects of single-ion anisotropy have been added to the Hamiltonian for an improved fit to the BLS data. The site-dependent anisotropy field is denoted as $SK_i/g\mu_B$, where the anisotropy coefficient can be taken as $K_i = K_{surf}$ at any surface site and $K_i = K_{bulk}$ elsewhere. We take $M_s = 0.069$ T, $= 9.3$ nm^2, $SK_{surf}/g\mu_B = -0.02$ T, and $SK_{bulk}/g\mu_B = 0$. We see a very good agreement between theory [58] and experiment [17], especially for the lowest SW frequency mode. In this plot the light scattering geometry was such that $q = 0$ for the longitudinal wave vector, but the transverse wave vector **Q** (in the x direction) could be varied in the experiments. Since the predicted SW frequencies do not depend on **Q** the theory lines are horizontal.

Figure 1.7b shows the comparison between the measured and calculated SW frequency dispersion [37] for a trilayer NW having the composition Py(30 nm)/Cu(10 nm)/Py(60 nm)/Cu(10 nm)/Py(30 nm) and width $W = 150$ nm. As a consequence of the lateral confinement within the stripes width, the BLS spectra of the stripes consist of several discrete peaks, whose frequencies do not change

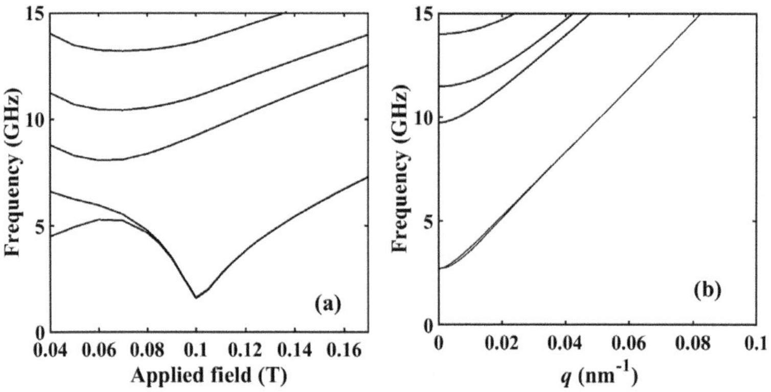

Figure 1.8 Lowest few SW branches of a rectangular NW stripe in a transverse applied field. The calculations are for a Permalloy NW with width 150 nm and thickness 5 nm, taking $a = 5$ nm: (a) frequencies plotted versus H_0 below and above the reorientation transition at $H_c \cong 0.10$ T for at fixed wavenumber $q = 0$ and (b) frequencies plotted versus q for a fixed field $\mu_0 H_0 = 0.11$ T above the transition.

with the wave vector **Q**. The lines are the results of calculations, and we have marked as continuous bold lines the modes expected to have larger BLS intensities. These intensity results are deduced from estimate of the mode probability amplitudes, as mentioned in the previous section.

Calculations have also been made for single NWs when the magnetic field is applied *transversely* (in the x direction) to the easy axis. Some numerical results are shown in Figs 1.8a and 1.8b. All spins are now canted in the direction of the applied field, but to differing degrees, and therefore the magnetization becomes non-uniform [56, 58]. At a large enough field value, corresponding to $H_0 > H_c$ with H_c denoting a critical reorientation field, the spins reorient to make the net magnetization align along the x axis. Figure 1.8a presents the SW frequencies versus the transverse applied field for a Permalloy NW with width $W = 150$ nm and thickness $d = 5$ nm. The main dip occurs at the critical field H_c. Below H_c competition between the Zeeman term and the dipole–dipole interaction leads to an overall decrease of the lowest SW frequencies with increasing

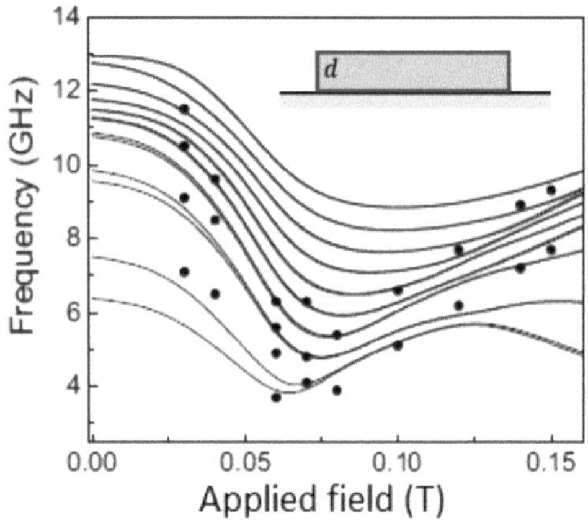

Figure 1.9 Comparison between BLS experiment (points) and theory (solid lines) for the SW frequencies of a rectangular Permalloy NW in a transverse applied field, following Ref. [69]. The inset shows a schematic of the geometry. See the text for further details.

field, while above H_c the SW frequencies increase with increasing field. The resulting dispersion relation for the same Permalloy NW at a fixed applied field (greater than the critical field) is shown in Fig. 1.8b. In contrast with the longitudinally magnetized NWs (Fig. 1.6) SW frequencies increase monotonically with q in this regime of the applied field.

Related BLS experimental results for isolated NW in a transverse field have been reported in Refs. [59, 68, 69], where comparisons between theory and experiment were included in the last two of these references. The main features observed for the magnetic field dependence of the lowest few SW frequencies, including the dip near the reorientation transition field, are well accounted for by theory. An example to illustrate this is shown in Fig. 1.9, which applies to Permalloy NWs with width 350 nm and thickness 30 nm, fabricated in an array with a lateral edge-to-edge separation of 55 nm [69]. This separation is sufficient to make the dipolar interwire interactions relatively small but not negligible. By an extension of the theory in Ref. [58] to include effects of the neighboring NWs, the theory lines in Fig. 1.9 could be deduced.

1.4 Hamiltonian Theory for Magnonic Crystals

A simple model of 1D MCs corresponds to a lateral (or side-by-side) arrays of magnetic NWs and nonmagnetic spacers on a nonmagnetic substrate. The NWs are typically arranged periodically, and so a new Bloch's theorem corresponding to the artificial periodicity length becomes applicable. Each NW in the array, which is labeled by an integer m (with $-\infty < m < \infty$), has N spins in its cross section. The NWs in this model are juxtaposed with a nonmagnetic spacer L in the xz plane. Thus, the artificial periodicity length of the MC in the x direction is $T = W + L$. The schematic cross section of a MC consisting of a periodic array of multilayer NWs is depicted in Fig. 1.10.

It is worthwhile to mention that MCs can be formed more generally of NWs with complicated shapes. In several recent works NWs with L-shaped cross sections were considered to study symmetry aspects of the magnetization dynamics. Lowering the symmetry of the system leads to an increase in the number of FMR peaks, since some degeneracies are removed. L-shaped NWs made of different material and different thicknesses of the layers [24, 30, 49] provide more information in FMR spectra than the L-shaped NWs made of single materials [28].

Another example of systems with a reduce symmetry was studied in Ref. [70], where the authors considered MCs formed by patterning, in terms of a periodic array of rectangular grooves, in one and both surfaces of ferromagnetic thin films. In particular, they

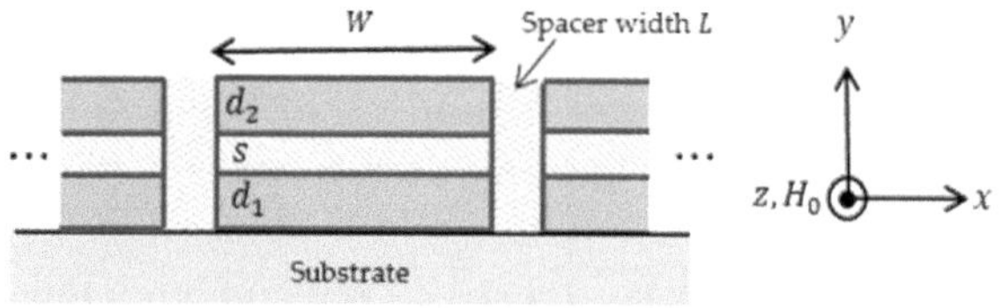

Figure 1.10 Cross-sectional representation of an MC consisting of a periodic array of multilayer stripes, for example, two magnetic layers of thicknesses d_1 and d_2 separated vertically by a spacer of thickness s. Each NW has a width of W in the x direction and is separated by nonmagnetic spacers of width L.

showed the dependence of the SW bands and gap on groove width and depth.

1.4.1 SW Frequencies

The procedure to generalize the microscopic dipole exchange theory for individual NWs to apply to MC arrays of NWs was provided in Refs. [70, 71]. In brief, it is necessary to incorporate the inter-NW interactions (due to dipolar and exchange terms, if any). By utilizing the periodicity property of the array and Bloch's theorem, the problem can again be recast in terms of solving a $2N \times 2N$ matrix problem, as before, where N again denotes the total number of spins in the cross section of a single NW. The exchange, dipolar and Zeeman contributions have the same form as in Eqs. 1.1–1.4, except that the long-range dipolar term is modified. If i and j indicate the position within any NW, while m and m' are labels for the NWs, the modified site separation vector takes the form

$$\mathbf{r}_{im,jm'} = \mathbf{r}_{im} - \mathbf{r}_{jm'} = \left(x_i - x_j + mT - m'T, y_i - y_j, z_i - z_j\right). \tag{1.12}$$

To proceed with the magnonic band and bandgap calculations, we generalize the method applied for NWs to MCs with macroscopic periodicity T in the z direction. Because of the periodicity property we may introduce a Bloch wave vector $\mathbf{Q}$, which will have Brillouin zone boundaries at multiples of π/T. Then we reformulate the exchange and dipole-dipole sums to include the inter- and intrastripe contributions, as mentioned above. In fact, the translational symmetry along the length of the NW (z direction) and the periodicity along the transverse direction (x direction) due to the MC lattice parameter a, lead us to use a double Fourier transformation and thus introduces the wave vectors $\mathbf{q}$ and $\mathbf{Q}$ respectively.

We still can use Eqs. 1.5–1.10, keeping in mind that the coefficients $A^{(2)}_{n,m}(\mathbf{q}, \mathbf{Q})$, $B^{(2)}_{n,m}(\mathbf{q}, \mathbf{Q})$ in Eq. 1.5, the dipolar terms $D^{(\alpha,\beta)}_{n,m}(\mathbf{q}, \mathbf{Q})$ in Eqs. 1.6 and 1.7, coefficients $\sigma_{n,l}(\mathbf{q}, \mathbf{Q})$ in Eq. 1.8, and the SW frequencies $\omega_l(\mathbf{q}, \mathbf{Q})$ in Eqs. 1.9 and 1.10 are now functions of both $\mathbf{q}$ and $\mathbf{Q}$. The SW bands arise from the $\mathbf{Q}$ dependence of $\omega_l(\mathbf{q}, \mathbf{Q})$ across the artificial Brillouin zone.

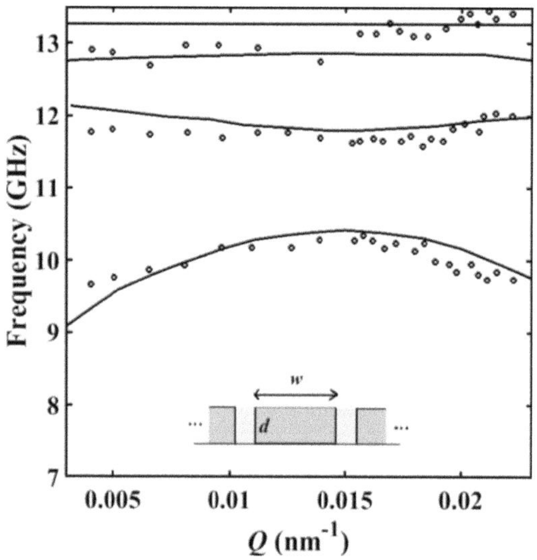

Figure 1.11 Comparison between the measured (points) and calculated (solid lines) SW frequencies as a function of Bloch wavenumber Q for an array of Permalloy stripes (each 175 nm $\times$20 nm) with 35 nm separation (the inset shows a schematic of the geometry). The assumed parameters are $\mu_0 H_0 = 0.05$ T, $D = 9.3$ T nm^2, $M_s = 0.0645$ T, and $g\mu_B = 29.5$ GHz/T.

1.4.2 Applications

We now present some numerical calculations based on the above theory for different MCs. We start with an MC consisting of an array of single-layer NWs. Figure 1.11 shows the comparison between the theory and BLS data obtained by Gubbiotti et al. [16, 17] for the MC of Permalloy stripes with width $W = 175$ nm and thickness $d = 20$ nm, separated by spacers of width $L = 35$ nm. We see a very good agreement between the theory and the experiment. It appears that the experimental data correspond to the lowest four bands in our theory, where the lowest two show significant dispersion and the upper two are relatively flat.

Next, we consider an array of layered Permalloy/Cu/Permalloy NWs (see Fig. 1.12) with width $W = 280$ nm, $d_1 = 30$ nm, $s = 10$ nm, $d_2 = 10$ nm, and spacing $L = 280$ nm at $H_0 = 500$ Oe. We show the SW dispersion relation by plotting the dependence of the lowest frequencies on the Bloch wavenumber Q in the x direction.

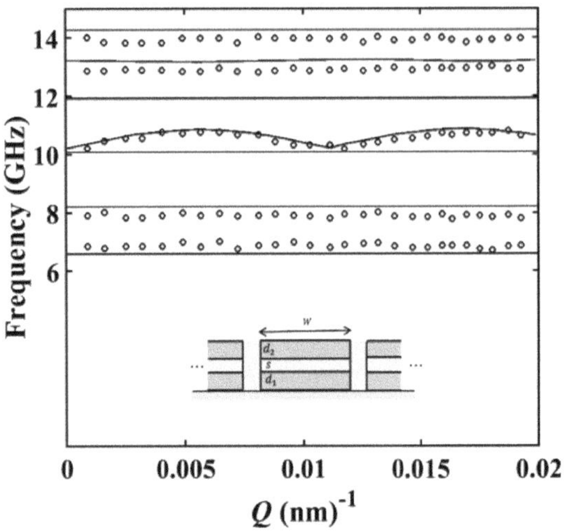

Figure 1.12 Comparison between the measured (circles) and calculated (solid lines) SW frequencies as a function of Bloch wavenumber Q for a Py(30 nm)/Cu(10 nm)/Py(10 nm) MC with separation L = 280 nm (the inset shows a schematic of the geometry) at H_0 = 500 Oe.

We see a very good agreement between theory and experiment for all the frequency modes. It can be seen that the third frequency mode (l = 3) shows a periodic frequency dispersion, which is because of the artificial periodicity of the NWs. The other frequency modes exhibit flat dispersion. This is different from the case of single-layer NW arrays, where we see that the lowest frequency mode has a dispersive behavior and the other modes are dispersionless [30].

In Fig. 1.12 the magnetizations in the two Py were in a P configuration along the easy (z) direction of the applied field. It was shown, however, in Ref. [6] that for an analogous Py(30 nm)/Cu(10 nm/Py(10 nm) MC, but with smaller lateral separation L = 80 nm, there are two stable magnetic states. Specifically, measurements of the Kerr hysteresis loop indicate that due to the different thickness of the two Py layers, there can be an AP state with oppositely directed NW magnetizations, as well as a P state for certain values of the applied field. The behavior for the SW bands in the two cases is significantly different, as seen from Fig. 1.13. For example, there are additional modes in the AP case that show dispersion (i.e., in the

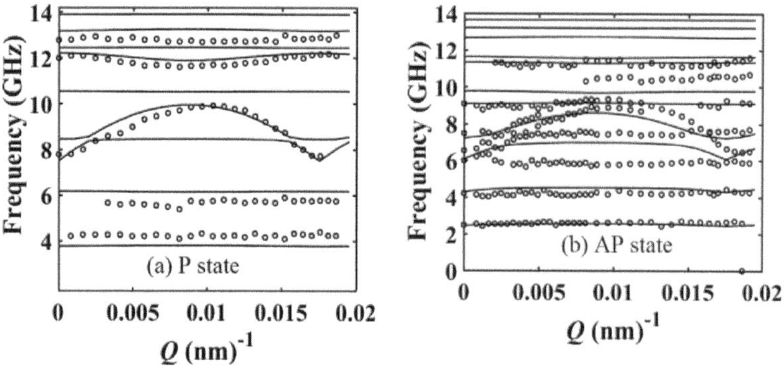

Figure 1.13 Same structure of the MC as depicted in Fig. 1.12 (see inset), but for the case when the lateral separation is reduced to $L = 80$ nm with an applied field of magnitude $H_0 = 150$ Oe. As explained in the text, there are two magnetic states corresponding to (a) parallel (P) and (b) antiparallel (AP) magnetizations in the two Permalloy layers.

approximate frequency range 6 to 10 GHz) compared to the P case, resulting in additional mode mixing in Fig. 1.13b).

As remarked earlier, by fabricating a multilayered structure with several different thicknesses of the layers, we achieve multiple reprogrammable magnetization configurations. Layered structures of this type are interesting as prototypes of multiple magnetic storage layers [72] and as nonvolatile vertical magnetic logic gates [73] where the knowledge of SW properties is important. Exploring the third dimension in MCs [74] is important in order to follow the latest trend in complementary metal-oxide semiconductor (CMOS) electronics, where expansion from 2D to 3D vertically integrated structures is being pursued [75].

1.5 Inclusion of Interlayer Exchange Coupling

Apart from the short-range direct exchange, there may be an indirect exchange that couples magnetic moments in some materials over relatively larger distances (e.g., several nanometers). In metals, when there is only weak or no direct overlap of the wave functions of neighboring electrons, this indirect exchange becomes important.

This interaction in metals acts through the conduction electrons and was considered by Ruderman and Kittel [76]. With later developments, the indirect coupling of magnetic moments via conduction electrons is now known as the RKKY interaction. Rare-earth metals also show rather weak direct exchange, and therefore the RKKY interaction contributes to magnetic ordering in these materials [7]. In some low-dimensional systems the RKKY coupling across a thin metallic layer between two other magnetic materials is of interest for device applications such as microwave resonators, magnetic filters, memory, and SW logic devices [9, 77, 78].

Additionally, there may be other exchange effects called bi-quadratic exchange. Typically, this type of interaction might be small in bulk materials, but it can become important at surfaces and interfaces [79, 80]. The name "biquadratic" is in distinction to the simpler interaction in Eq. 1.1, which is sometimes called "bilinear."

1.5.1 SW Frequencies

The interlayer RKKY interaction can be expressed as

$$H_{\mathrm{R}} = \frac{1}{2}\sum_{in,jm} J_{\mathrm{R}}\mathbf{S}_{in}\cdot\mathbf{S}_{jm}, \tag{1.13}$$

where J_{R} is the RKKY exchange term that couples the closest sites across the interface. The exchange J_{R} arising via polarization effects in the cloud of conduction electron of the Ru spacer. In general, J_{R} may be oscillatory with variation of the thickness s and extend over several nanometers. With this notation when $J_{\mathrm{R}} > 0$, an antiferromagnetic coupling (AP state) is favored and $J_{\mathrm{R}} < 0$ leads to a ferromagnetic ordering (P state).

The biquadratic exchange can by described by [8, 81, 82]

$$H_{\mathrm{BQ}} = \sum_{in,jm} J_{\mathrm{BQ}}(\mathbf{S}_{in}\cdot\mathbf{S}_{jm})^2. \tag{1.14}$$

This may be small in bulk materials but can be much larger across some thin metallic layers. We conclude that is important for Ru in the present NW studies. If $J_{\mathrm{BQ}} > 0$, which is usually the case, then this term in the Hamiltonian favors a magnetic state with $\mathbf{S}_{in}\cdot\mathbf{S}_{jm} = 0$ or $\mathbf{S}_{in}\perp\mathbf{S}_{jm}$ Various explanations for the enhanced role of biquadratic exchange at interfaces have been put forward, including interface roughness and the presence of interface impurities.

For simplicity, we assume J_{BQ} couples only a site i in the interface of one NW to the corresponding (opposite) site i in the interface layer of the other NW. We note that the competition between the RKKY coupling and other magnetic interactions, such as biquadratic and bilinear exchange interactions, dipole–dipole interactions and single-ion anisotropy lead to overall ferromagnetic and antiferromagnetic ordering and also spin-flop state, depending on the applied magnetic field.

As an example, we consider a trilayer NW consisting of two Permally layers separated by a nonmagnetic Ru spacer layer. From our studies on trilayer complete films [83] and wide stripes [34] and narrower NWs [35], it is known that interlayer exchange coupling through the Ru spacer layer can occurs via RKKY exchange and biquadratic exchange.

As before, the system is modeled by a lattice of effective spins with cell size a less than or comparable to the exchange correlation length (about 5 or 6 nm) in Permalloy. The Hamiltonian includes RKKY coupling (Eq. 1.13) and biquadratic coupling term (Eq. 1.14), plus exchange terms within each stripe (Eq. 1.1), dipole–dipole terms within and between stripes (Eq. 1.2), Zeeman energy (Eq. 1.4), and weak single-ion anisotropy, as well as shape anisotropy.

For the dynamics a linearized Hamiltonian is diagonalized in a matrix method to determine the SW modes, as described in Section 1.2.1. The low-temperature equilibrium spin configuration is found by solving for the canting angles (deduced by aligning each spin parallel to its self-consistent internal effective field). An energy minimization can be used to select the antiferromagnetic, spin-flop, or ferromagnetic state, as appropriate [35].

Next, a transformation to boson operators is made using these local axes defined with respect to the orientation of each spin. Then a transformation to boson operators is made using these local axes defined with respect to the orientation of each spin and an operator expansion of the Hamiltonian is made ($H = H^{(2)} + H^{(3)} + H^{(4)} + \ldots$). Finally, to obtain the linear SW spectrum, the Hamiltonian $H^{(2)}$ is solved.

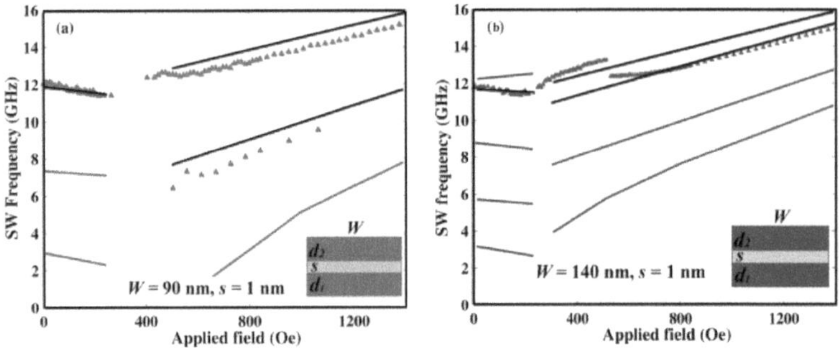

Figure 1.14 Theory (solid lines) and experiment (triangles) for the lowest NW modes magnonic branches showing the SW frequencies versus the applied field H_0 for the Py/Ru (s)/Py for (a) W = 90 and s = 1 nm and (b) W = 140 and s = 1 nm. The inset shows a schematic of the geometry for both cases.

1.5.2 Applications

In this section we present some numerical and experimental results for trilayer NWs consisting of two Permalloy layers separated by a Ru spacer layer. In particular, we show the results for the systems with the Ru spacer layer s = 1 nm, where the widths of the stripes are W = 90 and 140 nm. The Ru spacer layer can induce interlayer exchange coupling (both RKKY and biquadratic exchange interactions). For these samples the biquadratic coupling plays an important role. It can be shown that the components of the interlayer exchange, J_R and J_{BQ}, both decrease in magnitude as s is increased [35].

In Fig. 1.14a the SW frequencies are shown as a function of the applied field H_0 for the case with s = 1 and W = 90 nm at zero wave vector ($\mathbf{q} = 0$). We have only shown the lowest modes correspond to the range of measured frequencies in the FMR experiment. The fit to the experimental data is seen to be excellent. The lower SW branch predicted by theory shows the soft-mode behavior. Figure 1.14b shows a similar plot for the system with W = 140 nm (s = 1 nm). There is an excellent agreement between theory and experiment. A transition field occurs at $H_c = {\sim}230$ Oe. The antiferromagnetic phase is favored at a smaller field $H_0 < H_c$, where the magnetizations of the two Permalloy NWs are parallel to each other. The spin-flop state occurs at the range $H_0 < H_c < 500$ Oe, and $H_0 > 500$ Oe corresponds to the ferromagnetic phase where the magnetization of the two NWs is AP.

From the comparison of these results with the results for a trilayer of complete films [83] and wider NWs [34], we see that the biquadratic exchange interaction becomes more pronounced due to the lateral edge effects. Single-ion anisotropy is small compared to shape anisotropy [35] and seems to be important for NWs with a smaller width: $W = 90$ nm ($\frac{SK_i}{g\mu_B} = \sim -150$ Oe).

Our present results suggest ways to tailor the FMR mode behavior in layered NWs, exploiting a fine control of interlayer coupling. Potential applications relate to reconfigurable vertical magnetic logic devices and microwave filters. Other experimental studies (e.g., by BLS) would be useful in giving more information about spatially quantized SWs. It would also be of interest to extend Green's function microscopic theory to deduce mode intensities.

1.6 Conclusion

In conclusion, we have studied some properties of collective SWs in structured composite NW arrays. A microscopic (Hamiltonian-based) theory is employed, including the exchange and magnetic dipole–dipole interaction terms, together with the Zeeman energy of the applied magnetic field and other interaction terms, as appropriate.

The results presented in this work represent a step forward in understanding the SW band structure in layered MCs, where the propagation of collective SW with Bloch characteristics (due to the lateral periodicity) can be controlled by additional layers that are present in the structure and by their relative magnetization orientation.

Acknowledgments

M.G.C. acknowledges support from the Natural Sciences and Engineering Research Council (NSERC) of Canada (grant RGPIN-

2017-04429). A.O.A. was supported by the National Research 10 Foundation, Prime Minister's Office, Singapore, under its Competitive Research Programme (CRP Award No. NRFCRP 10-2012-03).

References

1. Ding, J. and Adeyeye, A. O. (2013). Binary ferromagnetic nanostructures: fabrication, static and dynamic properties, *Adv. Funct. Mater.*, **23**, 1648–1691.
2. Tacchi, S., Madami, M., Gubbiotti, G., Carlotti, G., Goolaup, S., Adeyeye, A. O., Singh, N. and Kostylev, M. P. (2010). Analysis of collective spin-wave modes at different points within the hysteresis loop of a one-dimensional magnonic crystal comprising alternative-width nanostripes, *Phys. Rev. B*, **82**, 184408 (9 pp).
3. Topp, J., Heitmann, D., Kostylev, M. P. and Grundler, D. (2010). Making a reconfigurable artificial crystal by ordering bistable magnetic nanowires, *Phys. Rev. Lett.*, **104**, 207205 (4 pp).
4. Krawczyk, M. and Grundler, D. (2014). Review and prospects of magnonic crystals and devices with reprogrammable band structure, *J. Phys.: Condens. Matter.*, **26**, 123202 (32 pp).
5. Vogel, M., Chumak, A. V., Waller, E. H., Langner, T., Vasyuchka, V. I., Hillebrands, B. and von Freymann, G. (2015). Optically reconfigurable magnetic materials, *Nat. Phys.*, **11**, 487–491.
6. Gubbiotti, G., Zhou, X., Haghshenasfard, Z., Cottam, M. G. and Adeyeye, A. O. (2018). Reprogrammable magnonic band structure of layered permalloy/Cu/permalloy nanowires, *Phys. Rev. B*, **97**, 134428 (6 pp).
7. Heinrich, B. and Bland, J. A. C., eds. (1994). *Ultrathin Magnetic Structures II* (Springer, Germany).
8. Cottam, M. G. and Tilley, D. R. (2005). *Introduction to Surface and Superlattice Excitations*, 2nd ed. (IoP Publishing, UK).
9. Kruglyak, V. V., Demokritov, S. O. and Grundler, D. (2010). Magnonics, *J. Phys. D: Appl. Phys.*, **43**, 264001 (14 pp).
10. Serga, A. A., Chumak, A. V. and Hillebrands, B. (2010). YIG magnonics, *J. Phys. D: Appl. Phys.*, **43**, 264002 (16 pp).
11. Khitun, A., Bao, M. and Wang, K. L. (2010). Magnonic logic circuits, *J. Phys. D: Appl. Phys.*, **43**, 264005 (10 pp).

12. Chumak, A. V., Neumann, T., Serga, A. A., Hillebrands, B. and Kostylev, M. (2009). A current-controlled, dynamic magnonic crystal, *J. Phys. D: Appl. Phys.*, **42**, 205005 (8 pp).
13. Lenk, B., Ulrichs, H., Garbs, F. and Münzenberg, M. (2011). The building blocks of magnonics, *Phys. Rep.*, **507**, 107–136.
14. Chumak, A. V., Serga, A. A. and Hillebrands, B. (2017). Magnonic crystals for data processing, *J. Phys. D: Appl. Phys.*, **50**, 244001 (20 pp).
15. Haldar, A., Kumar, D. and. Adeyeye, A. O. (2016). A reconfigurable waveguide for energy-efficient transmission and local manipulation of information in a nanomagnetic device, *Nat. Nanotechnol.*, **11**, 437–443.
16. Gubbiotti, G., Tacchi, S., Madami, M., Carlotti, G., Adeyeye, A. O. and Kostylev, M. (2010). Brillouin light scattering studies of planar metallic magnonic crystals, *J. Phys. D: Appl. Phys.*, **43**, 264003 (13 pp).
17. Gubbiotti, G., Tacchi, S., Carlotti, G., Vavassori, P., Singh, N., Goolaup, S., Adeyeye, A. O., Stashkevich, A. and Kostylev, M. (2005). Magnetostatic interaction in arrays of nanometric permalloy wires: a magneto-optic Kerr effect and a Brillouin light scattering study, *Phys. Rev. B*, **72**, 224413 (7 pp).
18. Gubbiotti, G., Tacchi, S., Carlotti, G., Singh, N., Goolaup, S., Adeyeye, A. O. and Kostylev, M. (2007). Collective spin modes in monodimensional magnonic crystals consisting of dipolarly coupled nanowires, *Appl. Phys. Lett.*, **90**, 092503 (3 pp).
19. Wang, Z. K., Zhang, V. L., Lim, H. S., Ng, S. C., Kuok, M. H., Jain, S. and Adeyeye, A. O. (2009). Observation of frequency band gaps in a one-dimensional nanostructured magnonic crystal, *Appl. Phys. Lett.*, **94**, 083112 (3 pp).
20. Wang, Z. K., Zhang, V. L., Lim, H. S., Ng, S. C., Kuok, M. H., Jain, S. and Adeyeye, A. O. (2010). Nanostructured magnonic crystals with size-tunable bandgaps, *ACS Nano*, **4**, 643–648.
21. Jain, S., Kostylev, M. and Adeyeye, A. O. (2010). Coupled oscillations in noncollinear microscale rectangular magnets, *Phys. Rev. B*, **82**, 214422 (9 pp).
22. Adeyeye, A. O. and Jain, S. (2011). Coupled periodic magnetic nanostructures, *J. Appl. Phys.*, **109**, 07B903 (6 pp).
23. Gubbiotti, G., Xiong, L. L., Montoncello, F., Giovannini, L. and Adeyeye, A. O. (2018). Spin wave dispersion and intensity correlation in width-modulated nanowire arrays: a Brillouin light scattering study, *J. Appl. Phys.*, **124**, 083903 (6 pp).

24. Gubbiotti, G., Xiong, L. L., Montoncello, F. and Adeyeye, A. O. (2017). Collective spin waves in arrays of permalloy nanowires with single-side periodically modulated width, *Appl. Phys. Lett.*, **111**, 192403 (4 pp).
25. Ciubotaru, F., Chumak, A. V., Grigoryeva, N. Yu., Serga, A. A. and Hillebrands, B. (2012). Magnonic band gap design by the edge modulation of micro-sized waveguides, *J. Phys. D: Appl. Phys.*, **45**, 255002 (6 pp).
26. Chumak, A. V., Vasyuchka, V. I., Serga, A. A. and Hillebrands, B. (2015). Magnon spintronics, *Nat. Phys.*, **11**, 453–461.
27. Piao, H., Shim, J., Yang, M., Pan, L. and Kim, D. (2015). Selective behavior of spin wave propagation in asymmetrically modulated ferromagnetic nanowires, *IEEE Trans. Magn.*, **51**, 1–6.
28. Gubbiotti, G., Kostylev, M., Tacchi, S., Madami, M., Carlotti, G., Ding, J., Adeyeye, A. O., Zighem, F., Stashkevich, A. A., Ivanov, E. and Samarin, S. (2014). Collective spin waves on a nanowire array with step-modulated thickness, *J. Phys. D: Appl. Phys.*, **47**, 105003 (8 pp).
29. Gubbiotti, G., Silvani, R., Tacchi, S., Madami, M., Carlotti, G., Yang, Z., Adeyeye, A. O. and Kostylev, M. (2017). Tailoring the spin waves band structure of 1D magnonic crystals consisting of L-shaped iron/permalloy nanowires, *J. Phys. D: Appl. Phys.*, **50**, 105002 (6 pp).
30. Gubbiotti, G., Tacchi, S., Madami, M., Carlotti, G., Yang, Z., Ding, J., Adeyeye, A. O. and Kostylev, M. (2016). Collective spin excitations in bicomponent magnonic crystals consisting of bilayer permalloy/Fe nanowires, *Phys. Rev. B*, **93**, 184411 (8 pp).
31. Kostylev, M., Yang, Z., Maksymov, I. S., Ding, J., Samarin, S. and Adeyeye, A. O. (2016). Microwave magnetic dynamics in ferromagnetic metallic nanostructures lacking inversion symmetry, *J. Appl. Phys.*, **119**, 103903 (13 pp).
32. Silvani, R., Kostylev, M., Adeyeye, A. O. and Gubbiotti, G. (2018). Spin wave filtering and guiding in permalloy/iron nanowires, *J. Magn. Magn. Mater.*, **450**, 51–59.
33. Gubbiotti, G., Kostylev, M., Sergeeva, N., Conti, M., Carlotti, G., Ono, T., Slavin, A. N. and Stashkevich, A. (2004). Brillouin light scattering investigation of magnetostatic modes in symmetric and asymmetric Ni Fe/ Cu/ Ni Fe trilayered wires, *Phys. Rev. B*, **70**, 224422 (8 pp).
34. Liu, X. M., Lupo, P., Cottam, M. G. and Adeyeye, A. O. (2015). Tuning of interlayer exchange coupling in Ni80Fe20/Ru/Ni80Fe20 nanowires, *J. Appl. Phys.*, **118**, 113902 (10 pp).

35. Lupo, P., Haghshenasfard, Z., Cottam, M. G. and Adeyeye, A. O. (2016). Ferromagnetic resonance study of interface coupling for spin waves in narrow NiFe/Ru/NiFe multilayer nanowires, *Phys. Rev. B*, **94**, 214431 (10 pp).
36. Xiong, L. L. and Adeyeye, A. O. (2018). Static and dynamic behavior of interlayer exchange coupled Ni80Fe20/Ru continuous films and nanowires, *J. Phys. D: Appl. Phys.*, **51**, 025004 (6 pp).
37. Gubbiotti, G., Tacchi, S., Nguyen, H. T., Madami, M, Carlotti, G., Nakano, K., Ono, T. and Cottam, M. G. (2013). Coupled spin waves in trilayer films and nanostripes of permalloy separated by nonmagnetic spacers: Brillouin light scattering and theory, *Phys. Rev. B*, **87**, 094406 (10 pp).
38. Tkachenko, V. S., Kuchko, A. N., Dvornik, M. and Kruglyak, V. V. (2012). Propagation and scattering of spin waves in curved magnonic waveguides, *Appl. Phys. Lett.*, **101**, 152402 (5 pp).
39. Xing, X., Yu, Y., Li, S. and Huang, X. (2013). How do spin waves pass through a bend? *Sci. Rep.*, **3**, 2958 (6 pp).
40. Zhi-xiong, L., Wang, M. N., Nie, Y. Z., Wang, D. W., Xia, Q. L., Tang, W., Zeng, Z. M. and Guo, G. H. (2016). Spin-wave propagation spectrum in magnetization-modulated cylindrical nanowires, *J. Magn. Magn. Mater.*, **414**, 49–54.
41. Van de Wiele, B., Hämäläinen, S. J., Baláž, P., Montoncello, F. and van Dijken, S. (2016). Tunable short-wavelength spin wave excitation from pinned magnetic domain walls, *Sci. Rep.*, **6**, 21330 (11 pp).
42. Banerjee, C., Choudhury, S., Sinha, J. and Barman, A. (2017). Pseudo-one-dimensional magnonic crystals for high-frequency nanoscale devices, *Phys. Rev. Appl.*, **8**, 014036 (8 pp).
43. Goolaup, S., Adeyeye, A. O., Singh, N. and Gubbiotti, G. (2007). Magnetization switching in alternating width nanowire arrays, *Phys. Rev. B*, **75**, 144430 (7 pp).
44. Topp, J., Mendach, S., Heitmann, D., Kostylev, M. and Grundler, D. (2011). Field- and geometry-controlled avoided crossings of spin-wave modes in reprogrammable magnonic crystals, *Phys. Rev. B*, **84**, 214413 (8 pp).
45. Zhang, V. L., Lim, H. S., Lin, C. S., Wang, Z. K., Ng, S. C., Kuok, M. H., Jain, S., Adeyeye, A. O. and Cottam, M. G. (2011). Ferromagnetic and antiferromagnetic spin-wave dispersions in a dipole-exchange coupled bi-component magnonic crystal, *Appl. Phys. Lett.*, **99**, 143118 (3 pp).
46. Kalinikos, A. B. (1981). Spectrum and linear excitation of spin waves in ferromagnetic films, *Sov. J. Phys.*, **24**, 718–731.

47. Cottam, M. G. (1994). *Linear and Nonlinear Spin Waves in Magnetic Films and Superlattices*, ed. Cottam, M. G. and Slavin, A. N., Chapter 1, Fundamentals of linear and nonlinear spin-wave processes in bulk and finite magnetic samples (World Scientific), pp. 1–88.
48. Gurevich, A. G. and Melkov, G. A. (1996). *Magnetization Oscillations and Waves* (CRC Press).
49. Kostylev, M., Ding, J., Ivanov, E., Samarin, S. and Adeyeye, A. O. (2014). Microwave magnetic dynamics in highly conducting magnetic nanostructures, *J. Appl. Phys.*, **115**, 173903 (7 pp).
50. Kostylev, M., Yang, Z., Maksymov, I. S., Ding, J., Samarin, S. and Adeyeye, A. O. (2016). Microwave magnetic dynamics in ferromagnetic metallic nanostructures lacking inversion symmetry, *J. Appl. Phys.*, **119**, 103903 (13 pp).
51. Kostylev, M. (2013). Non-reciprocity of dipole-exchange spin waves in thin ferromagnetic films, *J. Appl. Phys.*, **113**, 053907 (8 pp).
52. Benson, H. and Mills, D. L. (1969). Spin waves in thin films; dipolar effects, *Phys. Rev.*, **178**, 839 (9 pp).
53. Costa Filho, R. N., Cottam, M. G. and Farias, G. A. (2000). Microscopic theory of dipole-exchange spin waves in ferromagnetic films: linear and nonlinear processes, *Phys. Rev. B*, **62**, 6545 (16 pp).
54. Nguyen, T. M. and Cottam, M. G. (2005). Microscopic theory of dipole-exchange spin-wave excitations in ferromagnetic nanowires, *Phys. Rev. B*, **71**, 094406 (11 pp).
55. Nguyen, T. M. and Cottam, M. G. (2005). Spectral intensities and frequencies of spin waves in ferromagnetic cylinders: application to nickel nanowires, *Phys. Rev. B*, **72**, 224415 (10 pp).
56. Haghshenasfard, Z. and Cottam, M. G. (2016). Spin-wave instabilities of ferromagnetic nanowire stripes under parallel pumping, *J. Phys.: Condens. Matter*, **28**, 186001 (12 pp).
57. Nguyen, H. T., Akbari-Sharbaf, A. and Cottam, M. G. (2011). Spin-wave damping in ferromagnetic stripes with inhomogeneous magnetization, *Phys. Rev. B*, **83**, 214423 (11 pp).
58. Nguyen, T. M., Nguyen, H. T. and Cottam, M. G. (2007). Dipole-exchange spin waves in ferromagnetic stripes with inhomogeneous magnetization, *Phys. Rev. B*, **76**, 134413 (9 pp).
59. Nguyen, H. T., Gubbiotti, G., Madami, M., Tacchi, S. and Cottam, M. G. (2009). Brillouin light scattering study of the spin dynamics in nanoscale Permalloy stripes: Theory and experiment, *Microelectron. J.*, **40**, 598–600.

60. Nguyen, H. T. and Cottam, M. G. (2011). The effect of lattice structure on dipole–exchange spin waves in ultrathin ferromagnetic films, *J. Phys.: Condens. Matter*, **23**, 126004 (9 pp).

61. Haghshenasfard, Z. and Cottam, M. G. (2018). Squeezed states of Damon–Eshbach-like modes in ferromagnetic films under microwave pumping, *IEEE Trans. Magn.*, **54**, 1300609 (9 pp).

62. Haghshenasfard, Z., Nguyen, H. T. and Cottam, M. G. (2015). Spin-wave instability theory for ferromagnetic nanostructures, *Acta Phys. Pol. A*, **127**, 192–197.

63. Haghshenasfard, Z., Nguyen, H. T. and Cottam, M. G. (2017). Suhl instabilities for spin waves in ferromagnetic nanostripes and ultrathin films, *J. Magn. Magn. Mater.*, **426**, 380–389.

64. Holstein, T. and Primakoff, H. (1940). Field dependence of the intrinsic domain magnetization of a ferromagnet, *Phys. Rev.*, **58**, 1098 (16 pp).

65. Damon, R. W. and Eshbach, J. R. (1961). Magnetostatic modes of a ferromagnet slab, *J. Phys. Chem. Solids*, **19**, 308–320.

66. Grünberg, P. (1980). Magnetostatic spinwave modes of a ferro-magnetic double layer, *J. Appl. Phys.*, **51**, 4338–4341.

67. Gubbiotti, G., Zhou, X., Haghshenasfard, Z., Cottam, M. G., Adeyeye, A. O. and Kostylev, M. (2019). Interplay between intra- and inter-nanowires dynamic dipolar interactions in the spin wave band structure of Py/Cu/Py nanowires. *Sci. Rep.*, **9**, 4617.

68. Gubbiotti, G., Carlotti, G., Ono, T. and Roussigné, Y. (2006). High frequency magnetic excitations in patterned NiFe/Cu/NiFe trilayered wires subjected to a transverse magnetic field, *J. Appl. Phys.*, **100**, 023906 (7 pp).

69. Tacchi, S., Madami, M., Gubbiotti, G., Carlotti, G., Goolaup, S., Adeyeye, A. O., Nguyen, H. T. and Cottam, M. G. (2009). Field dependence of collective spin modes in transversely magnetized stripes with homogeneous and alternating width, *J. Appl. Phys.*, **105**, 07C102 (3 pp).

70. Nguyen, H. T. and Cottam, M. G. (2012). Microscopic dipole-exchange theory for magnonic crystals: application to ferromagnetic films with patterned surfaces, *J. Appl. Phys.*, **111**, 07D122 (3 pp).

71. Nguyen, H. T. and Cottam, M. G. (2011). Microscopic dipole-exchange theory for planar nanostriped magnonic crystals, *J. Phys. D*, **44**, 315001 (7 pp).

72. Albrecht, M., Hu, G., Moser, A., Hellwig, O. and Terris, B. D. (2005). Magnetic dot arrays with multiple storage layers, *J. Appl. Phys.*, **97**, 103910 (5 pp).

73. Butler, J., Shachar, M., Lee, B., Garcia, D., Hu, B., Hong, J., Amos, N. and Khizroev, S. (2014). Reconfigurable and non-volatile vertical magnetic logic gates, *J. Appl. Phys.*, **115**, 163903 (5 pp).
74. Popov, P. A., Sharaevskaya, A. Yu., Beginin, E. N., Sadovnikov, A. V., Stognij, A. I., Kalyabin, D. V. and Nikitov, S. A. (2019). Spin wave propagation in three-dimensional magnonic crystals and coupled structures, *J. Magn. Magn. Mater.*, **476**, 423–427.
75. Zanjani, S. M. M., Holt, M., Sadeghi, M. M., Rahimi, S. and Akinwande, D. (2017). 3D integrated monolayer graphene–Si CMOS RF gas sensor platform, *npj 2D Mater. Appl.*, **1**, 36 (9 pp).
76. Ruderman, M. A. and Kittel, C. (1954). Indirect exchange coupling of nuclear magnetic moments by conduction electrons, *Phys. Rev.*, **96**, 99–102.
77. Kostylev, M. P., Serga, A. A., Schneider, T., Leven, B. and Hillebrands, B. (2005). Spin-wave logic gates, *Appl. Phys. Lett.*, **87**, 153501 (3 pp).
78. Neusser, S. and Grundler, D. (2009). Magnonics: spin waves on the nanoscale, *Adv. Mater.*, **21**, 2927 (6 pp).
79. Rührig, M., Schäfer, R., Hubert, A., Mosler, R., Wolf, J. A., Demokritov, S. and Grünberg, P. (1991). Domain observations on Fe-Cr-Fe layered structures. Evidence for a biquadratic coupling effect, *Phys. Status Solidi A*, **125**, 635–656.
80. Azevedo, A., Chesman, C., Rezende, S. M., de Aguiar, F. M., Bian, X. and Parkin, S. S. P. (1996). Biquadratic exchange coupling in sputtered (100) Fe/Cr/Fe, *Phys. Rev. Lett.*, **76**, 4837–4840.
81. White, R. M. (2007). *Quantum Theory of Magnetism* (Springer).
82. Cottam, M. G. (1994). *Linear and Nonlinear Spin Waves in Magnetic Films and Superlattices*, ed. Rezende, S. M., Azevedo, A. and de Aguiar, F. M., Chapter 7, Spin-wave instabilities, auto-oscillations, chaos and control of chaos in YIG spheres (World Scientific), pp. 335–374.
83. Liu, X. M., Nguyen, H. T., Ding, J., Cottam, M. G. and Adeyeye, A. O. (2015). Interlayer coupling in Ni80Fe20/Ru/Ni80Fe20 multilayer films: ferromagnetic resonance experiments and theory, *Phys. Rev. B*, **90**, 064428, (11 pp).

Chapter 2

From 2D Planar Magnonic Crystals to 3D Magnonic Crystals

P. Graczyk,[a] P. Gruszecki,[b] S. Mamica,[b] J. W. Kłos,[b] M. Krawczyk,[b] and G. Gubbiotti[c]

[a]*Institute of Molecular Physics, Polish Academy of Sciences, 60-179 Poznan, Poland*
[b]*Faculty of Physics, Adam Mickiewicz University in Poznan, Umultowska 85, Poznan, 61-614 Poland,*
[c]*Istituto Offcina dei Materiali del CNR (CNR-IOM), Sede Secondaria di Perugia, c/o Dipartimento di Fisica e Geologia, Università di Perugia, I-06123 Perugia, Italy.*
krawczyk@amu.edu.pl

2.1 Introduction

Magnetic films with a periodic pattern were considered for controlling spin waves (SWs) already from 1976 [45, 52]. That time, the investigations were limited to relatively large structures, in which the dipolar interactions prevailed over the exchange interaction. The transmission of waves was molded by periodic distribution of the saturation magnetization, realized by ion implantation [16], a regular lattice of etched grooves in a magnetic dielectric, periodic modulation of metallic stripes or dots on a top of the ferromagnetic film [42, 43], or periodic perturbation of the effective magnetic

Three-Dimensional Magnonics: Layered, Micro- and Nanostructures
Edited by Gianluca Gubbiotti

ISBN 978-981-4800-73-0 (Hardcover), 978-0-429-29915-5 (eBook)
www.jennystanford.com

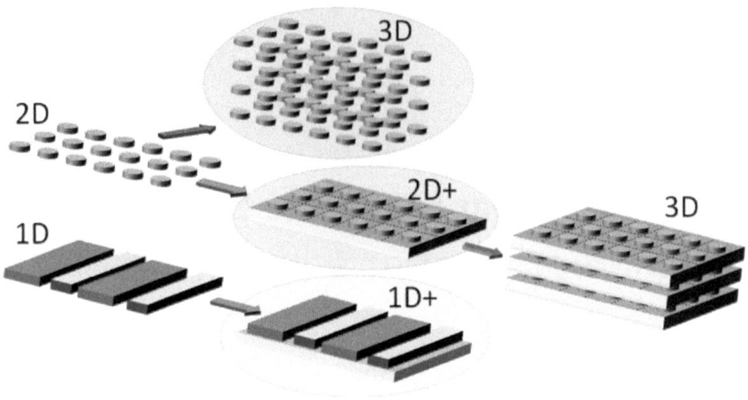

Figure 2.1 Schematic representation of the periodic structures and their compositions considered in the chapter. From planar 2D crystals by periodic repetition along the perpendicular direction we can form a 3D crystal, or by combining a 2D planar crystal with other planar structure, we can form a 2D+ structure. Similarly, a 1D crystal combined with other planar structures will form a 1D+ crystal.

field. The discovery of photonic crystals in 1987 [60] renewed the interest in periodic structures, inspiring new studies and providing abundant new physics also in magnetic materials. The concept of magnonic crystals (MCs) was proposed as an SW counterpart of photonic crystals [15, 25, 46, 57]. The MC can be formed from a magnetic material with periodic distribution of the constituent materials or periodic modulation of some magnetic parameters (e.g., saturation magnetization or the magnetocrystalline anisotropy) or other parameters relevant to the propagation of SWs, such as the external magnetic field, geometry, mechanical stress, temperature, or a surrounding of the homogeneous ferromagnetic medium.

According to the Bloch theorem in a medium with discrete spatial periodicity, the eigensolutions of the wave equation, regardless of its type, form a band structure in the frequency–wave vector space [5]. Thus, the qualitative understanding of the formation of the SW band structure can be inferred from the classical textbooks of solid-state physics [1, 17]. Nevertheless, in most cases, details of the dispersion relation can only be derived from numerical calculations. That is because SWs have a complex dispersion relation, and their curvature depends on the direction and magnitude of the wave vector, the relative strength of the short-range exchange interactions to the long-range dipolar interaction, the external shape of the sample, the magnetocrystalline anisotropy, the magnitude of the applied static magnetic field, and its orientation with respect to the direction of propagation. This makes MCs an intriguing subject for scientific studies in the field of magnonics [6, 8, 22, 28].

MCs with periodicity in 1D and 2D have been widely studied in various arrangements [6, 22, 53, 55]. Recently increased attention is being devoted to structures that have periodicity in the film plane but are also inhomogeneous along the perpendicular direction [4, 13, 34–36, 63]. The inhomogeneity along the out-of-plane direction can significantly influence the SW propagation in the periodic landscape of the film. We will call the systems, which exhibit such property, as 1D+ or 2D+ MCs. Such planar structures can be regarded as an intermediate step toward 3D MCs. Due to the exploitation of the 3D space for controlled propagation of SWs, they are interesting for research and promising for technological application in digital and analog signal processing, performed in large-scale integrated systems.

In the following sections, we will discuss the results for selected 1D+ and 2D+ structures investigated theoretically and experimentally, which we consider as promising toward the formation of full 3D MCs. The 1D and 2D magnonic systems have usually the form of a layer with an in-plane arrangement of constituent elements. These 2D planar structures are uniform in the third (out-of-plane) direction. The structuralization in the third direction can by introduced to 2D structures by shifting out-of-plane the elements (dots, stripes) forming the multicomponent structure being initially planar. This approach allows observing the gradual transition from a 2D to a 2D+ structure. The other strategy is to combine two or more 2D systems in the stack, as shown in Fig. 2.1. If the mentioned stack structure is formed by a large number of repetitions of the 2D+ system, then we obtain a 3D structure that will have the form of a 3D MC.

In the next section (Section 2.2.1), we will discuss the results of theoretical studies of SW dynamics in 2D+ MCs, where the ferromagnetic inclusions (dots) were lifted up with respect to the layer of the ferromagnetic matrix. Then (in Section 2.2.2) we will present the theoretical and experimental results for the hybrid 1D+ and 2D+ structures where the 1D or 2D MCs are coupled to the homogeneous ferromagnetic layer. In Section 2.3 we will get an overview of the theoretical results on the magnonic band structure in 3D MCs. Finally, in Section 2.4, we will conclude with prospects for further development toward 3D MCs.

2.2 Control of Spin-Wave Dynamics in 1D+ and 2D+ Magnonic Crystals

SW dynamics can be tailored by structural factors and further tuned by an external bias. The extension of planar 1D or 2D MCs along the third dimension can be realized by a relative shift of their components (the inclusions and the matrix) or by adding the uniform subsystem (the layer). These geometrical changes offer an additional degree of freedom for shaping the dispersion relation of SWs. Due to the complex geometry of 1D+ and 2D+ MCs, different shapes of the constituent elements, the switching of static magnetic configuration in a uniform and a periodic component of these hybrid structures happens at different fields and thus can be used for additional control of the SW dynamics in 1D+ and 2D+ MCs.

2.2.1 Bicomponent 2D+ Magnonic Crystals with Nonmagnetic Spacers

The typical geometry of a 2D MC is a planar system in the form of an antidot lattice (ADL) or a bicomponent MC, homogeneous in the direction normal to the plane of the structure. For the bicomponent system, the interesting issue is the possibility of tuning of the coupling between the constituent elements of the structure: matrix and inclusions. This coupling can be changed when the inclusions will be partially lifted up from the layer of the matrix, which leads to inhomogeneity in the third dimension and the formation of 2D+ MC. We have considered in Ref. [30] two scenarios: (i) with the inclusions in direct contact with the matrix that sustains the exchange coupling and (ii) the case with the inclusions separated from the matrix by

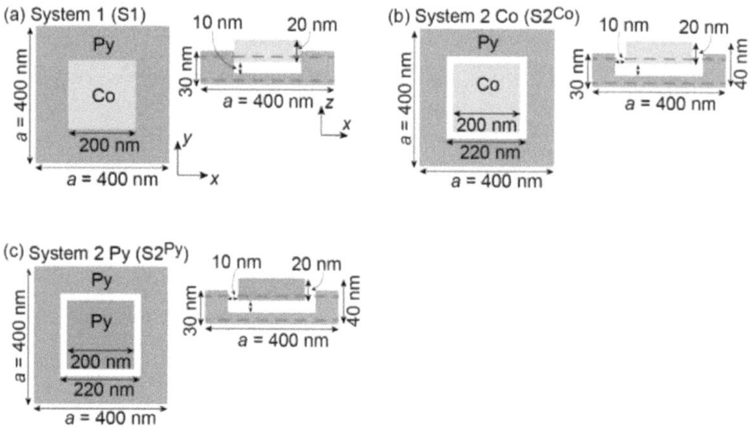

Figure 2.2 Schema (top and side views) of primitive cells of 2D+ MCs under study: (a) S1, (b) $S2^{Co}$, and (c) $S2^{Py}$. Reprinted (figure) from Ref. [30]. Copyright (2015) by the American Physical Society.

nonmagnetic spacers breaking the exchange interaction between the components of the structure (see Fig. 2.2a,b). The introduction of the nonmagnetic spacers increases the demagnetizing effects at the interfaces and enhances the presence of SW edge modes localized in close vicinity of the interfaces.

The considered structures consist of a 30-nm-thick Py film with an array of 20-nm-deep square cavities with the lattice constant of $a = 400$ nm. In the bottom of each grove, a 10-nm-thick nonmagnetic spacer is located, and on the top of each, a 20-nm-thick square dot of 200-nm-long sides is placed. In the study three types of 2D+ MCs were considered: (a) System 1 (S1): Co dots in direct contact with the Py matrix only at the lateral faces (b) System 2-Co ($S2^{Co}$): A similar MC as S1 but with the additional 10 nm nonmagnetic spacer around the lateral faces of dots (c) System 2-Py ($S2^{Py}$): Equivalent of $S2^{Co}$ but with Py dots. Schemes of the structures are presented in Fig. 2.2.

The influence of nonmagnetic spacers (between inclusions and matrix) on the SW spectra was studied by means of the dynamical matrix method (DMM) [11, 64]. The DMM is a finite difference micromagnetic approach used to solve the eigenvalue problem for the Landau–Lifshitz (LL) equation on a static magnetic configuration

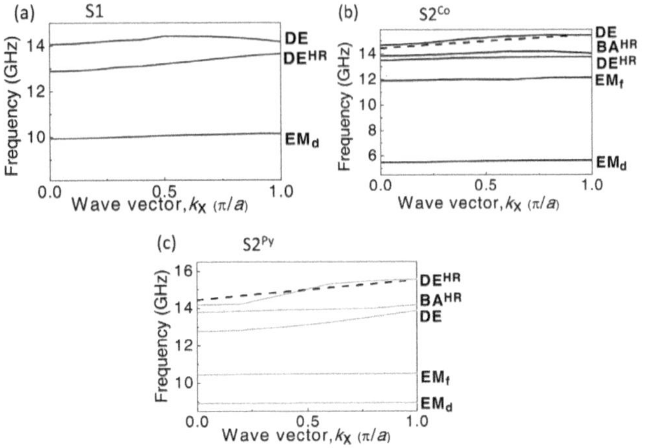

Figure 2.3 Dispersion relations in the first Brillouin zone (FBZ) along the direction perpendicular to the external magnetic field for (a) S1, (b) $S2^{Co}$, and (c) $S2^{Py}$. The black dashed line corresponds to the analytically derived dispersion relation of the DE mode in a homogeneous Py film of 10 nm thickness. Reprinted (figure) from Ref. [30]. Copyright (2015) by the American Physical Society.

obtained by means of micromagnetic simulations (e.g., OOMMF [9]). Using that method, by a single calculation a full SW spectrum (the SW modes frequencies and eigenvectors of any symmetry) can be computed directly in the frequency domain.

The full SW dispersion relations along the x direction obtained by means of DMM for all three structures (i.e., uniformly magnetized by the external magnetic field of value 2000 Oe along the y axis S1, $S2^{Co}$, and $S2^{Py}$) are presented in Fig. 2.3. In our analysis, we focus only on its low-frequency part within the first Brillouin zone (FBZ).

For the system S1 (Fig. 2.3a), we can see consecutively three main dispersion lines: end (edge) mode of the dot (EM_d), Damon–Eshbach-like (DE-like) mode, and DE-like mode in horizontal rows (DE^{HR}). We call these states DE-like modes due to their similarity to the DE mode in thin films. The nodal planes of DE-like modes (calculated at the center of the BZ) are parallel to the local static magnetization as it is known for typical DE modes [54] in planar MC. The introduction of the nonmagnetic spacer below Co dots results in the appearance of the end(edge) mode and the different SW amplitude distribution in the vicinity of the Co/Py interface of DE and DE^{HR} modes differs the S1 and the Co/Py MCs studied in Refs. [54, 64].

For the system $S2^{Co}$ (Fig. 2.3b), the incorporation of the nonmagnetic spacer at the sides of the Co dot causes the creation of two new bands corresponding to (i) the end (edge) mode of the Py film (EM_f) localized at the film's border and (ii) the backward-like mode (BA^{HR}) mainly concentrated in the horizontal rows that exhibits nodal planes perpendicular to the local static magnetization. The frequencies of the BA^{HR} mode are higher than the frequency of DE^{HR}, and the frequencies of DE^{HR} and DE modes are slightly increased with respect to corresponding modes in S1. The introduction of the spacer at the dots' edges results in a significant decrease with respect to S1 of the frequency of EM_d. The frequency of the new EM_f is a few GHz higher than the frequency of EM_d.

In the $S2^{Py}$ (Fig. 2.3c), the replacement of dots material, Co by Py, causes that the frequency of DE and EM_f bands drop down. In this scenario, the frequency of the DE mode is lower than the frequency of the DE^{HR} and the frequency of EM_f is much closer to that of EM_d. The frequencies of BA^{HR} are lower than the ones of DE^{HR}. In the spectra of the $S2^{Py}$ the sequence of frequencies of the subsequent modes is different (i.e., DE, DE^{HR}, and BA^{HR}) than in the case of the previous systems. In particular, the frequencies of the DE mode are lower than the frequencies of the DE^{HR} mode, similarly to the case of antidot lattices (ADLs) (one-component MCs) reported in Ref. [65]. Although $S2^{Py}$ is a 2D+ MC consisting only from one ferromagnetic material, it preserves features characteristic for 2D+ bicomponent MCs.

The main result of the study on the influence of the nonmagnetic spacer around the dots embedded in ADLs on SW dynamics reported in Ref. [30] is that such kind of 2D+ bicomponent MCs can be generalized to a separate kind of MCs different than 2D MCs. The nonmagnetic spacer breaks the exchange interaction between the magnetic materials of the matrix and the dots and, what seems to be more important, modifies the static effective magnetic field. That modification of the field is mainly at the dot edges. In consequence, two different edge modes appear in a single structure, that is, end modes of the film and of the dots. Moreover, employing an MC consisting only of a single material (both matrix and dots made from the same material, for example, Py), a one-component MC can be designed preserving the main features of the bicomponent counterpart. Interplay with the dot material and nonmagnetic spacer dimensions allows to further tailor the SW spectra of MCs in different ways.

2.2.2 Hybrid 1D+ or 2D+ Magnonic Crystals

By coupling the planar MC to the uniform ferromagnetic plane, we can achieve additional functionalities that do not exist in 1D or 2D MCs. In such hybrid 2D+ structures (see Fig. 2.1), the SW propagation in a uniform ferromagnetic film can be modified by dynamic magnetic interactions with the periodic component. Or on the other hand, the planar MC can induce the periodic magnetic landscape in the uniformly magnetized magnetic layer by the stray static magnetic fields, which also modifies SWs. Moreover, the mutual interactions (both static and dynamic) between both subsystems can also exist. For purely dynamic dipolar coupling between the both subsystems of the structure, we can easily exploit the reprogrammability feature. That is, the relative orientation of static magnetization between both subsystems can be switched from antiparallel (AP) to parallel (P) alignment or vice versa with the change of the in-plane bias magnetic field, which, in turn, allows to control the SW dynamics. Additionally, due to dynamic interaction between subsystems, the SW induced in one part of the system can be transferred to the second component of the structure.

2.2.2.1 Reprogrammability of Magnonic Band Structure in 2D+

Bandgap engineering plays a fundamental role in electronics, photonics, and also magnonics. Magnonics offers large flexibility in tuning the environmental properties for its carriers. The classical way to engineer band structure and bandgaps is a patterning of the thin films [14, 22, 39], which, however, is very receptive for defects and changes of the magnetic properties near the edges, which, in turn, can affect SW propagation in an uncontrollable manner [20, 44]. An alternative way is ion implantation, which modifies the material properties relevant for SW propagation [3, 59] and allows also to create the fine patterns suitable for the formation of magnonic bandgaps [40]. Although it does not etch the film, the crystallographic structure and composition are significantly affected, which usually is associated with increased damping. The other interesting idea for the formation of the magnonic band structure is based on the exploitation of the vertical dynamic coupling of propagating SW in a homogeneous film with patterned elements placed in its proximity. In Ref. [2] the control of the SW propagation in a ferromagnetic stripe by a change of the magnetization orientation in the bar placed above it has been demonstrated. This mechanism was also exploited to excite and to control the phase and amplitude of propagating SWs. This idea was further extended to an array of ferromagnetic nanodots deposited over the ferromagnetic film. In this system, the magnetization dynamics pumped in the array of dots by the microwave field has been used to induce and detect short-wavelength SWs propagating in a homogeneous film [63].

The formation of the magnonic band structure in a homogeneous Py film due to dynamic coupling with the magnetization oscillation in the array of stripes (1D+ MC) and dots (2D+ MC) deposited near the film surface has been demonstrated in two of our recent papers [10, 30]. In Ref. [30] we have shown that the vertical coupling between the array of Ni stripes and the Py thin film deposited below is sufficient for the formation of the SW band structure; thus the system represents a 1D+ MC according to our definitions (Fig. 2.1). In that study, the magnetic field, used to saturate the ferromagnetic materials, was aligned along the stripes. It prevented the coupling between the Py thin film and Ni stripes by the static demagnetizing fields. In Ref. [10], we investigated the hybrid structure composed of a Py thin film and the 2D array of the dots deposited on its top, which is an example of a 2D+ MC. The studied structure consisted of a Py thin film (thickness 20 nm) with the rectangular lattice of Py elliptical nanodots (with semiaxes 100 and 300 nm) 10 nm thick. The dots were separated by a 10-nm-thick nonmagnetic Pt spacer. The scanning electron microscopy (SEM) image of the structure is

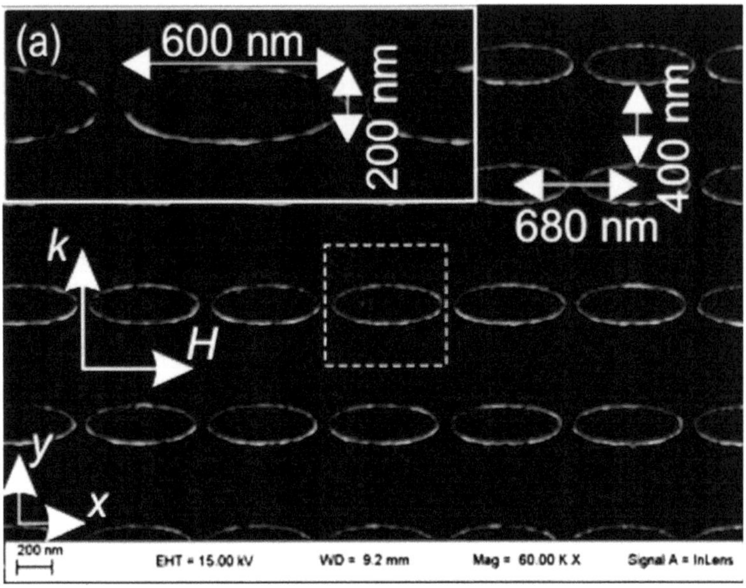

Figure 2.4 SEM image of Py elliptical dots deposited on the Py thin film. The direction of the applied magnetic field **H** and of the wave vector **k**, together with the Cartesian axes, are also represented by the arrows. Dashed white line marks the unit cell of the structure. Reprinted (figure) from Ref. [10]. Copyright (2018) by the American Physical Society.

shown in Fig. 2.4. Interestingly, the elliptical shape of the nanodots allows controlling relative magnetization orientation between the film and ellipses. This property is explored here to demonstrate the reprogrammability of the dipolarly induced magnonic band structure in the homogeneous Py film.

By the change of the magnetic field magnitude along the long axis of the ellipses, we can control the relative magnetization direction between nanodots and the layer, from P to AP. The hysteresis loop measured with the magneto-optical Kerr effect (MOKE) being shown in Fig. 2.5 confirms this hypothesis. The magnetization remains at the AP orientation for **H** covering the plateau visible in Fig. 2.5, that is, between −2 and −22 mT, when the field is decreased from large positive values. When the magnetic field is changed from the negative orientation, at the same range of **H**, the magnetizations are in the P state. In the following section, we will investigate

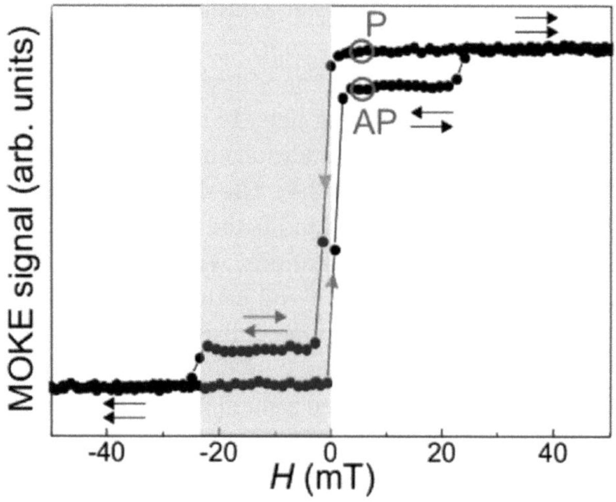

Figure 2.5 MOKE loop measured with the magnetic field along the long axis of the ellipses (x direction). Red circles mark the 5 mT field at which the SW dispersion in P and AP states was measured and is shown in Fig. 2.6a and Fig. 2.6b, respectively. Reprinted (figure) from Ref. [10]. Copyright (2018) by the American Physical Society.

experimentally and numerically the SW band structure at a 5 mT field in the P and AP configurations.

The dispersion relation of SWs was measured with Brillouin light scattering (BLS) spectroscopy from thermally excited SWs in the back-scattering configuration by focusing a monochromatic laser beam of wavelength 532 nm on the sample surface through a camera objective of numerical aperture (NA) 0.24. The scattered light was analyzed in frequency by a (3 + 3)-tandem Fabry–Perot interferometer. The SW dispersion was then measured at fixed $\mathbf{H} =$ 5 mT by sweeping the wave vector in the direction perpendicular to $\mathbf{H}$ in the range between 0 and 2×10^7 rad/m, which allows to map the SW dispersion along the y direction up to the third BZ of the reciprocal space in the DE configuration, it is when the wave vector is perpendicular to the saturation magnetization direction.

To explain the measured spectra and identify SW excitations, we solved the linearized LL equation with coupled Maxwell's equations in magnetostatic approximation on the dynamic components of the magnetization vector (m_y, m_z) in a frequency domain with damping neglected. We assumed full magnetic saturation and considered the effective magnetic field to be the sum of the direct current (DC) bias magnetic field **H** applied along the x direction, the exchange field, and the dynamic demagnetizing field. In calculations, we assumed the Bloch periodic boundary conditions on the borders of the unit cell in the plane (x, y) (see Fig. 2.4). The definition of the exchange and demagnetizing field terms and details of the computation can be found in Ref. [37]. In calculations, we assumed the exchange constant $A = 1.3 \times 10^{-11}$ J/m and saturation magnetization $M_S = 860$ kA/m, for which the calculated dispersion relation of SWs in the reference sample (20-nm-thick Py film) fits very well with the experimental one. We also added a small magnetic anisotropy term of value 10 mT parallel to the **H** in nanodots to fit confined mode frequencies with the experimental data. We solved the LL equations in 3D space using the finite element method for the in-plane unit cell and with a large free space above and below the Py along the z axis. From solutions of the LL equation, we found frequencies of the SWs for the successive wavenumbers k along the y axis. We also obtained the spatial distribution of the SW amplitude (m_y, m_z), and from the m_z we calculated the BLS intensity of the SWs [18].

The measured and calculated magnonic band structures for P and AP magnetic configurations at $\mu_0 H = 5$ mT are shown in Figs. 2.6a and 2.6b, respectively. These band structures are apparently different, which proves that SW spectrum of the considered system can be controlled by an external magnetic field. There are two bandgaps, one formed at the BZ border and the second at the smaller wavenumbers. The first is due to periodicity, while the second is due to the interaction of the propagating wave of the continuous film with standing resonant modes of the nano-ellipses. In both cases, the predominant role in the formation of the magnonic band structure and opening bandgaps is played by a vertical dynamic coupling between the propagating wave and magnetization oscillations in the nanodots.

In Fig. 2.6a the DE band is not experimentally detected at $k \approx 0$, it is due to the broad elastic peak. In measurements, the first standing mode in nanodots extends over the whole range of k with a slight frequency variation from 4.2 to 4.7 GHz. In simulations, this mode has a similar frequency; however, its intensity outside the

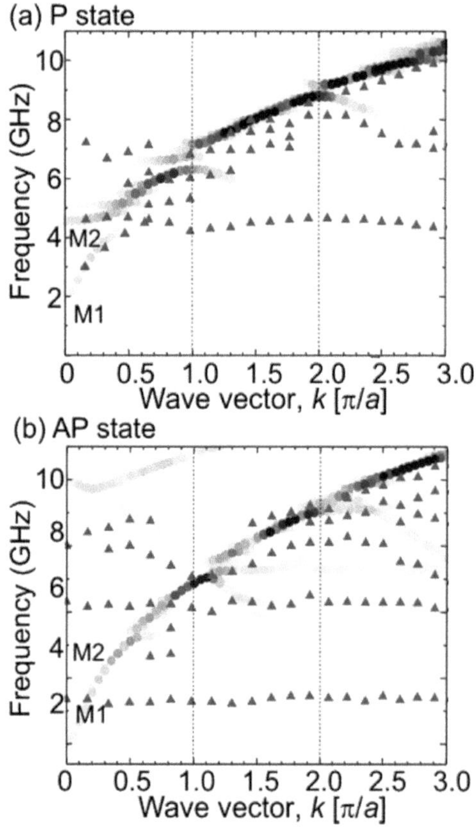

Figure 2.6 Dispersion relation of SWs propagating perpendicular to the magnetic field of value 5 mT aligned along the x axis for (a) parallel (P) and (b) antiparallel (AP) orientation of the static magnetization in nanodots and the film. Triangles mark the measured BLS data; dots show the numerical results. The color intensity of the numerical points indicates the calculated BLS intensity. Horizontal dashed lines mark the BZ boundary and center. Reprinted (figure) from Ref. [10]. Copyright (2018) by the American Physical Society.

FBZ is small. This discrepancy may be attributed to the approximate formula used for BLS intensity calculation, which does not take into account features of the confined modes. At higher k values the DE mode dominates in the spectra. The profile of the DE mode and of the first standing mode in the nanodot are shown in Fig. 2.7a

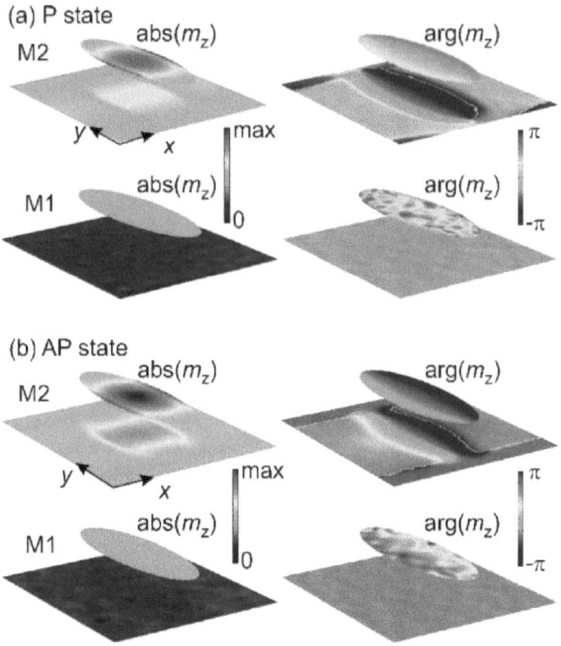

Figure 2.7 Calculated amplitude $|m_z|$ and phase arg(m_z) for SWs in the P (a) and AP (b) states of static magnetization. The calculations are for the DE wave (M1) and fundamental excitation in nanodots (M2) marked in Fig. 2.6 at $k = 0$. Reprinted (figure) from Ref. [10]. Copyright (2018) by the American Physical Society.

for the P state. The DE mode (M1) has amplitude concentrated mainly in the film, while the standing mode of the nanodot (M2) forces oscillations also in the Py film. In the film, the amplitude is weaker than in the nanodot, but oscillations in the film have the phase shifted by π with respect to the oscillation in nanodots. In the AP state (Fig. 2.6b), the two nondispersive lines are visible in BLS measurements through the whole k range, which is attributed to the standing modes in nanodots. The lowest is measured at around 3.4 GHz and the second at 6.3 GHz, both in agreement with calculations. However, their calculated intensity is low and they are not visible in the figure. In the experiment the DE mode becomes visible only from the second BZ, where the agreement with simulations is satisfactory.

In summary, the spectra for AP and P configurations, as well as the frequency of the standing modes in the nanodots, are significantly different, which confirms the reprogrammability in the investigated structure. Interestingly, in the AP state the bandgaps are much narrower than in the P state. The change of the standing waves frequency with respect to the P state shall affect also the mode M1 by the change of the frequencies at which hybridizations and additional bandgaps occur. Unfortunately, the confirmation of this effect is found in calculations but is invisible in BLS due to the low intensity of the modes.

2.2.2.2 Magnonic directional couplers in hybrid 1D+ magnonic crystals

We have shown above that the hybridizations of the DE mode of the thin film with the standing modes in elliptical nanodots (Fig. 2.6) opens an additional bandgap for the DE mode. This means that the DE mode excited at the frequency of the bandgap transfers the energy to the standing mode of the ellipses along the path of the propagation in the film. If instead of the hybridization with the standing mode, the DE mode will hybridize with another propagating mode, this opens the possibility to exchange the energy between two propagating modes and transfer the energy in the vertical direction.

The phenomenon of coupling and SW energy transfer between two identical uniform ferromagnetic films has already been discussed [49]. The effect was experimentally shown two years ago when the side coupling between two magnonic waveguides was investigated by the BLS technique [48]. Recently, the application of that effect, widely used in photonics for many years [56], has been proposed for signal processing with two coupled magnonic waveguides [58]. In this case, the resonant interaction between SWs propagates in the identical waveguides. If the layers are different then the SWs are decoupled due to the mismatch of wave vectors and each SW becomes confined to a single layer. However, the existence of the other layer still affects the dispersion of the SWs at some range of frequencies. We can couple them again by introducing the periodicity of the structure. Folding of the bands (effect of the

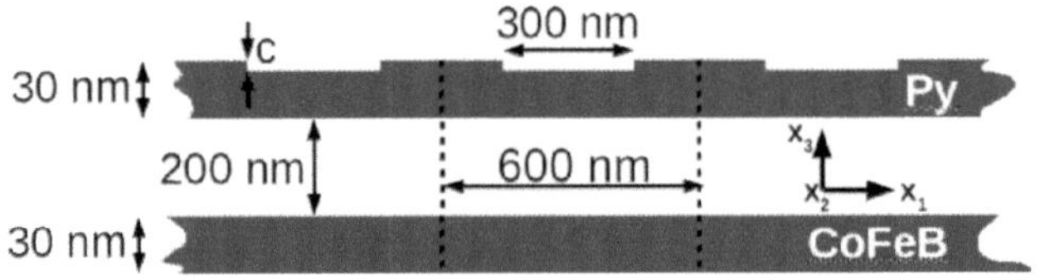

Figure 2.8 Scheme of the bilayer system composed of dipolarly coupled CoFeB and a Py film with the periodically corrugated surface. Dashed lines indicate the unit cell. https://creativecommons.org/licenses/by/3.0/.

periodicity) to the FBZ will result in multiple crossings of the bands that originate from different layers. Then, by tuning the distance between layers, we can achieve relatively strong hybridization between the SW bands. The crossing bands can have the same or opposite sign of the group velocity. In the first case, we call them codirectional couplings and in the second contradirectional couplings. To sum up, the criteria of efficient power exchange between different ferromagnetic media (waveguides or layers) are the following: (1) resonance, that is, equality of frequencies; (2) phase matching, that is, equality of wave vectors; and (3) strong interaction between layers.

We will consider the bilayer 1D+ system (Fig. 2.8) consisting of a CoFeB 30-nm-thick homogeneous film dipolar coupled to a 30-nm-thick Py patterned film. The Py film is patterned with a regular array of grooves of 300 nm width and a depth of either 5 nm (Py5) or 15 nm (Py15). The grating has a lattice constant 600 nm. The periodically patterned Py layer can itself be formally considered as a 1D+ system where Py stripes are exchanged and the dipolar coupled to the Py underlayer. Such dimensions make the structure feasible for fabrication and SW detection achievable by conventional techniques, such as BLS, X-ray magnetic circular dichroism (XMCD) [12], or SW spectroscopy [7, 29]. We assume that the layers are infinite along the x_2 coordinate. The separation between Co and Py layers has been optimized for sufficient strong interaction at the points of lowest band crossings and its value was fixed to 200 nm. To study coupling in the bilayered structure the Landau–Lifshitz–Gilbert equation has been solved either in the frequency domain with the approach described in the previous section or in the time domain with micromagnetic simulations.

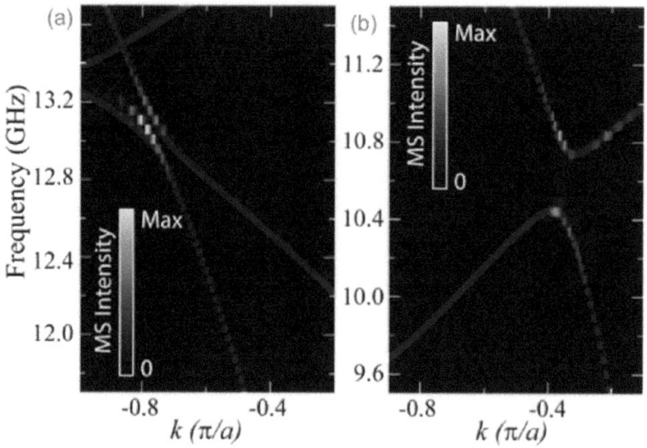

Figure 2.9 Dispersion relations at the vicinity of analyzed crossings for (a) Py5 (5-nm-deep groves)/CoFeB and (b) Py15/CoFeB bilayers. Dots result from finite element simulations, and they are superimposed on the results of the micromagnetic simulations. https://creativecommons.org/licenses/by/3.0/.

The branches that originate from the dispersions of two different films anticross at some k points. Figure 2.9a shows the hybridizations of DE modes from Py and CoFeB layers at 13 GHz for codirectional coupling and Fig. 2.9b at 10.6 GHz for contradirectional coupling.

To demonstrate the transfer of energy from one layer to the second, we put the line source of the radiofrequency magnetic field in time domain simulations at $x_1 = 28$ μm in the CoFeB film (see dashed horizontal line in Fig. 2.10 and Ref. 2.11). The CoFeB film exists along the whole x_1 range, while Py spreads out only from $x_1 = 0$ to $x_1 = 25$ μm.

To demonstrate codirectional coupling the corrugation of the Py film (grooves' depth) is select as $c = 5$ nm (Py5). Figure 2.10a shows the magnetization component m_{x1} after 15 ns from the excitation. The amplitude of the SW that goes to the left is decreasing from the maximum value at $x_1 > 25$ μm to almost zero at $x_1 = 14.5$ μm. Its wavelength is 1.51 μm. At the same distance, the SW in Py appears to reach a maximum amplitude at $x_1 = 14.5$ μm. Therefore, the energy of the wave from CoFeB has been vertically transformed

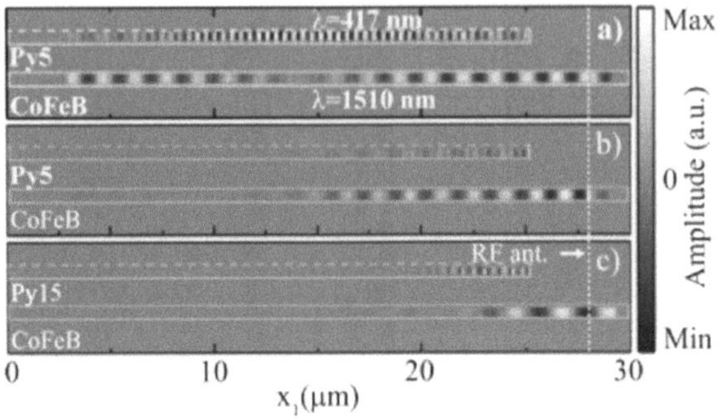

Figure 2.10 Spatial distribution of the dynamical magnetization component m_{x1} after 15 ns of sinusoidal excitation by antenna (vertical dashed line) for (a) Py5/CoFeB without damping, (b) Py5/CoFeB with damping, and (c) Py15/CoFeB with damping. Lines indicate boundaries of the layers; the dashed line indicates the periodic surface of the Py layer. https://creativecommons.org/licenses/by/3.0/.

into the energy of the wave in Py5 completely at a distance of 10.4 μm. The wave induced in Py has a wavelength of 417 nm, that is, almost 4 times shorter than that which excites it. When the damping is introduced, the effect of energy exchange between layers is strongly suppressed. As shown in Fig. 2.10b, the short wave in Py5 is still excited, but neither a clear maximum amplitude in Py nor a corresponding minimum in CoFeB is visible. The damping length for the Py film is of about 2–3 μm; thus it is much smaller than the length at which transfer from CoFeB occurs. At a given distance, the Py film loses more energy than it gets from CoFeB and full transfer of energy and its accumulation cannot take place. We recover accumulation of energy in the Py layer by increasing the coupling coefficient between layers. For this purpose, we increase the corrugation depth of Py to $c = 15$ nm (Py15). The time-dependent simulations are presented in Fig. 2.10c. The wave in CoFeB vanishes much faster with the distance because it transfers the energy to Py in a shorter time. The excited wave in Py grows, and the amplitude reaches the maximum value at $x_1 = 23$ μm. Therefore, the energy is clearly accumulated in Py despite damping.

The contradirectional coupling between SWs in a bilayered system has been demonstrated for the same CoFeB film and Py film but with 15-nm-deep grooves (denoted as Py15). The coupling strength, in this case, is 0.476 μm^{-1}, which gives a transfer length of 3.34 μm. We performed time domain simulations with the 10.6 GHz excitation at the source line in the system similar to that discussed above for codirectional coupling. The difference is that we introduced the fragment of the homogeneous Py layer at the x_1 between 20 and 25 μm with strong damping close to the right edge. This allows us to observe the wave going out of the periodic structure toward the right direction. Figure 2.11a shows the results of simulations after 5 ns of excitation. Clearly, the wave of a wavelength of about 700 nm is excited in Py15 by the SW of wavelength 3.85 μm in CoFeB. However, the amplitudes of both waves decay with the distance. The distance at which amplitude decreases by a factor of $1/e$ is about 4 μm. At the output (in homogeneous part of Py) we observe the wave that propagates to the right, as shown in Fig. 2.11a. Therefore, the energy of the wave in CoFeB that propagates to the left has been vertically transferred to the energy of the wave in Py, which propagates to the right.

With the damping introduced (see Fig. 2.11b), the amplitudes of both waves in Py15 and CoFeB decrease. However, the transfer of SWs from the CoFeB to the Py15 film is still visible.

2.3 3D Magnonic Crystals

The 2D+ MCs, discussed in the previous section, are planar structures with out-of-plane inhomogeneity. For these systems, the in-plane dimensions (e.g., period of the structure) are much larger that out-of-plane dimensions (e.g., thicknesses of layers). The 3D MCs are periodic in all three dimensions and are considered bulk structures where all three dimensions are in the comparable range of sizes.

The reports on SW dynamics of 3D MCs are quite rare in the literature so far. The most challenging problem related to experimental studies on 3D MCs is the fabrication of a high-quality structure with periodicity along three directions in the nanoscale. Due to the lack of experimental papers, there are just a few theoretical works that have studied such systems. However,

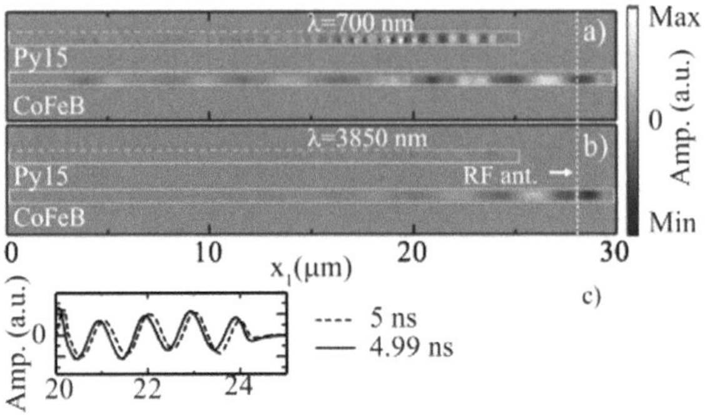

Figure 2.11 Spatial distribution of the m_{x1} component after 5 ns of sinusoidal excitation at 10.6 GHz (antenna is marked with the dashed vertical line) in (a) Py15/CoFeB without damping and (b) Py15/CoFeB with damping. Lines indicate boundaries of the layers; the dashed line indicates the periodic surface of the Py15 layer. (c) The m_{x1} component of the dynamical magnetization in the two consecutive time intervals in the region of uniform Py layer, which demonstrates the direction of propagation toward the positive x_1 axis. https://creativecommons.org/licenses/by/3.0/.

presented results are interesting in terms of new phenomena and promising for prospective applications. Below, we summarize the main findings related to the complete (omnidirectional) magnonic bandgap existence in the case of 3D MCs. We also recapitulate our results on the subject of magnetoferritine (mFT)-based 3D MCs—thinkable candidates for experimental realization. But let us start with the description of the theoretical method used.

2.3.1 The Model

In the classical continuous-medium approach the dynamics of the space- and time-dependent magnetization vector $\vec{M} = \vec{M}(\vec{r}, t)$ can be described by the LL equation:

$$\frac{\partial \vec{M}}{\partial t} = -|\gamma|\,\mu_0 \vec{M} \times \vec{H}_{\text{eff}}, \tag{2.1}$$

where γ is the gyromagnetic ratio and μ_0 is the vacuum permeability. In the majority of MC studies the damping is neglected as it is in Eq. 2.1. The vector $\vec{H}_{\text{eff}}$ is the effective magnetic field that contains few different components such as magnetostatic field, exchange field, magnetic anisotropy, and static external magnetic field $\vec{H}$. Different formulations of the effective field components are discussed, for example, in Ref. [27]. The usual assumption is that $\vec{H}$ is strong enough to saturate the magnetization within the entire sample, which allows to use linear approximation. In this case, in the Cartesian coordinate system with the x axis oriented along the external field, the magnetization vector can be expressed as $\vec{M} = [M_S, m_y, m_z]$, where M_S is the saturation magnetization and m_y and m_z are two dynamic components.

One of the approaches in magnonics to solve the LL equation in MCs is the plane wave method (PWM), which transforms the linearized LL equation together with Maxwell equations to reciprocal space, that is, the space of the wave vector $\vec{k}$. The method involves two transformations that require ideal periodicity and allow to consider only the unit cell with the periodic boundary conditions. To the dynamic functions, such as dynamic components of magnetization or the demagnetizing field, Bloch's theorem applies. Material parameters, for example, saturation magnetization M_S or exchange stiffness constant A, as periodic in real space are Fourier-expanded. The final set of equations is equivalent to the algebraic eigenvalue problem with the block matrix of the following form:

$$\hat{M} = \begin{bmatrix} \hat{M}_{yy} & \hat{M}_{yz} \\ \hat{M}_{zy} & \hat{M}_{zz} \end{bmatrix}_{2N\times 2N}, \tag{2.2}$$

where N is the number of plane waves used in the Fourier/Bloch expansion.

The particular expression for each element of the matrix 2.2 depends on the specific problem. For the first time the PWM was used for bicomponent 2D MCs treated as infinite in the third direction with the external field parallel to the rods [57]. The method was adapted to finite-thickness thin-film MCs with the external field aligned in plane (parallel to the surface) [21, 24] and out of plane (normal to the surface) [44]. The 2D MC with a metal overlayer was taken into account in Ref. [50]. In the case of ADL the problem arises due to the absence of the magnetic material inside antidots, which makes the LL equation undefined in the area of the antidots. The problem was solved by introducing an artificial magnetic material with saturation magnetization much smaller than in the material of the film. To eliminate unphysical modes excited in this material one can shift these modes up to high frequencies by introducing strong uniaxial magnetic anisotropy inside the antidots [20, 38]. The full 3D version of the PWM was derived in Ref. [26] and extended in Ref. [32].

The final step of the method is numerical diagonalization of the matrix 2.2. Calculated eigenvalues give information about SW frequencies, while eigenvectors $\vec{m}_{\vec{k}}(\vec{G})$ are coefficients of the Bloch expansion of the dynamic magnetization component. Eigenvectors let one calculate the SW profile, that is, the spatial distribution of the dynamic magnetization for a given mode and $\vec{k}$. Additionally, eigenvectors allow to estimate the intensity of the Brillouin light scattering (BLS) signal. In the first approximation, BLS intensity is proportional to the square modulus of the fundamental harmonics of the dynamic magnetization $\vec{m}_{\vec{k}}(\vec{G} = 0)$.

2.3.2 Complete Magnonic Bandgaps

An example of a 3D MC is shown in Fig. 2.12, where the face-centered cubic (fcc) lattice is considered (see panel b). The SW spectrum shown in Fig. 2.12a is obtained for magnetite/maghemite spheres of diameter 8 nm embedded in the cobalt matrix. The lattice constant is 14 nm and the material parameters are given in Table 2.1. All calculations in 3D MCs are done for the external field 100 mT strong enough to saturate the sample. The spectrum is plotted along the high-symmetry path in the FBZ shown in Fig. 2.12c. There are two complete magnonic gaps in this case: the first gap, between two lowest modes, around 300 GHz wide, and the second gap, between the 5th and 6th modes, around 70 GHz wide.

The mechanism of the opening of both gaps as well as the influence of structural and material parameters on their existence were studied for the very first time in the theoretical work by Krawczyk and Puszkarski [26]. Both gaps were found to open between the modes excited in different materials, that is, between one mode excited in spheres and another one excited in the matrix. This finding suggests that magnetic properties of the constituent

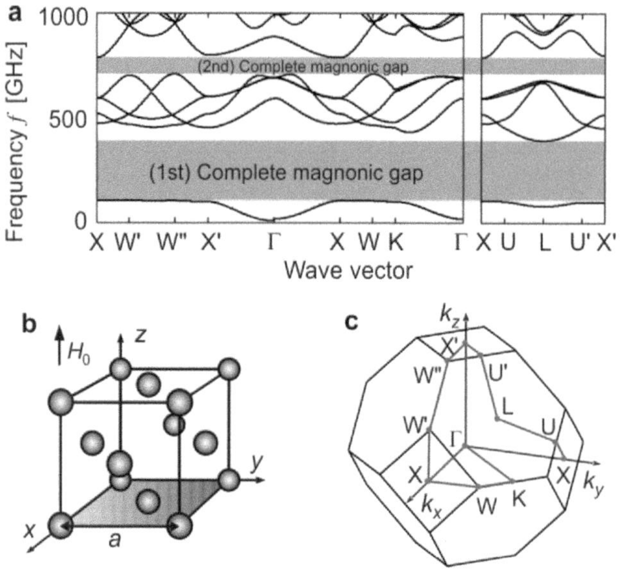

Figure 2.12 (a) The SW spectrum of the MC consists of magnetite/maghemite spheres with diameter 8 nm embedded in an iron matrix and arranged in an fcc lattice with lattice constant 14 nm. The spectrum is plotted along the high-symmetry path in the FBZ shown in (c). Shaded areas represent complete magnonic gaps. (b) Schematic plot of the fcc structure with coordinating system used. (c) The FBZ for the fcc lattice with the high-symmetry path marked by gray lines and high-symmetry points marked by full dots.

materials should be crucial for the occurrence and width of such gaps. The main results of Ref. [26] concerning the influence of material parameters are shown in Fig. 2.13a, where we combine the position and width of the first gap in its dependence on the saturation magnetization contrast ($M_{S,A/B} = M_{S,A}/M_{S,B}$) for MCs based on three cubic lattices: simple cubic (sc), body-centered cubic (bcc), and fcc. The lattice constant is 10 nm, and the filling fraction $ff = 0.2$ (the filling fraction is defined as the ratio of the volume of inclusions in the elementary cell to the cell volume; thus the sphere radius is 3.63 nm for sc, 2.88 nm for bcc, and 2.29 nm for fcc). The contrast is controlled by changing saturation magnetization in spheres, $M_{S,A}$. The magnetization in the matrix is fixed at its value for

Table 2.1 Values of material parameters: Spontaneous magnetization M_S, exchange stiffness constant A, and exchange length λ_{ex}, in the materials considered for 3D MCs [32]

Material:	Fe	Co	Py	YIG	mFT[a]
M_S [kA/m]	1752	1390	810	194	346
A [J/m] 10^{-11}	2.1	2.8	1.1	0.4	1.0
λ_{ex} [nm]	3.30	4.80	5.17	13.01	11.53

[a]Values based on [19].

YIG and both exchange stiffness constants are set to iron. Increasing of the contrast results in widening of the gap and there is a threshold value below which the gap does not occur: around 2.8 in fcc, 3.2 in bcc, and 4.8 in sc. If the gap exists its width is highest for fcc and smallest in sc, which is consistent with threshold values. The existence of such a threshold is found also for the hexagonal (sh) lattice [23].

In Fig. 2.13b we combine the first gap for three regular lattices versus the structural parameter, namely the filling fraction. The MC consists of Fe spheres embedded in a YIG matrix and the lattice constant is 10 nm. Calculations are done for the ff ranges from 0, that is, homogeneous YIG, up to its maximum for the particular lattice, that is, the MC with spheres touching each other (dashed lines in Fig. 2.13b). For all lattices the best conditions for the gap existence are for the intermediate values of the ff. And again, the widest gap occurs for the fcc lattice, while the narrowest occurs for the sc lattice.

Both features, that is, the existence of the contrast threshold and the fcc lattice as the most convenient for the gap opening, have the same origin, which we explain in Ref. [32]. Complete gaps are *indirect* gaps spanned between frequencies obtained for different directions of the SW propagation. Therefore, their width strongly depends on the difference between the length of the wave vector at the FBZ border for different directions—a smaller difference gives a wider gap. In other words, the more the FBZ shape is close to the sphere the better are conditions for the complete gap opening and the lowest is the minimal contrast of the material parameters necessary for its occurring.

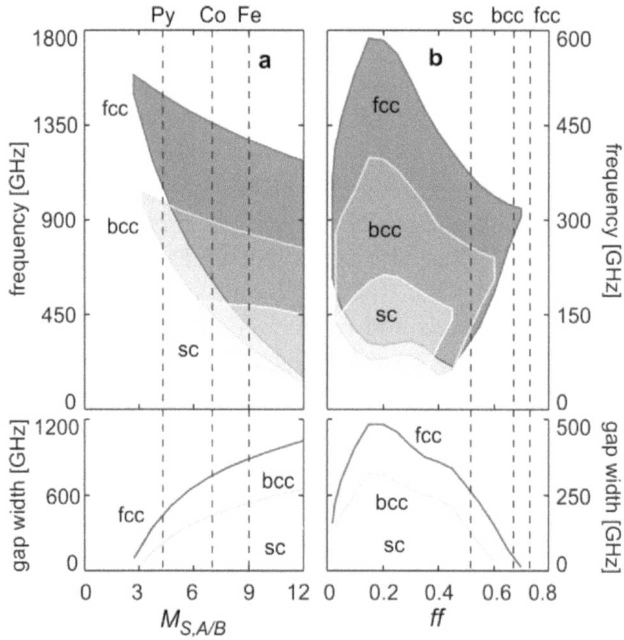

Figure 2.13 Lowest gap in 3D MCs for three cubic lattices vs. (a) the saturation magnetization contrast $M_{S,A/B}$ and (b) the ff. Dashed lines in (a) depict the magnetization contrast for particular materials of spheres and in (b) the maximum of the ff, that is, for touching spheres. Below each graph the dependence of the gaps width is shown.

2.3.3 Perspectives toward Experimental Realization

Another result presented in Ref. [26] is that the wide complete gap appears in the SW spectrum of 3D MCs with the lattice constant in the range of a dozen or so nanometers. This makes such MCs pretty tough to fabricate. In Refs. [32] and [31] we studied mFT crystals [19] as candidates for a 3D MCs base. mFT is a biomimetic nanoparticle (NP) produced from the natural ferritine used to store iron in living organisms [33]. The main part of the ferritine is a spherical protein cage, called apoferritine, the external diameter of which is 12 nm and the internal one 8 nm, carrying an inside ferrihydrite core. Since the natural core is nonmagnetic it has to be removed to obtain empty apoferritine ready to be filled with magnetic material. Typical mFT is loaded with magnetite/maghemite ($Fe_3O_4/\gamma - Fe_2O_3$) having a total magnetic moment of the order of $10^4\mu_B$ [19], but also other materials can be used [61, 62], which gives a possibility to produce NPs with the same size but different total magnetic moments (different saturation magnetization). Such mFT NPs have a natural tendency toward self-organization, which is used in the protein crystallization technique for the production of highly ordered 3D fcc crystals of external size up to 0.4 mm [41]. The lattice constant of mFT crystals as prepared is around 18.5 nm, but it shrinks while drying down to 14 nm [19]. (For more details and references please see Ref. [32].) Due to their properties mFT crystals appear to be good candidates for 3D MC fabrication; however, introducing the matrix, that is, filling spaces between mFT NPs with a magnetic material, is definitely a difficult task. The technique that seems to be very promising in this question is atomic layer deposition (ALD) [47].

The main results obtained for mFT-based MCs with three materials, cobalt, iron, and permalloy, considered as a matrix, are combined in Fig. 2.14. In Fig. 2.14a the dependence of the lowest gap on the lattice constant is shown for typical mFT NPs, that is, filled with magnetite/maghemite. The behavior of the gap originates in the interplay of the exchange and dipolar interactions, as explained in Ref. [32]. Even if the maximum of the gap width is below the lattice constant values available for mFT crystals there is still a 300–350 GHz wide complete gap in the SW spectrum for the Co or Fe matrix at $a = 14$ nm, that is, the lattice constant typical for dried mFT crystals. The gap could be even wider if we use different materials to fill apoferritine shells, as shown in Fig. 2.14b. Here the lattice constant is fixed at 14 nm. The result is qualitatively the same as in Fig. 2.13a—the stronger contrast gives a wider gap and below some threshold value the gap does not occur. Please notice that now the saturation magnetization is lower in spheres than in the matrix, so the contrast is lower than 1.0. In Fig. 2.14b its value grows from right to left toward 1.0, which means it becomes stronger from left to right.

Despite technical problems concerning fabrication, mFT crystals appear promising candidates for a 3D MC base having an fcc structure, which is the most convenient for complete gap opening, and a lattice constant in the range of a dozen of nanometers.

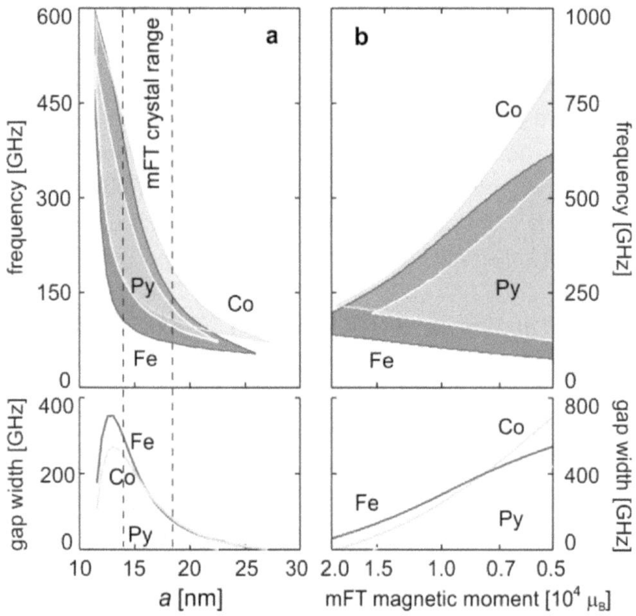

Figure 2.14 Lowest gap in mFT MCs for three matrix materials vs. (a) the lattice constant a and (b) the mFT magnetic moment. In (a) the magnetic moment of inclusions is fixed at the typical value for mFT NPs ($10^4\mu_B$). In (b) the lattice constant of MCs is fixed at 14 nm (dried mFT crystal). Below each graph the dependence of the gap width is shown.

Possibilities for the adjusting of the lattice constant as well as filling apoferritines with different magnetic materials give an additional opportunity for the tuning of the magnonic spectrum, including the complete gap width and its position.

2.4 Conclusions and Prospects

The 1D+ and 2D+ MCs are planar systems with in-plain periodicity where the inhomogeneity in out-of-plane direction plays a crucial role in tailoring of SW propagation. The 1D+ and 2D+ MCs can be built in many different ways. We discussed in detail 2D+ MCs formed by shifting the constituent elements of the structure in the out-of-plane direction and hybrid 1D+ and 2D+ MCs where the additional magnetic layer was stacked to 1D or 2D MCs. Such structures, being feasible for fabrication with current technology, are nontrivial extensions of the conventional planar 1D and 2D MCs into the third dimension. Being aware of difficulties and challenges in the fabrication of 3D structures, we discussed theoretical results for 3D MCs, which are bulk structures with periodicity in 3D. The investigations of 3D MCs were focused on bicomponent structures composed of spherical magnetic inclusions embedded in a magnetic matrix.

The main finding concerning 1D+ and 2D+ MCs discussed in this chapter can be summarized as follows:

- By shifting the inclusions of 2D bicomponent MCs in the out-of-plane direction we can mold the exchange and dipolar coupling between inclusions and the matrix. The additional effect is the modification of the demagnetizing field at the edges of inclusions and the matrix, which mainly affects the edge modes in the system.
- The magnetic layer coupled dipolarly to the planar 1D (or 2D) MC forms the hybrid 1D+ (or 2D+) system were the static and dynamic interaction between both subsystems can evoke new effects, absent in uncoupled components. (i) The SW spectrum can be reconfigurable when the static magnetic configuration is switched from antiparallel to parallel alignment of the magnetization between both subsystems. (ii) The magnetic layer provides additional coupling within the MC. (iii) The SW propagating in one subsystem can resonantly (at the selected frequency and wavenumber) excite the copropagating (or counterpropagating) SW in the second subsystem of the hybrid 1D+ or 2D+ MC.

The research on 3D MCs was concentrated on the calculation of the magnonic band structure and influence of the material and geometrical parameters on the magnonic bandgaps. The main outcomes of these studies concern the optimization of structural parameters (filling fraction, lattice type) and material parameters (contrast of magnetization saturation) for maximization of the width of the magnonic bandgap. In particular, we found that:

- The widest magnonic gap is observed for the inclusions arranged in the fcc lattice and the lattice constant range 10–20 nm.
- The gap can be open only when the contrast of magnetization saturation between the matrix and inclusions exceeds the threshold value—the smallest for an fcc lattice.

According to these results, we have proposed self-organized magnetoferritine crystals as a base for 3D MC fabrication. Having an fcc lattice and a 14–18 nm lattice constant after filling with a magnetic matrix, they can be good candidates for 3D MCs with a highly tunable spectrum of exchange SWs.

Acknowledgments

The research has received funding from the European Union's Horizon 2020 research and innovation program under the Marie Skłodowska-Curie grant agreement no. 644348, from the Polish Ministry of Science and Higher Education resources for science in 2017–2019 granted for the realization of an international cofinanced project (W28/H2020/2017), and the National Science Centre Poland for Grant No. 2015/17/B/ST3/00118 and UMO-2016/21/B/ST3/00452. The simulations were partially performed at the Poznan Supercomputing and Networking Center (Grant No. 209).

References

1. Ashcroft, N. W. and Mermin, N. D. (1976). *Solid State Physics* (Harcourt College, Orlando, FL).
2. Au, Y., Ahmad, E., Dmytriiev, O., Dvornik, M., Davison, T. and Kruglyak, V. (2012). Resonant microwave-to-spin-wave transducer, *Appl. Phys. Lett.*, **100**(18), 182404.
3. Bali, R., Wintz, S., Meutzner, F., Hubner, R., Boucher, R., Unal, A. A., Valencia, S., Neudert, A., Potzger, K., Bauch, J., et al. (2014). Printing nearly-discrete magnetic patterns using chemical disorder induced ferromagnetism, *Nano Lett.*, **14**(2), 435–441.

4. Beginin, E. N., Sadovnikov, A. V., Sharaevskaya, A. Y., Stognij, A. I. and Nikitov, S. A. (2018). Spin wave steering in three-dimensional magnonic networks, *Appl. Phys. Lett.*, **112**(12), 122404, doi:10.1063/1.5023138.
5. Bloch, F. (1928). Quantum mechanics of electrons in crystal lattices, *Z. Phys.*, **52**, 0555–600.
6. Chumak, A., Serga, A. and Hillebrands, B. (2017). Magnonic crystals for data processing, *J. Phys. D: Appl. Phys.*, **50**(24), 244001.
7. Ciubotaru, F., Devolder, T., Manfrini, M., Adelmann, C. and Radu, I. P. (2016). All electrical propagating spin wave spectroscopy with broadband wavevector capability, *Appl. Phys. Lett.*, **109**(1), 012403, doi:10.1063/1.4955030.
8. Demokritov, S. O. and Slavin, A. N. (2012). *Magnonics: From Fundamentals to Applications*, Vol. 125 (Springer Science & Business Media).
9. Donahue, M. J. (1999). OOMMF user's guide, version 1.0, Tech. rep.
10. Graczyk, P., Krawczyk, M., Dhuey, S., Yang, W.-G., Schmidt, H. and Gubbiotti, G. (2018). Magnonic band gap and mode hybridization in continuous permalloy films induced by vertical dynamic coupling with an array of permalloy ellipses, *Phys. Rev. B*, **98**(17), 174420.
11. Grimsditch, M., Giovannini, L., Montoncello, F., Nizzoli, F., Leaf, G. K. and Kaper, H. G. (2004). Magnetic normal modes in ferromagnetic nanoparticles: a dynamical matrix approach, *Phys. Rev. B*, **70**(5), 054409.
12. Gross, F., Trager, N., Frster, J., Weigand, M., Schtz, G. and Grfe, J. (2019). Nanoscale detection of spin wave deection angles in permalloy, *Appl. Phys. Lett.*, **114**(1), 012406, doi:10.1063/1.5074169.
13. Gubbiotti, G., Tacchi, S., Madami, M., Carlotti, G., Jain, S., Adeyeye, A. O. and Kostylev, M. P. (2012). Collective spin waves in a bicomponent two-dimensional magnonic crystal, *Appl. Phys. Lett.*, **100**(16), 162407, doi:10.1063/1.4704659.
14. Gubbiotti, G., Zhou, X., Haghshenasfard, Z., Cottam, M. G. and Adeyeye, A. O. (2018). Reprogrammable magnonic band structure of layered permalloy/cu/permalloy nanowires, *Phys. Rev. B*, **97**, 134428.
15. Gulyaev, Y. V. and Nikitov, A. (2001). Magnonic crystals and spin waves in periodic structures, in *Doklady Physics*, Vol. 46 (Springer), pp. 687–689.
16. Hartemann, P. (1987). Reection of magnetostatic forward volume waves by ion-implanted periodic arrays, *J. Appl. Phys.*, **62**(5), 2111–2115.
17. Joannopoulos, J. D., Johnson, S. G., Winn, J. N. and Meade, R. D. (2011). *Photonic Crystals: Molding the Flow of Light* (Princeton University Press).

18. Jorzick, J., Demokritov, S., Mathieu, C., Hillebrands, B., Bartenlian, B., Chappert, C., Rousseaux, F. and Slavin, A. (1999). Brillouin light scattering from quantized spin waves in micron-size magnetic wires, *Phys. Rev. B*, **60**(22), 15194.
19. Kasyutich, O., Sarua, A. and Schwarzacher, W. (2008). Bioengineered magnetic crystals, *J. Phys. D*, **41**, 134022.
20. Kłos, J. W., Kumar, D., Romero-Vivas, J., Fangohr, H., Franchin, M., Krawczyk, M. and Barman, A. (2012a). Effect of magnetization pinning on the spectrum of spin waves in magnonic antidote waveguides, *Phys. Rev. B*, **86**(18), 184433.
21. Kłos, J. W., Sokolovskyy, M. L., Mamica, S. and Krawczyk, M. (2012b). The impact of the lattice symmetry and the inclusion shape on the spectrum of 2d magnonic crystals, *J. Appl. Phys.*, **111**, 123910.
22. Krawczyk, M. and Grundler, D. (2014). Review and prospects of magnonic crystals and devices with reprogrammable band structure, *J. Phys.: Condens. Matter*, **26**(12), 123202.
23. Krawczyk, M., Kłos, J., Sokolovskyy, M. L. and Mamica, S. (2010). Materials optimization of the magnonic gap in three-dimensional magnonic crystals with spheres in hexagonal structure, *J. Appl. Phys.*, **108**, 093909.
24. Krawczyk, M., Mamica, S., Mruczkiewicz, M., Kłos, J. W., Tacchi, S., Madami, M., Gubbiotti, G., Duerr, G. and Grundler, D. (2013). Magnonic band structures in two-dimensional bi-component magnonic crystals with in-plane magnetization, *J. Phys.D*, **46**, 495003.
25. Krawczyk, M. and Puszkarski, H. (1998). Magnonic spectra of ferromagnetic composites versus magnetization contrast, *Acta Phys. Pol. A*, **93**(5), 805–810.
26. Krawczyk, M. and Puszkarski, H. (2008). Plane-wave theory of threedimensional magnonic crystals, *Phys. Rev. B*, **77**, 054437.
27. Krawczyk, M., Sokolovskyy, M., Kłos, J. W. and Mamica, S. (2012). On the formulation of the exchange field in the Landau-Lifshitz equation for spin-wave calculation in magnonic crystals, *Adv. Condens. Matter Phys.*, **2012**, 764783.
28. Kruglyak, V., Demokritov, S. and Grundler, D. (2010). Magnonics, *J. Phys. D: Appl. Phys.*, **43**(26), 264001.
29. Krysztofik, A., Głowiński, H., Kuświk, P., Zietek, S., Coy, L. E., Rychły, J. N., Jurga, S., Stobiecki, T. W. and Dubowik, J. (2017). Characterization of spin wave propagation in (1 1 1) YIG thin films with large anisotropy, *J. Phys. D: Appl. Phys.*, **50**(23), 235004.

30. Malago, P., Giovannini, L., Zivieri, R., Gruszecki, P. and Krawczyk, M. (2015). Spin-wave dynamics in permalloy/cobalt magnonic crystals in the presence of a nonmagnetic spacer, *Phys. Rev. B*, **92**(6), 064416.
31. Mamica, S. (2013). Tailoring of the partial magnonic gap in three-dimensional magnetoferritinbased magnonic crystals, *J. Appl. Phys.*, **114**, 043912.
32. Mamica, S., Krawczyk, M., Sokolovskyy, M. L. and Romero-Vivas, J. (2012). Large magnonic band gaps and spectra evolution in three-dimensional magnonic crystals based on magnetoferritin nanoparticles, *Phys. Rev. B*, **86**, 144402.
33. Meldrum, F. C., Heywood, B. R. and Mann, S. (1992). Magnetoferritin: in vitro synthesis of a novel magnetic protein, *Science*, **257**, 522–523.
34. Morozova, M., Romanenko, D., Matveev, O., Grishin, S., Sharaevskii, Y. and Nikitov, S. (2018). Suppression of periodic spatial power transfer in a layered structure based on ferromagnetic films, *J. Magn. Magn. Mater.*, **466**, 119–124, doi:https://doi.org/10.1016/j.jmmm.2018.06.07.
35. Morozova, M. A., Sharaevskaya, A. Y., Sadovnikov, A. V., Grishin, S. V., Romanenko, D. V., Beginin, E. N., Sharaevskii, Y. P. and Nikitov, S. A. (2016). Band gap formation and control in coupled periodic ferromagnetic structures, *J. Appl. Phys.*, **120**(22), 223901, doi:10.1063/1.4971410.
36. Mruczkiewicz, M., Graczyk, P., Lupo, P., Adeyeye, A., Gubbiotti, G. and Krawczyk, M. (2017). Spin-wave nonreciprocity and magnonic band structure in a thin permalloy film induced by dynamical coupling with an array of Ni stripes, *Phys. Rev. B*, **96**(10), 104411.
37. Mruczkiewicz, M., Krawczyk, M., Sakharov, V., Khivintsev, Y. V., Filimonov, Y. A. and Nikitov, S. (2013). Standing spin waves in magnonic crystals, *J. Appl. Phys.*, **113**(9), 093908.
38. Neusser, S., Bauer, H. G., Duerr, G., Huber, R., Mamica, S., Woltersdorf, G., Krawczyk, M., Back, C. H. and Grundler, D. (2011). Tunable metamaterial response of a Ni80Fe20 antidot lattice for spin waves, *Phys. Rev. B*, **84**, 184411.
39. Nikitov, S. A., Kalyabin, D. V., Lisenkov, I. V., Slavin, A., Barabanenkov, Y. N., Osokin, S. A., Sadovnikov, A. V., Beginin, E. N., Morozova, M. A., Filimonov, Y. A., et al. (2015). Magnonics: a new research area in spintronics and spin wave electronics, *Phys. Usp.*, **58**(10), 1002.
40. Obry, B., Pirro, P., Brächer, T., Chumak, A. V., Osten, J., Ciubotaru, F., Serga, A. A., Fassbender, J. and Hillebrands, B. (2013). A micro-structured ion-implanted magnonic crystal, *Appl. Phys. Lett.*, **102**(20), 202403.

41. Okuda, M., Eloi, J.-C., Jones, S. E. W., Sarua, A., Richardson, R. M. and Schwarzacher, W. (2012). Fe_3O_4 nanoparticles: proteinmediated crystalline magnetic superstructures, *Nanotechnology*, **23**, 415601.
42. Owens, J., Collins, J., Smith Jr, C. and Chiang, I. (1977). Oblique incidence of magnetostatic waves on a metallic grating, *Appl. Phys. Lett.*, **31**(11), 781–783.
43. Owens, J., Smith, C., Lee, S. and Collins, J. (1978). Magnetostatic wave propagation through periodic metallic gratings, *IEEE Trans. Magn.*, **14**(5), 820–825.
44. Pal, S., Kłos, J., Das, K., Hellwig, O., Gruszecki, P., Krawczyk, M. and Barman, A. (2014). Optically induced spin wave dynamics in [Co/Pd] 8 antidot lattices with perpendicular magnetic anisotropy, *Appl. Phys. Lett.*, **105**(16), 162408.
45. Parekh, J. and Tuan, H. (1977). Magnetostatic surface wave reectivity of a shallow groove on a yigfilm, *Appl. Phys. Lett.*, **30**(12), 667–669.
46. Puszkarski, H. and Krawczyk, M. (2003). Magnonic crystals—the magnetic counterpart of photonic crystals, in *Solid State Phenomena*, Vol. 94 (Trans Tech Publ), pp. 125–134.
47. Ritala, M. and Leskela, M. (1999). Atomic layer epitaxy - a valuable tool for nanotechnology? *Nanotechnology*, **10**, 19.
48. Sadovnikov, A., Beginin, E., Sheshukova, S., Romanenko, D., Sharaevskii, Y. P. and Nikitov, S. (2015). Directional multimode coupler for planar magnonics: side-coupled magnetic stripes, *Appl. Phys. Lett.*, **107**(20), 202405.
49. Sasaki, H. and Mikoshiba, N. (1979). Directional coupling of magnetostatic surface waves in layered magnetic thin films, *Electron. Lett.*, **15**(6), 172–174.
50. Sokolovskyy, M., Kłos, J. W., Mamica, S. and Krawczyk, M. (2012). Calculation of the spin-wave spectra in planar magnonic crystals with metallic overlayers, *J. Appl. Phys.*, **111**, 07C515.
51. Stancil, D. D. and Prabhakar, A. (2009). *Spin Waves* (Springer).
52. Sykes, C., Adam, J. and Collins, J. (1976). Magnetostatic wave propagation in a periodic structure, *Appl. Phys. Lett.*, **29**(6), 388–391.
53. Tacchi, S., Botters, B., Madami, M., Kłos, J. W., Sokolovskyy, M. L., Krawczyk, M., Gubbiotti, G., Carlotti, G., Adeyeye, A. O., Neusser, S. and Grundler, D. (2012a). Mode conversion from quantized to propagating spin waves in a rhombic antidot lattice supporting spin wave nanochannels, *Phys. Rev. B*, **86**, 014417, doi:10.1103/PhysRevB.86.014417.

54. Tacchi, S., Duerr, G., Kłos, J., Madami, M., Neusser, S., Gubbiotti, G., Carlotti, G., Krawczyk, M. and Grundler, D. (2012b). Forbidden band gaps in the spin-wave spectrum of a two-dimensional bicomponent magnonic crystal, *Phys. Rev. Lett.*, **109**(13), 137202.
55. Tacchi, S., Gruszecki, P., Madami, M., Carlotti, G., Kłos, J. W., Krawczyk, M., Adeyeye, A. and Gubbiotti, G. (2015). Universal dependence of the spin wave band structure on the geometrical characteristics of two-dimensional magnonic crystals, *Sci. Rep.*, **5**, 10367, doi:10.1038/srep10367.
56. Tien, P. (1977). Integrated optics and new wave phenomena in optical waveguides, *Rev. Mod. Phys.*, **49**(2), 361.
57. Vasseur, J., Dobrzynski, L., Djafari-Rouhani, B. and Puszkarski, H. (1996). Magnon band structure of periodic composites, *Phys. Rev. B*, **54**(2), 1043.
58. Wang, Q., Pirro, P., Verba, R., Slavin, A., Hillebrands, B. and Chumak, A. V. (2018). Reconfigurable nanoscale spin-wave directional coupler, *Sci. Adv.*, **4**(1), e1701517.
59. Wawro, A., Kurant, Z., Jakubowski, M., Tekielak, M., Pietruczik, A., Böttger, R. and Maziewski, A. (2018). Magnetic properties of coupled co/mo/co structures tailored by ion irradiation, *Phys. Rev. Appl.*, **9**(1), 014029.
60. Yablonovitch, E. (1987). Inhibited spontaneous emission in solid-state physics and electronics, *Phys. Rev. Lett.*, **58**(20), 2059.
61. Yamashita, I., Iwahori, K. and Kumagai, S. (2010). Ferritin in the field of nanodevices, *Biochim. Biophys. Acta*, **1800**, 846.
62. Yoshimura, H. (2006). Protein-assisted nanoparticle synthesis, *Colloids Surf., A*, **282–283**, 464.
63. Yu, H., Duerr, G., Huber, R., Bahr, M., Schwarze, T., Brandl, F. and Grundler, D. (2013). Omnidirectional spin-wave nanograting coupler, *Nat. Commun.*, **4**, 2702.
64. Zivieri, R., Malagò, P. and Giovannini, L. (2014). Band structure of collective modes and effective properties of binary magnonic crystals, *Photonics Nanostruct. Fundam. Appl.*, **12**(5), 398–418.
65. Zivieri, R., Tacchi, S., Montoncello, F., Giovannini, L., Nizzoli, F., Madami, M., Gubbiotti, G., Carlotti, G., Neusser, S., Duerr, G., et al. (2012). Bragg diffraction of spin waves from a two-dimensional antidot lattice, *Phys. Rev. B*, **85**(1), 012403.

Chapter 3

3D Magnonic Crystals

E. N. Beginin,[a] D. V. Kalyabin,[b,c] P. A. Popov,[b,c] A. V. Sadovnikov,[a] A. Yu. Sharaevskaya,[b] A. I. Stognij,[d] and S. A. Nikitov[a,b,c]

[a]*Saratov State University, 112 Bol'shaya Kazach'ya, Saratov, 410012, Russia*
[b]*Kotelnikov Institute of Radio-engineering and Electronics of RAS, 11-7 Mokhovaya st., Moscow, 125009, Russia*
[c]*Moscow Institute of Physics and Technology, 9 Instituskij per., Dolgoprudny, 141700, Moscow Region, Russia*
[d]*Scientific and Practical Materials Research Centre, National Academy of Sciences of Belarus, 220072, Minsk, Belarus*
nikitov@cplire.ru

3.1 Introduction

In 1965 INTEL engineer Gordon Moore proposed the empirical regularity what was later known as Moore's law. According to this law the number of transistors based on one chip with 1 sq. inch is doubled each year [1, 2]. Therefore, the following rule was established pursuant to which the computers' calculation power will be increased (and the cost will be reduced) exponentially approximately every two years. The last 10 years, due to essential technological developments, the transistors' size reach its physical limit, what led to the necessity of Moore's law corrections or even

Three-Dimensional Magnonics: Layered, Micro- and Nanostructures
Edited by Gianluca Gubbiotti

ISBN 978-981-4800-73-0 (Hardcover), 978-0-429-29915-5 (eBook)
www.jennystanford.com

on its refusal [2]. On the other hand, conventional semiconductor complementary metal-oxide semiconductor (CMOS) technology becomes essentially 3D (vertically integrated) instead of 2D. This comes from the fact that the size of the transistors approaches units of nanometers and future technologies tend to be developed in 3D topologies. In particular, one of the possible solutions for transistors integration with CMOS technology is 3D integration with through holes in silicon (through-silicon vias [TSVs]) (see, for example, Refs. [3, 4]). Such vertical constructions combine quality of CMOS technology (i.e., minimal consumed power) and vertical integration of computing devices for achievement of essential information density.

About 40 years ago for development of an element base for microwave electronics it was suggested to use technology of growth of high-quality magnetic films of yttrium iron garnet (YIG, $Y_3Fe_5O_{12}$) [5, 6]. In these films the signal carriers for information transmission and processing are magnetostatic spin waves (MSWs). To excite and receive propagating MSWs in such films the microstrip transducers or strip lines are usually used. The saturation magnetization of YIG is 1750 G, and this defines the basis for microwave frequency devices based on MSWs and YIG films. Besides, to be in a single domain state, these films should be magnetized to saturation; therefore they should be placed into an external magnetic field. Both the value of the external magnetic field and the saturation magnetization define the frequency range of the work of MSW microwave devices. Typically, this range starts at units of gigahertz and depending on the value of the field lasts up to tens of gigahertz. In such a way, a number of microwave devices were developed, including delay lines, frequency transmission and stop filters, resonators, power dividers, demultiplexers, nonlinear devices, and many others [5–7]. The velocities of propagating MSWs by a few orders of value (up to 6) are less than electromagnetic waves' velocity in YIG, what makes possible to develop effective delay lines for information carriers. Small losses of the wave power at propagation in YIG films relatively far (distance up to cm) permit to develop devices with the quality factor up to 10,000. The following technology achievements also permit to develop element bases, not only using single YIG films, but also using narrow directed waveguides and coupled waveguides where coupling can be lateral or vertical [8–10]. Besides, the growth of the nanometer-thick YIG films and high-quality Permalloy ($Ni_{80}Fe_{20}$) films permits to increase working frequencies a few times and decrease lateral sizes dozens of times (up to a few micrometers). In such a way a new investigation field in microwave electronics has been established, namely magnonics [11–14]. Magnonics is a field of electronics that studies physical properties of magnetic micro- and nanostructures, properties of propagating spin waves (SWs), and possibility of SW use for development of signal transmission and processing devices and networks based on novel physical principles.

Summarizing the previous discussions one can conclude that magnetic micro- and nanostructures are extensively investigated due to their potential as candidates for future spintronic and magnonic devices. The main carriers of signals used for information processing in such structures are magnons [15] and, in particular, MSWs. These structures are also possible candidates for beyond-CMOS technologies, as they are free of Joule heating and, respectively, free of power loss associated with traditional electronics [16–31a,b]. The majority of works related to theoretical and experimental studies of MSWs are based on planar geometries in which ferromagnetic films, or structures containing such films, are subjected to the external saturating magnetic field. Thus, during the last few years, a number of sections of MSW investigations in the field of magnonics were established. Among them are artificial magnetic materials with properties periodically varied in space (magnonic crystals [MCs]) [32, 33], magnonic circuits [34, 35], edge magnons [36], coupled structures [9, 10, 25, 28, 29, 31a], narrow and irregular waveguides [17, 27], etc. The main objects of investigations are YIG films and Permalloy films, and coupled structures also containing pure Ni, Co, and other magnetic elements, Heusler alloys, etc. Due to its low damping, YIG allows propagation of MSWs over long distances, which opens up the possibility for the fabrication of functional magnonic networks [26]. SW transport can also be manipulated by the presence of lateral boundaries, which leads to non-uniformity of the internal magnetic field. Typical geometries of films and structures are thin films, coupled films, and narrow waveguides with a shape of stripes or even curved waveguides. The structures with etched holes and/or stripes were developed to study

SW edge modes. Mainly, the structures mentioned above can be used as SW conduits and filters, delay lines and phase shifters, data-buffering elements, and magnonic logic gates. However, if magnonics wants to remain on the same technological platform, it should also be extended toward three dimensions. Previous works have explored the idea that magnetic parameters of the media can be spatially changed. For example, magnetic permeability tensors can be spatially dispersive due to insertion of the magnetic particles of the material into the matrix with different magnetization, as well as 3D MCs can be fabricated with magnetic particles as the centers of a 3D lattice in a host magnetic medium [37, 38]. Edge magnons can possibly exist in a magnetic matrix medium with embedded cylinders of another magnetic material. The vertically integrated 3D magnonic structures open the perspective to increase in the density of magnonic and spintronic elements (interferometers, switches, logic elements, spin valves, data buses, etc.) for the fabrication of scalable and configurable magnonic networks [39]. 3D magnonic architecture [40–44] can provide both frequency and spatial multiplexing and demultiplexing in three dimensions due to the reduction of the total length of connections between functional elements and the simultaneous implementation of spatial separation of SW beams for different frequencies of carrier signals.

The chapter is organized in the following way: Section 3.1 describes the technology of creating 3D structures based on YIG grown onto structured nonmagnetic a gadolinium-gallium-garnet (GGG, $Gd_3Ga_5O_{12}$) substrate. Section 3.2 is devoted to ferromagnetic resonance (FMR) investigations of meander-type structures. In Section 3.3 we study surface MSW propagation in 3D structures, and in the subsequent Section 3.4 we describe volume MSW propagation in 3D structures. The following Section 3.5 contains results of MSW propagation in coupled 3D structures, and the Conclusion section is devoted to a discussion of possible use of obtained results for development of components on the basis of new physical principles.

3.1.1 Technology of 3D Magnonic Crystal Development

Figure 3.1 demonstrates the sketch of the proposed for investigation structure, which can be treated as 3D for propagating SWs. Typically,

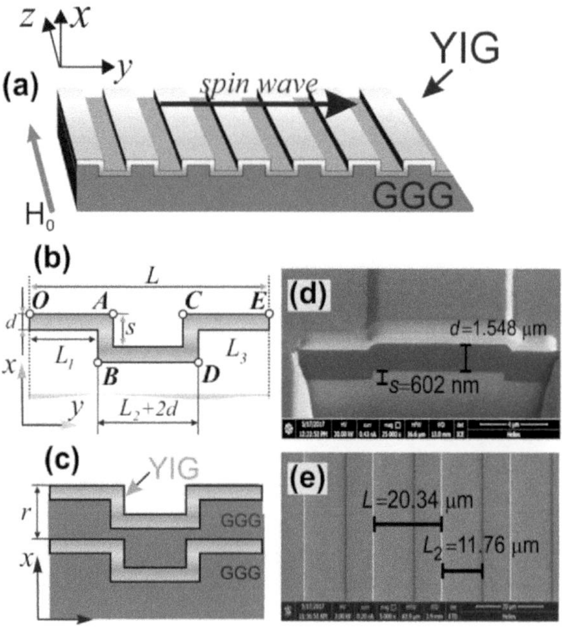

Figure 3.1 (a) Schematic view of the meander magnonic waveguide, (b) one period of the meander waveguide, (c) symmetric type of a coupled MC with parallel shift *r*, and (d and e) SEM images of the cross section and surface of the fabricated structure.

the SW waveguide structure consists of a nonmagnetic substrate at which the magnetic film is grown. Few methods exist to grow magnetic films, for example, liquid-phase epitaxy, sputtering, and laser ablation. The typical substrate for YIG films is GGG. To fabricate a 3D magnonic structure, we suggest to process the surface of the GGG (111) substrate by etching the single grooves or periodic structures of them. On the top of such a processed substrate, the YIG is then sputtered by the means of ion beam sputtering with fast annealing. Further, at the surface of the YIG film, the second (third, etc.) films can be grown with different or same magnetizations. This can be achieved by making YIG impure with ions of Ca, Sc, Bi, etc., or even with the variation of the substrate temperature. In such a way, a 3D meander type (Fig. 3.1b) or multilayer (Fig. 3.1c) magnonic structure can be fabricated. The main advantage

Figure 3.2 Cross-section of the double-layer structure of a YIG (840 nm)/GGG (860 nm)/LPE YIG film.

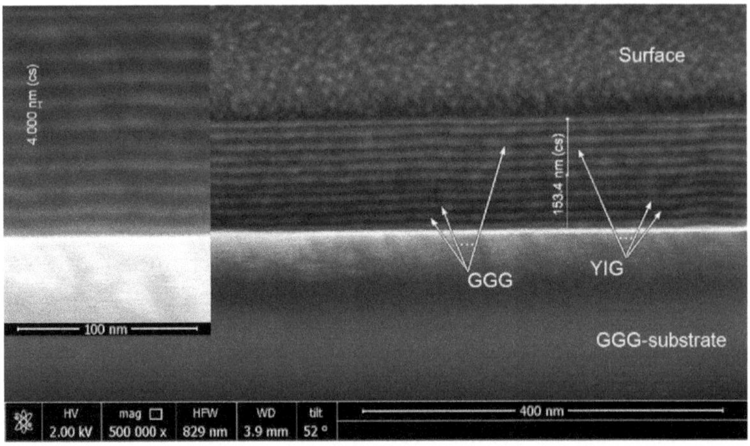

Figure 3.3 Multilayered structure of 12 YIG layers (each 4 nm thick) interlayered by 12 GGG layers (each 4 nm thick) grown on a single-crystal GGG substrate.

of the anisotropic ion beam sputtering (IBS) over, for example, pulsed laser deposition is the uniform surface of hundreds of square centimeters and predefined YIG thickness in the wide range of 0.1–1 μm. In addition to the above, both deposition on the cold substrate and vacuum fast annealing provide the flatness and appropriate interface quality with the possibility of multilayer fabrication with the sequential repetitions of the fabrication stages. Moreover, this method also enables the sputtering of GGG on the YIG substrate. The details about the growth and the standard characterization of the structural properties of our YIG films can be found in Ref. [33]. By ion beam cutting using the FIB Helios NanoLab 600 Station (FEI Company, USA), the scanning electron microscopy (SEM) image of the cross section of the fabricated structure of the meander SW waveguide was obtained (Fig. 3.1d). The surface of the 3D magnonic structure is shown in Fig. 3.1e. The structure of YIG films was determined by X-ray diffraction (XRD CuKa) and by energy dispersive X-ray spectroscopy. The XRD spectrum confirms the existence of the YIG phase in the sample. It is found that YIG films have a polycrystallic state but a homogeneous phase structure.

In Fig. 3.2 we also present double-layer structure of a YIG (840 nm)/GGG (860 nm)/LPE, where LPE means the YIG substrate grown by liquid-phase epitaxy (LPE). This last film has a single-crystal structure. To get this structure grown by IBS we used an 8-μm-thick LPE-grown YIG film as a substrate and single-crystal YIG and GGG targets. Further, this structure was annealed in vacuum with a small pressure of 1 Pa at 820°C–850°C for 5–10 minutes in an atmosphere containing 80% N_2 and 20% O_2. According to Fig. 3.2 the film growth and its crystallization at these external conditions are quite perfect and interphase boundaries are parallel, which is required for applications in optics and microwave electronics.

In Fig. 3.3 we present a multilayered structure containing 12 YIG layers 4 nm thick each alternating with 12 GGG layers 12 nm thick each grown on a single-crystal GGG substrate. Annealing was done at above-described conditions. The layer thickness was controlled by a fixed time of deposition of each matter. Finally, a structure with parallel interfaces and without pinholes and other defects was obtained.

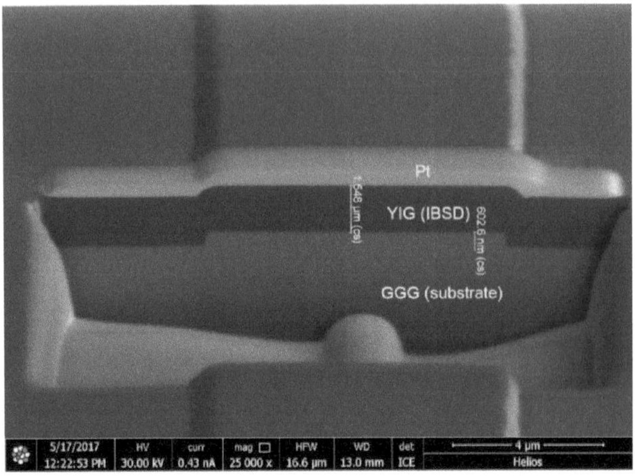

Figure 3.4 YIG film (1.5 µm thick) deposited on a GGG substrate with surface relief 600 nm.

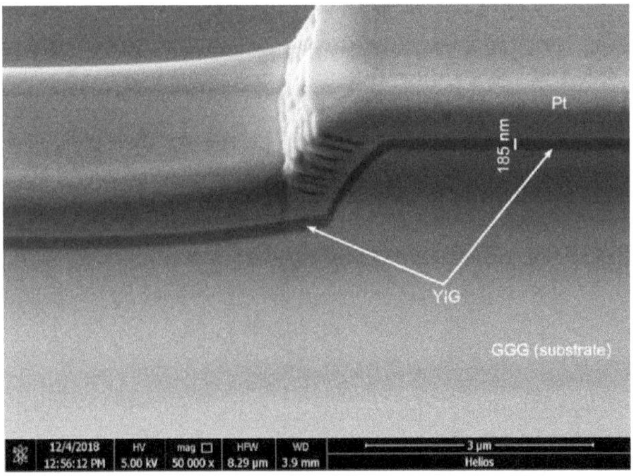

Figure 3.5 YIG film (185 nm thick) conformally covering a laterally processed GGG (1D) substrate with period 5 µm and thickness 2 µm.

In Fig. 3.4 we show a 1.5-µm-thick YIG film deposited onto a 1D relief with height 600 nm at a GGG substrate. It can be seen that the YIG film is conformally filling all the relief and keeps its shape after annealing. Such structures, in fact, are 1D MCs and can serve as substrates for further deposition of 3D magnonic structures.

In Fig. 3.5 a YIG film with thickness 185 nm is grown onto a 1D MC substrate having a surface relief with thickness 2 μm.

3.2 Ferromagnetic Resonance in a Meander-Type Ferromagnetic Film

Prior to discussing SW propagation in meander-type ferromagnetic films and 3D structures consisting of such films, let us consider properties of FMR that can be excited in such films. FMR is widely used to study and measure material parameters of magnetic films and structures and field-frequency characteristics of SW oscillations. The investigated structure is shown in Fig. 3.1a,b and is introduced in a previous section. However, to simplify the analysis of obtained results and their interpretation the sketch of this structure with cross sections is shown in the inset of Fig. 3.6. Numerical investigations of FMR were done using the micromagnetic calculation method based on the MuMax3 program [45]. As a ferromagnetic layer we considered the isotropic YIG layer with thickness $d = 1$ μm width of the horizontal segment $w = 21$ μm, and the following material parameters: saturation magnetization $M_s = 139$ G, Gilbert damping constant $\alpha = 10^{-3}$, and non-uniform exchange constant $A_{ex} = 3.614 \times 10^{-2}$ cm^2. Numerical calculations were provided for a meander area with size $L \times s \times \Delta z$ (at fixed values $L = 40$ μm and $\Delta z = 5$ μm) and at different values of height s. At the boundaries of this fixed area at the axes x and z the periodic boundary conditions (PBCs) are applied. The external magnetic field $\vec{H}_0$ was oriented in a meander plane along either axis x (transverse magnetization) or along axis z (longitudinal magnetization). To study magnetization oscillations we need to subject the meander structure in uniform transverse pumping by a dynamic magnetic field $\vec{h}(t)$ ($\vec{h}(t) \perp \vec{H}_0$) with time dependence $h(t) = h_0 \sin c(2\pi f_0(t - t_0))$, where $h_0 \ll \left|\vec{H}_0\right|$ is the amplitude of the dynamic field, and f_0 and t_0 are the frequency spectrum boundary and time shift of the pumping pulse, respectively. The characteristic frequencies of the SW excitations during pumping action were defined by the spectral power density $W(f)$ obtained

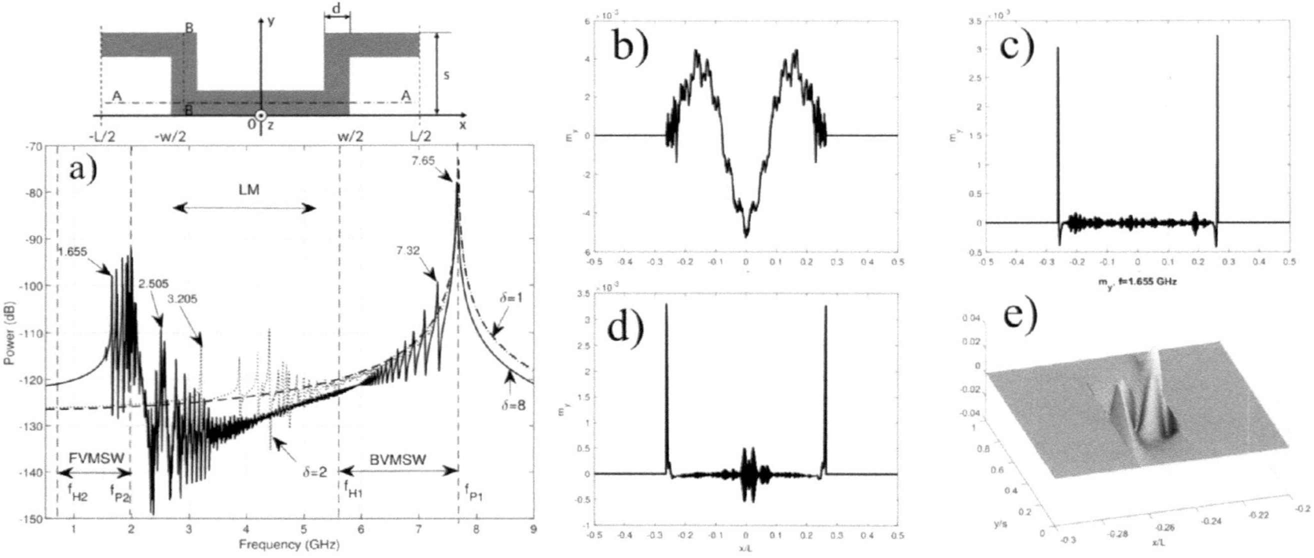

Figure 3.6 Dependence of characteristic frequencies of the transversely magnetized meander (the cross section of the structure is shown in the inset map) on the meander height (a). Magnetization profiles of m_y in section A-A and spatial distributions at various frequencies (b) $-f = 7.32$ GHz, $\delta = 8$; (c) $-f = 3.205$ GHz, $\delta = 2$; (d) $-f = 2.505$ GHz, $\delta = 8$; (e) $-f = 1.655$ GHz, $\delta = 8$; arrows indicate the frequencies of magnetization distributions.

at Fourier transform of the time realization of the function $\langle \vec{m}(t) \rangle = \langle \vec{M}(t) \rangle / M_s$, where $\langle \vec{M}(t) \rangle$ is the mean value of the magnetization vector in the calculation area (segment) $L \times s \times \Delta z$ in a time moment t.

3.2.1 Transversely Magnetized Meander

At first, we will present results of numerical calculations of FMR at the transverse pump of the magnetic system by a linearly polarized magnetic field pulse with amplitude $h_y(t)$ with limit frequency $f_0 = 10$ GHz and time shift $t_0 = 20$ ns. A saturating external magnetic field was directed along the x axis ($H_{0x} = H_0 = 2$kOe). Figure 3.6a presents the frequency spectrum and values of some characteristic frequencies of SW excitations, depending on the value of normalized height $\delta = s/d$ of the meander. The case $\delta = 1$ corresponds to the FMR in the tangentially magnetized uniform YIG film, which is shown to be a reference film. The spectrum has two regions of characteristic frequencies. The first frequency region of the existence of a backward-volume magnetostatic spin wave (BVMSW) propagating along the x axis has the boundary frequencies $f_{H1} = \gamma H_0$, $f_{P1} = \sqrt{f_{H1}(f_{H1} + f_M)}$, where $\gamma = 2.8$ GHz/kOe is the gyromagnetic ratio and $f_M = \gamma 4\pi M_s$ corresponds to the frequency domain of the existence of BVMSW propagating along the x axis. The second region with boundary frequencies $f_{H2} = \gamma(H_0 - 4\pi M_s)$, $f_{P2} = \sqrt{f_{H2}(f_{H2} + f_M)}$ corresponds to the case $\delta = \infty$ and the frequency domain of existence of a forward-volume magnetostatic spin wave (FVMSW) propagating along the y axis in a normally magnetized ferromagnetic layer with thickness d and that is infinite in the (y, z) plane.

We note that the resonant frequencies exist for any value of $\delta > 1$ and are weakly dependent on height δ of the meander in the frequency domain of the BVMSW. The oscillations of the magnetization are localized in horizontal segments at these frequencies. Hence the distributions of the magnetization m_y components correspond to standing waves. Such waves are formed by the BVMSW multiple reflections from the boundaries of the horizontal segments of width w. Quasi-sinusoidal standing waves in the segments are formed when conditions for the wavenumbers k_n of the BVMSW are met: $n = 1, 3, 5 \ldots$, where n is a number of half-wavelengths in the horizontal segment of w. Wavenumbers k_n and their corresponding frequencies f_n are related to each other by the dispersion relation $f_n(k_n)$ for the BVMSW, and these results were presented, for example, in Ref. [46]. Furthermore, the modes with only odd n are excited at a uniform pumping of the meander by a linearly polarized dynamic field $h_y(t)$.

Figure 3.1b presents the magnetization profile $m_y(x)$ along section A-A for frequency 7.32 GHz and for meander height $\delta = 8$. The weak dependence of the resonance frequencies f_n on the height of the meander follows from the results of solution of a static problem. For this case it is found that the distributions of internal magnetic fields in the horizontal segments do not depend on each other. We note that the observed magnetization distributions in horizontal segments at resonant frequencies are superpositions of resonant and nonresonant wave processes with significantly different wavenumbers k. The resonant wave processes with $k \sim 1/w$ correspond to the BVMSW. Nonresonant wave processes with $k \gg 1/w$ correspond to dipole exchange waves propagating along the x axis with propagation constants $k(x)$ depending on the longitudinal coordinate x. The specific form of this dependence is determined by the nature of the distribution of local static internal magnetic fields within the segment. The waves of this type can be effectively excited by uniform pumping in gradient static magnetic fields in the vicinity of the turning points of $k(x_0) \approx 0$, (see, for example, Ref. [47]).

Let us consider the frequency domain (Fig. 3.7a) between the BVMSW and the FVMSW (designated by the letters LM) of the spectrum of SW excitations in the meander. The resonant frequencies correspond to the localized oscillation modes (LM) of the magnetization near the edges of the segments in this region, where the internal magnetic field $H_x(x)$ rapidly decreases. The localization of the modes is due to the effects of "spin wave walls" [48]. The main features of the SW spectra in the region LM for rectangular magnetic elements and systems of periodic magnetic strips are studied in detail in a number of works, for example, Refs. [49, 50]. Then the symmetric modes with odd numbers are effectively excited with a uniform pumping field. Figure 3.1c shows the $m_y(x)$ profile of the lowest mode at the height of the meander $\delta = 2$. Furthermore, the resonant frequencies of the modes are shifted to smaller values with an increase of the meander height δ. Figure 3.1d shows the magnetization profile of the lowest mode at the resonant frequency $f = 2.505$ GHz at $\delta = 8$.

Now we consider the distribution of the magnetization in the meander in the region of the resonant frequencies of FVMSWs with height equal to $\delta = 8$. The internal field in the vertical segments of the meander with a big height becomes significantly inhomogeneous, and the distributions $m_y(x, y)$, $m_z(x, y)$ at the resonant frequencies cannot be represented as combinations of trigonometric function of the meander. We assumed that standing waves in this case are formed by superposition of quasi-FVMSWs that will pass into the common FVMSW with trigonometric distributions over spatial coordinates only in the limit case of an infinitely high meander at $\delta = \infty$. The regions of the resonant frequencies of the meander only partially overlap with the region of FVMSWs in the case of a meander with a finite height.

Figure 3.1e presents the distribution $m_y(x, y)$ for a frequency of 1.655 GHz. Magnetization oscillations occur in only vertical segments and have a complex spatial profile. The following conclusions can be drawn from micromagnetic calculations of FMR spectra in a meander magnetized along the x axis. Namely, for $\delta > 1$ in the FMR spectrum there are always resonant frequencies in the BVMSW region with localized magnetization oscillations in horizontal segments. The resonant frequencies of localized modes are formed at the edges of the horizontal segments. Then, also low-frequency oscillations in the FMR spectrum can be generated by quasi-FVMSWs in the vertical segments at $\delta > 2$.

3.2.2 Longitudinally Magnetized Meander

In the case of a longitudinally magnetized meander (the magnetic field is directed along the z axis), the internal magnetic fields in the segments are weakly inhomogeneous. As it has been shown in Refs. [42, 43], the magnetostatic surface spin waves (MSSWs) propagating along the x axis are weakly reflected at the regions of junctions between the vertical and horizontal segments. Transmitted and reflected MSSWs form quasi-standing waves in the

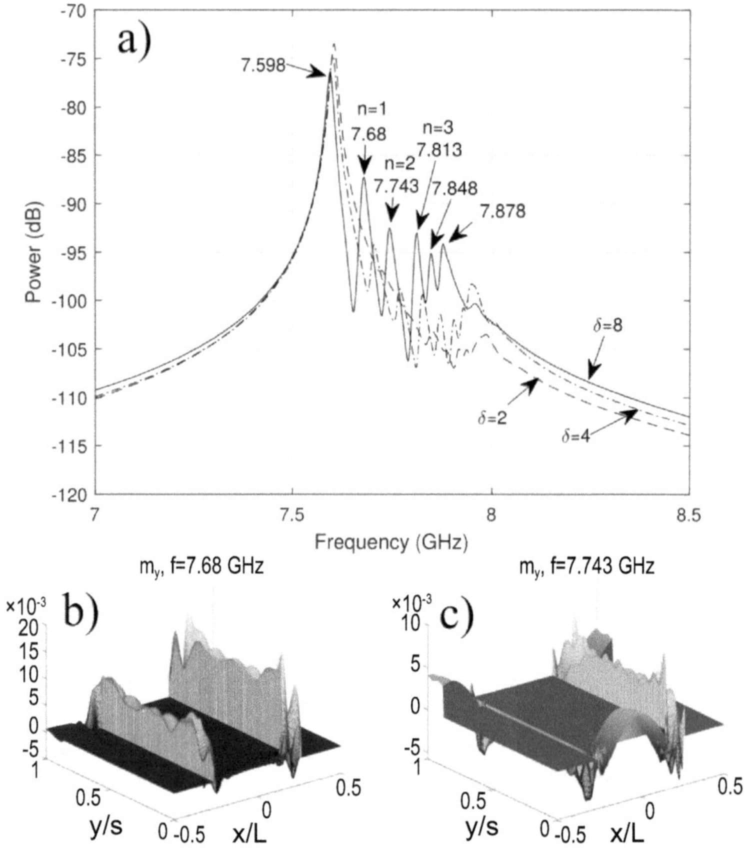

Figure 3.7 (a) Dependence of the characteristic frequencies of the longitudinally magnetized meander on meander height δ. Spatial distributions of m_y at different frequencies for $\delta = 8$, (b) $- f = 7.68$ GHz, $n = 1$; (c) $- f = 7.743$ GHz, $n = 2$.

entire meander, and the resonance frequencies and corresponding spatial distributions of the magnetization are primarily determined by the PBCs for the computational domain and by the overlapping integrals of the uniform pumping field and MSSW fields. The spectrum can be presented by the discrete set of frequencies characterized by an index $n = 1, 2, 3\ldots$ (Fig. 3.7a) for a meander with a sufficiently big height equal to $\delta > 2$. It is followed from the results of calculations of the spatial distribution of magnetization for various resonant frequencies (Fig. 3.7b–d) that the odd indices n correspond to common-mode oscillations of magnetization in vertical segments and the even indices correspond to antiphase-mode oscillations, and $(n - 1)$ half-waves of MSSWs fit in horizontal segments.

3.3 Magnetostatic Surface Waves in 3D Structures

In this section, we discuss the idea how magnonics can be essentially 3D. We present the concept of 3D magnonic structures. We demonstrate results of investigations of the meander magnonic waveguide fabricated on the structured substrate. We study SW propagation regimes in isolated and coupled meander magnonic structures using the finite element method (FEM), transfer matrix method (TMM), and micromagnetic simulations. We analyze and explain the features of formation of magnonic bandgaps in such structures. Using micromagnetic simulation, the potential for realizing highly efficient, SW computing concepts based on 3D structures is demonstrated.

The schematic of the meander-type magnonic waveguide is shown in Fig. 3.1a. The structure is placed in the external magnetic field parallel to the z axis, ferromagnetic film thickness d, structure period L, and groove depth s. The excited wave propagates along the positive direction of the y axis. This geometry corresponds to the case of Damon–Eshbach MSSW [51]. Reaching point A the wave turns down along the negative direction of the x axis, as shown in Fig. 3.1b. And again in section AB this is the MSSW propagating toward the corner at point B. Reaching point B the wave turns to the positive y direction. Assuming that the thickness of the YIG film is the same in each part of the meander waveguide (OA, AB, BD, DC, and CE), the structure of the wave will be the same in each part. However, the essential thing is that the wave propagates not in a plane but in two dimensions. In particular, the 3D structure of the MC can be realized if we narrow this structure in the z direction. At the points of turnouts (A, B, C, D) the wave is reflected and transmitted at the corners of coupled parts of the waveguide. The role of reflections and transmission will be discussed later in this chapter.

Next, we present the results of numerical calculations of wave dispersion and magnetic potential distribution for MSW propagation in a meander-type structure of different depths of the grooves (parts AB, CD, etc.). The calculations were done using FEM [52] (calculated using COMSOL Multiphysics software from COMSOL, Inc.). Figure 3.1b shows the primitive cell of the magnonic structure. The periodicity is implied to be along the y axis. Using the Bloch theorem the magnetostatic potential can be written as $\varphi(x, y, z) = \Psi(x, y, z)e^{-jk_s y}$, where $\Psi(x, y, z) = \Psi(x, y + L, z)$ is the periodic function of the spatial coordinate y, k_s is the Bloch wavenumber, and $L = L_1 + L_2 + L_3 + 2d$ is the period of the 3D MC. The PBCs were taken at planes $y = 0$ and $y = L$, which are denoted with dotted vertical lines in Fig. 3.1b. At the surfaces of the YIG film the electrodynamic boundary conditions were applied (continuity of the magnetostatic potential and the normal component of the magnetic induction). Figure 3.1a demonstrates the results of calculations of the dispersion characteristics of the MSW for different depths of the grooves (s = 1, 3, 6 and 9 μm). The following parameters were taken for numerical calculations: YIG film thickness d = 1 μm, $L_1 = L_3$ = 9 μm, L_2 = 20 μm, and L = 40 μm; YIG film saturation magnetization $4\pi M_0$ = 1750 G; and external magnetic field H_0 = 250 Oe. If $s = 0$, then the periodicity along the structure is absent and the dispersion characteristic coincides with the dispersion of the MSSW in a film with thickness d (black curve in Fig. 3.1a denoted as MSSW). If $s \neq 0$ the structure becomes periodic in the y direction, and as the meander depth increases the Bloch wave phase and group velocities $v_s = \omega/k$ and $v_{sg} = d\omega/dk$ become smaller. The phase shift of the Bloch wave after propagation through the one period is $\theta = k_s(f)L$. On the other hand the same phase change $\theta = k(f)l$ should be for the MSSW propagating in an elementary cell of length $l = L + 2s$. Thus, the relation between the wavenumber of MSSW k and Bloch wavenumber is read as $k_s(f) = k(f)l/L$. Therefore, the Bloch wavenumber k_s increases and the phase velocity is decreased with an increase of the meander parameter s.

It can be seen from Fig. 3.8a that the bandgaps are clearly distinguished at the frequencies of Bragg resonances when the

MSW wavelength is equal to two periods of the structure. The first bandgap appears at $k_b = 785\ \text{cm}^{-1}$, and the bandgaps of the next order with $k > k_b$ are shown by vertical dashed lines. Two main conclusions can be made on the basis of the results presented in Fig. 3.8a. First, as the depth of the groove is increased the width of the bandgap is also increased. The MSSW wavelength respecting to the first Bragg resonance is about 75 μm and the depth $s = 1$ μm with respect to the case when the YIG surface is slightly modulated only, and there is no direct propagation of the wave into the depth of the YIG film. On the other hand for depth $s = 9$ μm and, especially, for the higher Bragg resonances the wavelengths become smaller than the depth of the groove and the MSW clearly propagates inside the film depth (x axis). Second, at fixed wavenumbers (or wavelengths) the dispersion curves are shifted in the region of lower frequencies due to a decrease of the MSSW phase velocity.

To describe properties of MSSW propagation in a 3D MC with finite number N of the periods, we used TMM [53]. A primitive cell of the 3D MC was "unwrapped" into the line of the horizontally coupled segments OA, AB, BD, DC, and CE of the ferromagnetic waveguide with thickness d. In such way the 3D MC was replaced by the equivalent 1D MC with uniform segments divided by narrow non-uniform regions (at the places of horizontal and vertical waveguide connections). This non-uniform segments were considered as being infinitely thin but having the phenomenological MSSW reflection coefficient ζ. Setting the amplitudes of SWs at the input and output of the 1D MC, the reflection coefficient ζ, and solving the matrix equation we obtain the transmission response of MSSW $T(f)$ (Fig. 3.8b). Calculation was performed for the 1D MC with $N = 20$ elementary cells, $s = 9$ μm, reflection coefficient $\zeta = 0.05$, dissipation parameter $\alpha = 10^{-4}$. and MC period $l = 58$ μm, which corresponds to the Bragg wavenumber $k_{bD} = 542\ \text{cm}^{-1}$. The transmission response has the forbidden zones at the Bragg frequencies f_1, f_2, f_3, and f_4, for which the Bragg wavenumbers are shown by points and arrows at dispersion characteristics in Fig. 3.8a. Figure 3.8c–h demonstrates the results of magnetostatic potential spatial distribution $\Psi(x, y)$ in the 3D MC for different meander depths and fixed wavenumbers $k_s = 100\ \text{cm}^{-1}$ ($k_b < 1000\ \text{cm}^{-1} < 2k_b$) and $k_s = 2700\ \text{cm}^{-1}$ ($3k_b < 2700\ \text{cm}^{-1} <$

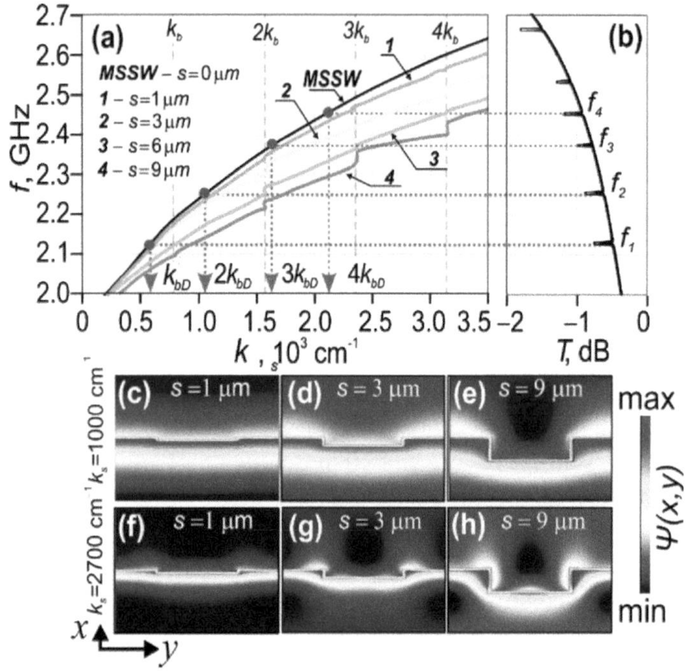

Figure 3.8 (a) Dispersion characteristics of MSSWs in the uniform film (denoted as MSSW) and Bloch wave dispersion in the MC at various meander depths s. The vertical dashed lines denote Bragg wavenumbers. (b) Transmission response for equivalent MC at $s = 9$ μm calculated with TMM. (c–h) Ψ profile in the primitive cell of the MC at various s.

$4k_b$). As parameter s is increased the potential distribution becomes periodically modulated with amplitude exponential decrease across the film thickness in horizontal and vertical MC segments.

To elucidate the SW (magnon) transport features across the meander SW waveguide we performed micromagnetic simulation using MuMax3 [45]. First, we calculated the transmission response and dispersion of the reference waveguide, which is presented as a uniform YIG film of thickness $d = 1$ μm. The transmission response and dispersion are shown in Fig. 3.9a with dashed blue and brown curves. The dispersion in this case corresponds to the MSSW dispersion [51]. Next, we consider $N = 40$ period length meander structure with the period of $L = 40$ μm, film thickness $d = 1$ μm,

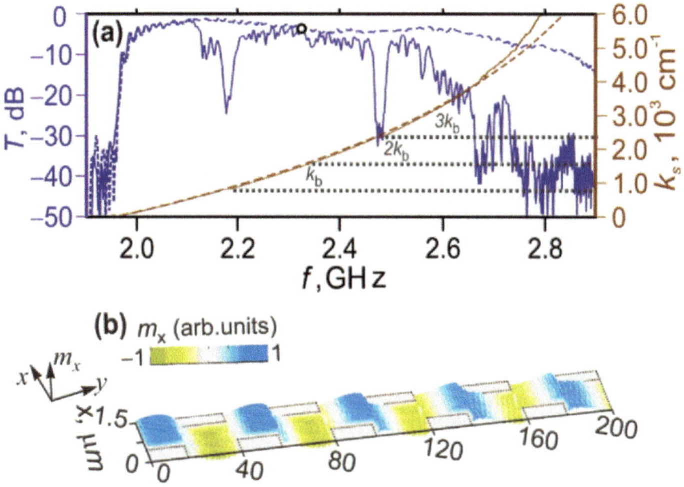

Figure 3.9 (a) Transmission response (blue curves) and dispersion characteristics (brown curves) for the reference structure (dashed curves) and MC (solid curves). (b) The spatial distribution of m_x at $f = 2.235$ GHz.

and $s = 1$ μm. It is assumed further that to excite a propagating SW in the waveguide, a sinusoidal magnetic field $b_x = b_0 \sin(2\pi f t)$ is applied over an area 5 μm in length, with the oscillation amplitude $b_0 = 0.1$ mT and microwave frequency $f = 2.235$ GHz. The dispersion and transmission response are demonstrated with the solid blue and brown curves in Fig. 3.9a. Spatial distribution of the m_x component of dynamic magnetization at $f = 2.235$ GHz is shown in Fig. 3.9b for the fragment of five periods.

The forbidden zone formation for SWs in the proposed 3D MC can be considered for a Bloch wave propagating along a periodic structure with a wavenumber k_b having a projection only on the y coordinate axis and for a partial wave propagating in horizontal (OA, BD, and CE) and vertical (AB and CD) segments with wavenumbers having, respectively, projections onto the horizontal y axis and vertical x axis. To form forbidden bands from the point of view of Bloch waves, it is necessary to satisfy the phase conditions on one period of the structure for the reflected wave (the reflected wave must coincide in phase with the incident wave for the realization of constructive interference). However, on the vertical segments of the

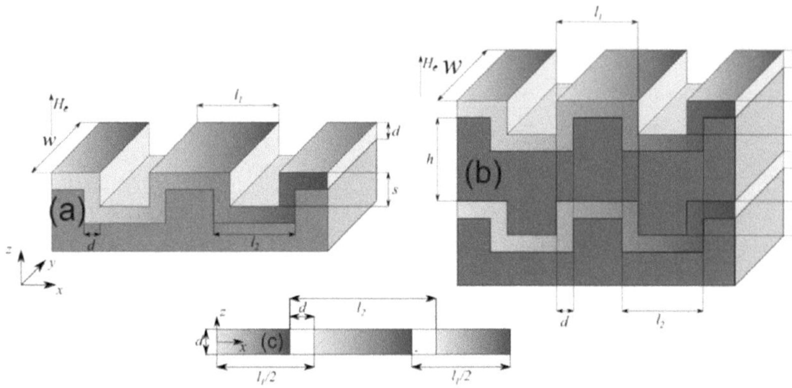

Figure 3.10 Schemes of (a) magnonic crystal with the meander configuration, (b) vertically coupled magnonic crystal with the meander configuration, and (c) equivalent ferromagnetic film.

periodic structure (AB and CD), the wave undergoes an additional phase shift depending on the length of the vertical segment s and the wavenumber. The forbidden bands in the 3D MC will be observed if the conditions of constructive interference in the vicinity of the nth Bragg frequency are satisfied for both Bloch and partial waves. Therefore, the situation is possible when for the first forbidden frequency zone ($n = 1$) for the length of the vertical segment s these conditions are met, and for the second zone ($n = 2$) these conditions are not satisfied at the same length of vertical section s.

3.4 Forward-Volume and Backward-Volume Magnetostatic Waves in 3D Structures

In the case of an external saturating magnetic field H_e (see Fig. 3.10a), applied along the x or z axis, volume MSWs can propagate: FVMSWs in normally magnetized segments and BVMSWs in segments magnetized along the direction of propagation. Dispersion characteristics of these waves have differences from MSSW dispersion and from each other. Thus, it is necessary to develop a theory of wave propagation that would take into account mechanisms of different wave type transformations.

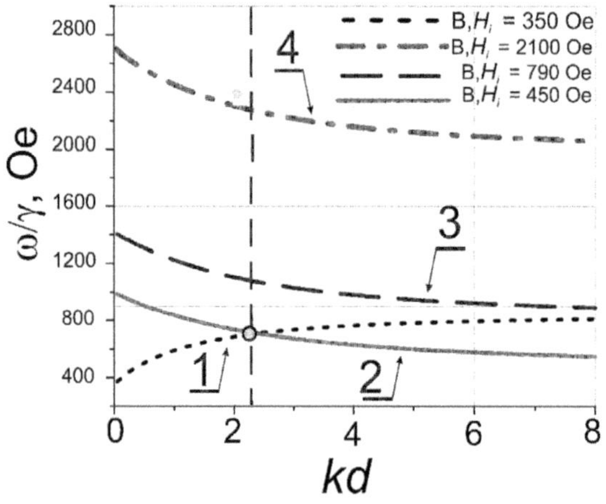

Figure 3.11 Dispersion curves for the first modes of (1) FVMSWs and (2–4) BVMSWs in infinite films with effective magnetic fields of different H_i at $4\pi M = 1750$ G.

It is known from Ref. [6, 52, 54, 55] that the FVMSW and BVMSW dispersion relations in infinite ferromagnetic films are determined by a single frequency interval provided that the film parameters (in particular, saturation magnetization and internal magnetic field) are identical. Figure 3.11 presents the dispersion relations for such waves in the presence of different effective internal fields.

In general, the demagnetization factors of the films are different in the presence of identical external magnetic fields. Thus, the wave spectra may be not overlapped. For a normally magnetized infinite plate, the internal field is represented as $H_i = H_e - 4\pi M$, where H_e is the external static magnetic field. Thus, the FVMSWs propagate in the frequency interval

$$\omega_{H_1} = \gamma H_e(1 - 4\pi M/H_e) < \omega < \omega_{\perp 1} = \gamma H_e\sqrt{1 - 4\pi M/H_e}.$$

For a tangentially magnetized plate, we have $H_i = H_e$. Therefore, the BVMSWs propagate in the frequency interval

$$\omega_{H_2} = \gamma H_e(1 - 4\pi M/H_e) < \omega < \omega_{\perp 2} = \gamma H_e\sqrt{1 + 4\pi M/H_e}.$$

It is seen that the intervals are not overlapped. The wave dispersion also strongly depends on the remaining transverse and longitudinal (with respect to wave propagation) sizes of the film in which the waves propagate. The reason for this is also related to the strong dependence of the internal magnetic field on the geometrical sizes of the sample. In addition, both waves exhibit a multimode character, since the wavenumber is quantized along the film thickness. In experiments, the first modes are predominantly excited and the remaining modes are insignificantly coupled to the first mode, especially in thin films, and, therefore, the coupling can be disregarded [51, 54]. For the structure under study, the periodicity along the x axis and the reflection of waves from the interfaces of the horizontal and vertical segments must lead to dispersion characteristics with the Bragg bandgaps at frequencies f_n, where the following conditions are satisfied: $k_s(f_n) = Nk_b$ ($N = 1, 2\ldots$ is the band number and k_b is the Bragg wavenumber). The bandgaps result from multiple reflections of waves from interfaces of vertical and horizontal segments of the structure. Such a scenario is possible if the wave structures inside segments are identical. The identity is reached when the dispersion curves of the waves that propagate in different segments are intersected.

In the structure under study (see Fig. 3.10), volume MSWs propagate as follows. An external magnetic field is exerted on the structure along the z or x axis. The waves are excited using the conventional microwave method and propagate along the x axis from the left- to the right-hand side. Regardless of the direction of the external magnetic field, an FVMSW must propagate in one segment and a BVMSW must propagate in another segment. If the conditions for coexistence of the frequency intervals of the waves are satisfied, the spatial structures of the waves are identical in the region of intersection of the dispersion curves. In this case, a wave (e.g., FVMSW) passes through a horizontal segment, turns downward and propagates as the BVMSW along the z axis, reaches the next segment, is transformed into the FVMSW, an so on. Such processes are possible provided that the dispersion curves of different volume waves are intersected.

To find dispersion relations of FVMSWs and BVMSWs, first we solve a static problem of the internal effective magnetic field at different points of the structure. Note that an element of period can be divided into two horizontal and two vertical segments in which directions of vector **M** are perpendicular and parallel to the segment, respectively. In the approximation $l_{1,2}$, w, $s \gg d$, the structure under study can be represented as a series of infinite films with different magnetizations. Under such conditions wave propagation is impeded, since a wave that propagates in one segment exhibits exponential decay in another one. However, variations in frequency s can be used to choose demagnetization factors that allow overlapped frequency intervals. When $s \approx d$ and $s \ll d$, the demagnetization factor in vertical elements becomes close to 2π and 4π, respectively. For the parameter $4\pi M = 1750$ G, which corresponds to the YIG film and an external field of $H_e = 2100$ Oe, we obtain $\omega_{H_1} = 350\gamma = 6.16$ rad/s and $\omega_{\perp 1} \approx 860\gamma = 15.14$ rad/s. For a demagnetization factor of 3π, we have $\omega H_1 = 790\gamma = 13.9$ rad/s for vertical segments, which belong to the frequency interval of the horizontal segments. Such demagnetization factor corresponds to $s < d$. Note that the internal fields are substantially non-uniform at the edges of elements at distances of about two minimum linear sizes.

To calculate dispersion relations for the propagation of volume waves in meander-type structures, we calculate the internal effective fields with the aid of micromagnetic simulation using Mumax3 [45]. The simulation was performed for structures with a saturation magnetization of $4\pi M = 1750$ G, which corresponds to YIG and the following parameters: external magentic field $H_e = 2100$ Oe, width of ferromagnetic film, $w = 40$ μm, and thickness $d = 1$ μm.

Figure 3.12 shows the distribution of the internal effective field over the film width for several configurations of the ferromagnetic structures (physically bounded waveguides). Figures 3.12a and 3.12b present distributions for the FVMSW and the BVMSW, respectively, and Fig. 3.13 shows the distribution for two (vertical and horizontal) coupled segments of the structure.

The analysis of the above dependencies shows that (i) a decrease in the waveguide length along the propagation direction leads to significant non-uniformity of the internal fields and (ii) field magnitudes exhibit significant variations so that the lengths at which the dispersion curves of different waves are intersected can be determined.

Figure 3.13 presents amplitude profiles of effective fields in the vicinity of interfaces of the segments that show noticeable non-

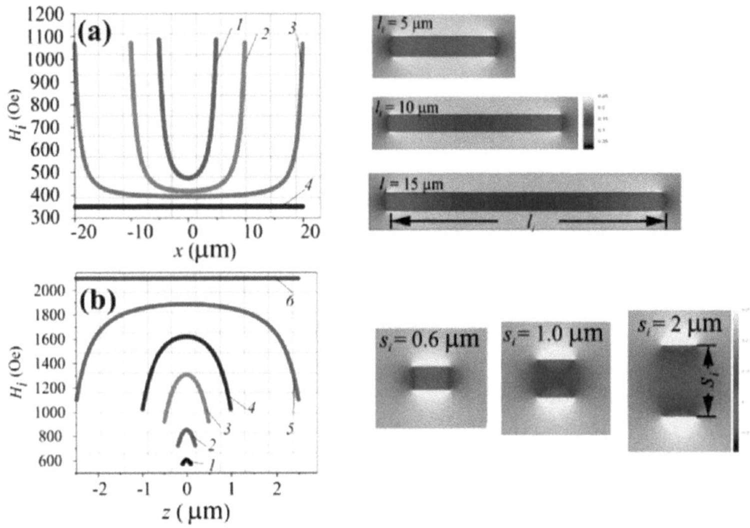

Figure 3.12 Results of simulation of the internal effective magnetic field: (a) (left-hand panel) distribution of the internal effective field and (right-hand panel) distribution of the z component of the field over length in normally magnetized plates with lengths of (1) 10, (2) 20, and (3) 40 μm and (4) infinite length; (b) (left-hand panel) distribution of the internal effective field and (right-hand panel) distribution of the z component of the field over height in tangentially magnetized plates with different heights. The parameters of the plates are $d = 1$ μm, $w = 40$ μm, $H_e = 2100$ Oe, and $4\pi M = 1750$ G.

uniformity of the field at such interfaces. Thus, the interval of geometrical sizes of the segments of the meander-type structure at which the FVMSW and BVMSW may simultaneously exist in different segments is relatively small. In particular, the lengths of vertical fragments must be less than the lengths of horizontal fragments (e.g., $s < d$ and $d \ll L$). Such a structure corresponds to the equivalent ferromagnetic film of Fig. 3.10c. We assume that the film with thickness d exhibits periodic (with respect to coordinate z) non-uniform internal field H_i. The regions of non-uniformity correspond to the regions of the vertical segments of the original structure and have identical widths.

To calculate the dispersion of magnetostatic waves, we change the structure of Fig. 3.10a by an equivalent structure with allowance

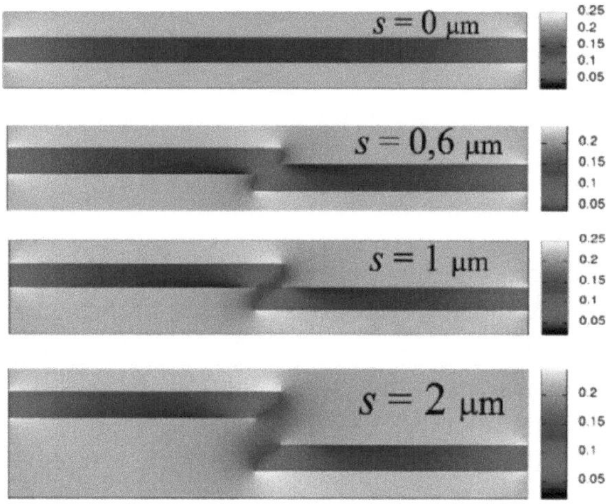

Figure 3.13 Distribution of the z component of the field over width of the cross section of a meander-type crystal in the vicinity of the step. The parameters of the plates are $d = 1$ μm, $w = 40$ μm, $H_e = 2100$ Oe, and $4\pi M = 1750$ G.

for the geometrical sizes at which the FVMSW and BVMSW may exist in different segments. We assume that the film with thickness d exhibits a periodic (with respect to coordinate x) non-uniform internal field H. The regions of non-uniformity correspond to the regions of vertical segments of the original structure, and the widths of the former and latter are approximately equal.

The plane wave expansion method was already applied [56, 57] to the planar periodic magnonic structures. Peculiarities of the PWE method in the case of 3D structures and equivalent structures (see Fig. 3.10c) were studied in detail [44]. To avoid cumbersome derivations of matrix equations we reveal here only milestones of this technique. On the basis of the periodicity with respect to the x axis, we employ the Bloch theorem and expand quantities Ψ and μ in terms of the Bloch harmonics [54]:

$$\Psi = \sum_n \Psi_n(z)\exp(i(k + b_n)x) = \sum_n \Psi_n(z)\exp(i(k_n)x),$$

$$\mu = \sum_m M_n(z)\exp(i(b_m)x),\ b_n = 2\pi n/(l_1 + l_2 - 2d),$$

$$M_m = 1/(l_1 + l_2 - 2d) \int_0^{l_1+l_2-2d} \mu(x) \exp(-ib_m x) dx,$$

$$C_m = 1/(l_1 + l_2 - 2d) \int_0^{l_1+l_2-2d} \exp(-ib_m x) dx,$$

$$1 = \sum_m C_m \exp(ib_m x).$$

Substituting these expressions into Maxwell's equations under magnetostatic approximation [6], renumerating summation indices, and limiting the expansion set by $2N + 1(-N \leq p \leq N)$, we obtain an eigenvalue problem $[M_1]\Psi(z) = \lambda^2[M_2]\Psi(z)$ in matrix representation, where $[M_1]_{pn} = -M_{p-n}k_n k_p$, $[M_2] = C_{p-n} = \delta_{pn}$. This equation is an eigenvalue problem, the solution to which yields a set of $2N + 1$ eigenvalues λ_s^2 and vectors $\Psi_s(z)$. The eigenvalues are real, since the matrices are Hermitian matrices. Introducing a set of $8N + 4$ amplitude coefficients (A_s, B_s, C_s, D_s) and applying boundary conditions of continuity of magnetostatic potential Ψ and the magnetic field normal component $b_z = (\overleftrightarrow{\mu}\vec{\nabla}(\Psi))_z$ we can rewrite boundary conditions in matrix form $[F(k, \omega)] \times \{A_s, B_s, C_s, D_s\}^T = 0$. $[F(k, \omega)]$ is an $4N + 8$ order matrix and its determinant should be equal to zero $det[F(k, \omega)] = 0$. For each fixed wavenumber k we obtain $2N + 1$ frequencies ω_s.

To consider the properties of waves that propagate in the above periodic inhomogeneous film, we calculate the transfer function of the structure. Figure 3.14a shows the frequency dependence of the transmission coefficient for a structure with a finite number of periods, and Fig. 3.14b presents the dispersion relations for the FVMSW and BVMSW that propagate in the meander-type ferromagnetic film. We clearly observe the formation of bandgaps in the wave spectra with the corresponding Bragg frequencies and, hence, wavenumbers. The bandgap decreases in width with an increase in the number of the Bragg resonance due to a decrease in the group velocity of waves and, hence, weaker interaction of the spatial harmonics of waves. With relatively high accuracy, the positions of amplitude minima (Fig. 3.14a) coincide with the positions of the Bragg bandgaps (BG1, BG2, and BG3) on the dispersion curves of the waves (Fig. 3.14b), so we obtain minimum transmission and maximum opacity for the corresponding frequency intervals.

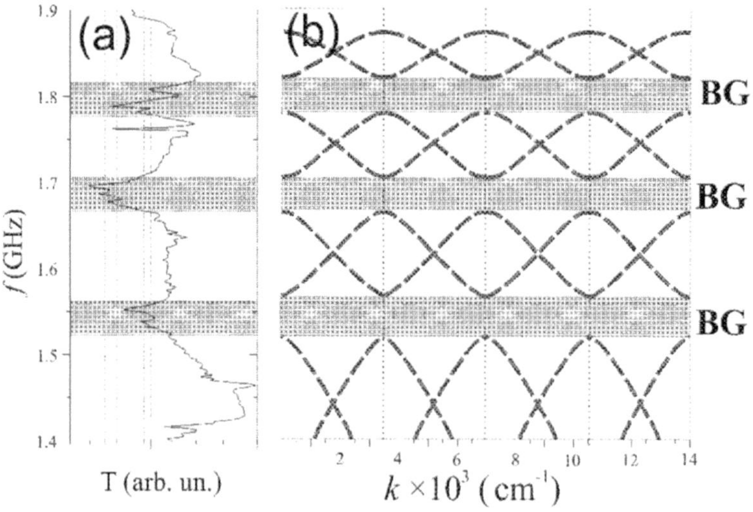

Figure 3.14 Calculated transmission coefficient (a) and analytical dispersion (b) for $s = 0.3$ μm, $l_1 = l_2 = 10$ μm, $d = 1$ μm, and $w = 40$ μm.

3.5 Magnetostatic Waves in Vertically Coupled Structures

3.5.1 Magnetostatic Surface Waves in Vertically Coupled Structures

Here, we consider coupled 3D structures (see Fig. 3.1c). The SW spectra in coupled ferromagnetic films [58–60] and coupled MCs [8, 61] are essentially more complicated that in a single film or even in a single MC. This is due to the fact that in such structures instead of only modes of the propagating wave in single films, the coupled modes appear having different dispersion.

We consider the meander-type ferromagnetic films coupled through the nonmagnetic intermediate layer with the smallest distance between the ferromagnetic meanders being equal to $r = 5$ μm. The results of calculations are presented in Fig. 3.15a (dispersion) and Fig. 3.15b–e (magnetostatic potential distribution), along with geometrical parameters at which the calculations were performed. To analyze the properties of MSW propagation in two coupled 3D MCs we use the reference model based on two coupled

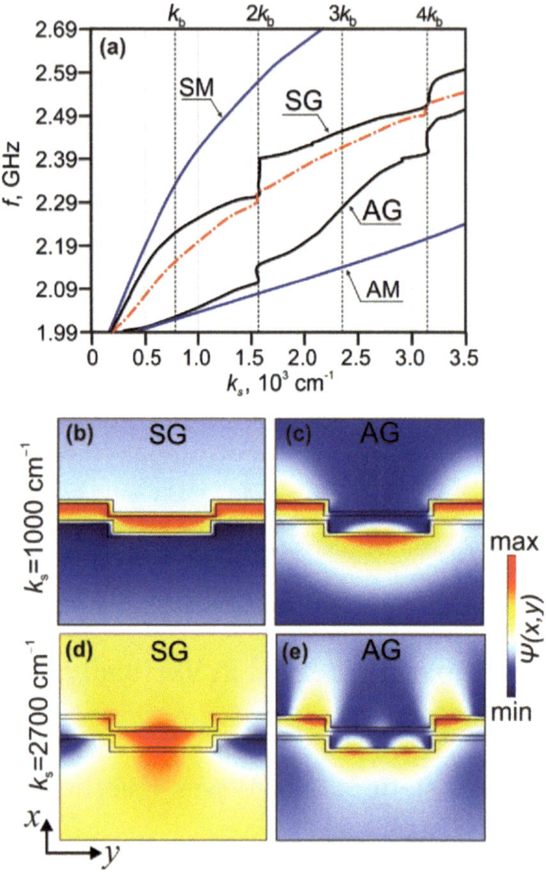

Figure 3.15 (a) Dispersion characteristics of MSWs in a coupled 3D MC structure (r = 5 μm, d = 1 μm, L = 40 μm, s = 3 μm), black solid curves; dispersion MSWs in a single 3D MC, red dashed-dotted curve; and dispersion of a reference-coupled uniform magnetic film, blue curves. Magnetostatic potential distribution for a coupled 3D MC at fixed Bloch wavenumbers $k_s = 1000$ cm^{-1} (b) and (c) and $k_s = 2700$ cm^{-1} (d) and (e). The vertical dashed line denotes the symmetry axes.

uniform YIG films separated by a dielectric nonmagnetic layer with thickness 5 μm. The coupling between films lead to splitting of the dispersion curve into two symmetric (SM) and antisymmetric (AM) modes [8] (see Fig. 3.15a). The dashed line in Fig. 3.15a denotes the dispersion curve for a single 3D MC with the meander paramater $s = 3$ μm. Similar to a single 3D MC the dispersion band structure is formed for both coupled modes, which we denote as SG and AG for the SM and AM, respectively. Positions of bands respect to the positions of a single 3D MC and, respectively, have intermediate values. The magnetostatic potentials distributions for coupled MCs were calculated for two fixed values of wavenumbers that are far from those for the magnonic bandgap. The geometrical structure of two coupled 3D MCs (shown in Fig. 3.1c) has symmetry only with respect to the vertical x axis. The potential distribution for the SM and AM conserves this symmetry type for all possible values of k_S. However, along the horizontal y axis coupled 3D MCs are nonsymmetrical, and this leads to special features of the magnetostatic amplitudes distributions. The symmetric mode SG for small wavenumbers ($k_S = 1000$ cm^{-1}) has a local minimum at the lower surface of the upper 3D MC at any values of the y coordinate. As the wavenumber (frequency) is increased ($k_S = 2700$ cm^{-1}) the potential amplitude distribution is defined also by the superposition of the propagating and reflected SG waves (SM). However, the interleaving of the local maximum and minimums for modes is observed (Fig. 3.15c). Localized SM and AM have quite a different character of field distribution where the local maxima and minima are interchanged. For both $k_S = 1000$ cm^{-1} and $k_S = 2700$ cm^{-1} in the upper MC there is maximum of Ψ in SG mode and minimum of Ψ for AG mode. However, the magnetostatic potential has maxima at points A and C for each mode. Finally, the developed model could be used to explain the nonreciprocal features of the dipolar SWs' propagation in the proposed class of structures. Thus, our results show that vertical SW coupling opens up the possibility of the utilization of 3D magnonic structures as SW-based building blocks for processing analog and digital signals via the manipulation of SW characteristics.

3.5.2 Forward-Volume and Backward-Volume Magnetostatic Waves in Vertically Coupled Structures

We also consider the propagation of magnetostatic SWs in two vertically coupled structures for different non-uniform fields (Fig. 3.10b).

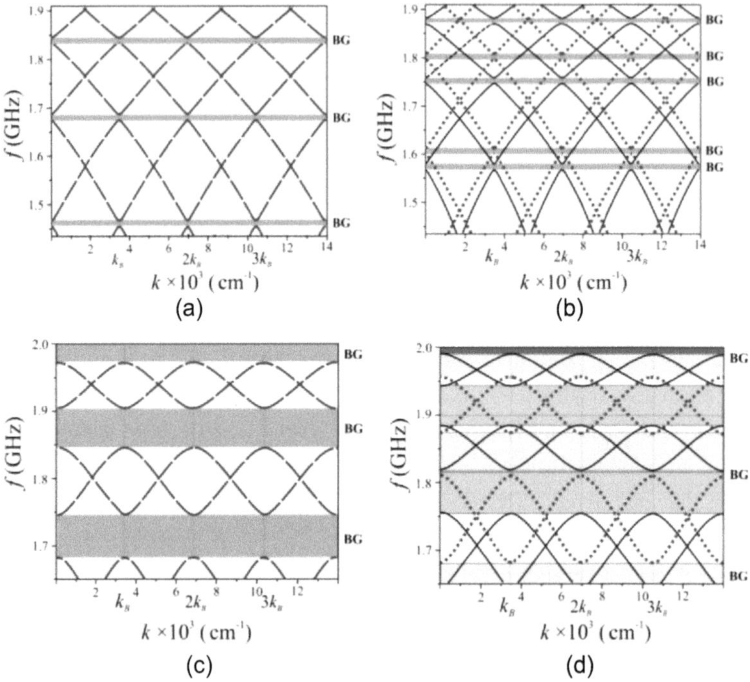

Figure 3.16 Dispersion curves for (a and c) single (dashed curves) and (b and d) coupled structures in the presence of non-uniform fields of H = (a and b) 420 Oe and (c and d) 550 Oe for the distance between the plates d = 2 μm: (dotted curves) fast and (solid curves) slow modes.

The spectra of SWs in vertically coupled ferromagnetic films are significantly more complicated in comparison with the spectra of single films. Such result is due to the fact that the spectra of single modes are supplemented with the spectra of coupled oscillations of the magnetic structures [60–62]. The coupled waves have more complicated dispersion, and the structure consisting of two meander-type ferromagnetic films separated by a dielectric layer exhibits FVMSWs and BVMSWs and, in addition, coupled waves of the two types. For the system under study, the thickness of the dielectric layer is 2 μm. In general, the coupling of two films leads to symmetric and antisymmetric coupled modes. Figures 3.16a and 3.16d show the dispersion relations of such coupled waves and their comparison with ones in single-layered structures (see Figs. 3.16a and 3.16d).

3.6 Conclusion

Magnetic thin films and heterostructures based on them are quite essential structures for applications in novel types of electronics, namely spintronics and magnonics. Especially this concerns the beyond-CMOS technology due to a possibility of these structures to vary their properties in a wide frequency range, their small radiation resistance, and their low power consumption. We suggested in this chapter a new idea of how magnonics can go in three dimensions (instead of one or two dimensions), thus staying on the same technological platform as conventional electronics. We presented theoretical investigations of FMR study of these structures, along with investigations of propagating SWs in simple 3D films or coupled heterostructures. In such way the proposed 3D MCs and heterostructures are building blocks of any complex integral magnonic networks for high-speed signal processing devices.

Acknowledgments

The research was carried out within the RFBR grant (project nos. 18-57-00006 and 18-37-00373), within the state assignment (theme Nanoferroid-3 No. 0030-2015-0152) and supported in part by the Government of the Russian Federation (agreement no. 074-02-2018-286) within the laboratory "Terahertz spintronics" of the Moscow Institute of Physics and Technology (State University).

References

1. Moore, G. E. (1965). *Cramming More Components Onto Integrated Circuits* (McGraw-Hill).
2. Semiconductor Industry Association (2015). International technology roadmap for semiconductors (itrs), URL http://www.itrs2.net/itrs-reports.html.

3. Lau, J. H. (2013). *Through-Silicon Vias for 3D Integration* (McGraw-Hill).
4. Burghartz, J. (ed.) (2011). *Ultra-thin Chip Technology and Applications* (Springer New York), doi:10.1007/978-1-4419-7276-7, URL https://doi.org/10.1007/978-1-4419-7276-7.
5. Ishak, W. (1988). Magnetostatic wave technology: a review, *Proc. IEEE*, **76**(2), 171–187, doi:10.1109/5.4393, URL https://doi.org/10.1109/5.4393.
6. Gurevich, A. G. and Melkov, G. A. (1996). *Magnetization, Oscillations and Waves* (CRC Press New York).
7. Sodha, M. S. and Srivastava, N. C. (1981). *Microwave Propagation in Ferrimagnetics* (Springer US), doi:10.1007/978-1-4757-5839-9, URL https://doi.org/10.1007/978-1-4757-5839-9.
8. Sadovnikov, A. V., Beginin, E. N., Sheshukova, S. E., Romanenko, D. V., Sharaevskii, Y. P. and Nikitov, S. A. (2015b). Directional multimode coupler for planar magnonics: side-coupled magnetic stripes, *Appl. Phys. Lett.*, **107**(20), 202405, doi:10.1063/1.4936207, URL https://doi.org/10.1063/1.4936207.
9. Sadovnikov, A., Grachev, A., Beginin, E., Sheshukova, S., Sharaevskii, Y. and Nikitov, S. (2017a). Voltage-controlled spin-wave coupling in adjacent ferromagnetic-ferroelectric heterostructures, *Phys. Rev. Appl.*, **7**(1), doi:10.1103/physrevapplied.7.014013, URL https://doi.org/10.1103/physrevapplied.7.014013.
10. Sadovnikov, A. V., Odintsov, S. A., Beginin, E. N., Sheshukova, S. E., Sharaevskii, Y. P. and Nikitov, S. A. (2017d). Toward nonlinear magnonics: intensity-dependent spin-wave switching in insulating side-coupled magnetic stripes, *Phys. Rev. B*, **96**(14), doi:10.1103/physrevb.96.144428, URL https://doi.org/10.1103/physrevb.96.144428.
11. Kruglyak, V. V., Demokritov, S. O. and Grundler, D. (2010). Magnonics, *J. Phys. D: Appl. Phys.*, **43**(26), 264001, doi:10.1088/0022-3727/43/26/264001, URL https://doi.org/10.1088%2F0022-3727%2F43%2F26%2F264001.
12. Nikitov, S. A., Kalyabin, D. V., Lisenkov, I. V., Slavin, A., Barabanenkov, Y. N., Osokin, S. A., Sadovnikov, A. V., Beginin, E. N., Morozova, M. A., Filimonov, Y. A., Khivintsev, Y. V., Vysotsky, S. L., Sakharov, V. K. and Pavlov, E. S. (2015). Magnonics: a new research area in spintronics and spin wave electronics, *Phys. Usp.*, **58**(10), 1002–1028, doi:10.3367/ufne.0185.201510m.1099, URL https://doi.org/10.3367/ufne.0185.201510m.1099.

13. Demokritov, S. O. (ed.) (2017). *Spin Wave Confinement: Propagating Waves* (Pan Stanford Publishing).
14. Sharaevsky, Y. P., Sadovnikov, A. V., Beginin, E. N., Sharaevskaya, A. Y., Sheshukova, S. E. and Nikitov, S. A. (2018). Functional magnetic metamaterials for spintronics, in *NanoScience and Technology* (Springer), pp. 221–245, doi: 10.1007/978-3-319-90481-8 11, URL https://doi.org/10.1007/978-3-319-90481-8_11.
15. Demokritov, S. O. (2018). *Topology in Magnetism*, Springer Series in Solid-State Sciences, Vol. 192, chap. Magnons (Springer).
16. Locatelli, N., Cros, V. and Grollier, J. (2014). Spin-torque building blocks, *Nat. Mater.*, **13**(1), 11–20, doi:10.1038/nmat3823, URL https://doi.org/10.1038/nmat3823.
17. Krawczyk, M. and Grundler, D. (2014). Review and prospects of magnonic crystals and devices with reprogrammable band structure, *J. Phys.: Condens. Matter*, **26**(12), 123202, doi:10.1088/0953-8984/26/12/123202, URL https://doi.org/10.1088/0953-8984/26/12/123202.
18. Vogel, M., Chumak, A. V., Waller, E. H., Langner, T., Vasyuchka, V. I., Hillebrands, B. and von Freymann, G. (2015). Optically reconfigurable magnetic materials, *Nat. Phys.*, **11**(6), 487–491, doi: 10.1038/nphys3325, URL https://doi.org/10.1038/nphys3325.
19. Nikitin, A. A., Ustinov, A. B., Semenov, A. A., Chumak, A. V., Serga, A. A., Vasyuchka, V. I., Lähderanta, E., Kalinikos, B. A. and Hillebrands, B. (2015). A spin-wave logic gate based on a widthmodulated dynamic magnonic crystal, *Appl. Phys. Lett.*, **106**(10), 102405, doi:10.1063/1.4914506, URL https://doi. org/10.1063/1.4914506.
20. Klingler, S., Pirro, P., Brächer, T., Leven, B., Hillebrands, B. and Chumak, A. V. (2015). Spin-wave logic devices based on isotropic forward volume magnetostatic waves, *Appl. Phys. Lett.*, **106**(21), 212406, doi:10.1063/1.4921850, URL https://doi.org/10.1063/1.4921850.
21. Demidov, V. E., Urazhdin, S., Zholud, A., Sadovnikov, A. V. and Demokritov, S. O. (2014). Nanoconstriction-based spin-hall nanooscillator, *Appl. Phys. Lett.*, **105**(17), 172410, doi:10.1063/1.4901027, URL https://doi.org/10.1063/1.4901027.
22. Demidov, V. E., Urazhdin, S., Zholud, A., Sadovnikov, A. V. and Demokritov, S. O. (2015a). Dipolar field-induced spin-wave waveguides for spin-torque magnonics, *Appl. Phys. Lett.*, **106**(2), 022403, doi:10.1063/1.4905869, URL https://doi.org/10.1063/1.4905869.

23. Demidov, V. E., Urazhdin, S., Zholud, A., Sadovnikov, A. V., Slavin, A. N. and Demokritov, S. O. (2015b). Spin-current nano-oscillator based on nonlocal spin injection, *Sci. Rep.*, **5**(1), doi:10.1038/srep08578, URL https://doi.org/10.1038/srep08578.

24. Khitun, A., Nikonov, D. E., Bao, M., Galatsis, K. and Wang, K. L. (2007). Efficiency of spin-wave bus for information transmission, *IEEE Trans. Electron Devices*, **54**(12), 3418–3421, doi: 10.1109/ted.2007.908898, URL https://doi.org/10.1109/ted.2007.908898.

25. Sadovnikov, A. V., Beginin, E. N., Bublikov, K. V., Grishin, S. V., Sheshukova, S. E., Sharaevskii, Y. P. and Nikitov, S. A. (2015a). Brillouin light scattering study of transverse mode coupling in confined yttrium iron garnet/barium strontium titanate multiferroic, *J. Appl. Phys.*, **118**(20), 203906, doi: 10.1063/1.4936320, URL https://doi.org/10.1063/1.4936320.

26. Sadovnikov, A. V., Beginin, E. N., Morozova, M. A., Sharaevskii, Y. P., Grishin, S. V., Sheshukova, S. E. and Nikitov, S. A. (2016). Nonlinear spin wave coupling in adjacent magnonic crystals, *Appl. Phys. Lett.*, **109**(4), 042407, doi:10.1063/1.4960195, URL https://doi.org/10.1063/1.4960195.

27. Sadovnikov, A. V., Davies, C. S., Kruglyak, V. V., Romanenko, D. V., Grishin, S. V., Beginin, E. N., Sharaevskii, Y. P. and Nikitov, S. A. (2017b). Spin wave propagation in a uniformly biased curved magnonic waveguide, *Phys. Rev. B*, **96**(6), doi:10.1103/ physrevb.96.060401, URL https://doi.org/10.1103/physrevb. 96.060401.

28. Sadovnikov, A. V., Grachev, A. A., Beginin, E. N., Odintsov, S. A., Sheshukova, S. E., Sharaevskii, Y. P., Serdobintsev, A. A., Mitin, D. M. and Nikitov, S. A. (2017c). Coupled spin waves in magnetic waveguides induced by elastic deformations in YIG–piezoelectric structures, *JETP Lett.*, **106**(7), 465– 469, doi:10.1134/s0021364017190110, URL https://doi.org/10.1134/s0021364017190110.

29. Sadovnikov, A., Grachev, A., Sheshukova, S., Sharaevskii, Y., Serdobintsev, A., Mitin, D. and Nikitov, S. (2018a). Magnon straintronics: reconfigurable spin-wave routing in strain-controlled bilateral magnetic stripes, *Phys. Rev. Lett.*, **120**(25), doi:10. 1103/physrevlett.120.257203, URL https://doi.org/10.1103/ physrevlett.120.257203.

30. Sadovnikov, A., Gubanov, V., Sheshukova, S., Sharaevskii, Y. and Nikitov, S. (2018b). Spin-wave drop filter based on asymmetric side-coupled magnonic crystals, *Phys. Rev. Appl*, **9**(5), doi:10.1103/physrevapplied .9.051002, URL https://doi.org/10.1103/physrevapplied.9.051002.

31. (a) Sadovnikov, A. V., Odintsov, S. A., Beginin, E. N., Grachev, A. A., Gubanov, V. A., Sheshukova, S. E., Sharaevskii, Y. P. and Nikitov, S. A. (2018c). Nonlinear spin wave effects in the system of lateral magnonic structures, *JETP Lett.*, **107**(1), 25–29, doi:10.1134/s0021364018 010113, URL https://doi.org/10.1134/s0021364018010113. (b) Sadovnikov, A. V., Davies, C. S., Grishin, S. V., Kruglyak, V. V., Romanenko, D. V., Sharaevskii, Y. P. and Nikitov, S. A. (2015c). Magnonic beam splitter: the building block of parallel magnonic circuitry, *Appl. Phys. Lett.*, **106**(19), 192406, doi: 10.1063/1.4921206, URL https://doi.org/ 10.1063/1.4921206.

32. Nikitov, S., Tailhades, P. and Tsai, C. (2001). Spin waves in periodic magnetic structures—magnonic crystals, *J. Magn. Magn. Mater.*, **236**(3), 320–330, doi:10. 1016/s0304-8853(01)00470-x, URL https://doi. org/10.1016/ s0304-8853(01)00470-x.

33. Chumak, A. V., Serga, A. A. and Hillebrands, B. (2017). Magnonic crystals for data processing, *J. Phys. D: Appl. Phys.*, **50**(24), 244001, doi:10.1088/1361-6463/aa6a65, URL https://doi.org/10.1088/1361-6463/aa6a65.

34. Beginin, E. N., Sadovnikov, A. V., Sharaevsky, Y. P. and Nikitov, S. A. (2014). Multimode surface magnetostatic wave propagation in irregular planar YIG waveguide, *Solid State Phenom.*, **215**, 389–393, doi: 10.4028/www. scientific. net/ssp.215.389, URL https://doi.org/10. 4028/www.scientific.net/ssp.215. 389.

35. Sheshukova, S. E., Beginin, E. N., Sadovnikov, A. V., Sharaevsky, Y. P. and Nikitov, S. A. (2014). Multimode propagation of magnetostatic waves in a width-modulated yttrium-iron-garnet waveguide, *IEEE Magn. Lett.*, **5**, 1–4, doi:10.1109/lmag.2014.2365431, URL https://doi.org/10.1109/lmag.2014.2365431.

36. Barabanenkov, Y., Osokin, S., Kalyabin, D. and Nikitov, S. (2015). Spin-wave bound modes in a circular array of magnetic inclusions embedded into a metallized ferromagnetic thin-film matrix, *Phys. Rev. B*, **91**(21), doi:10.1103/physrevb.91.214419, URL https://doi. org/10.1103/physrevb.91.214419.

37. Krawczyk, M. and Puszkarski, H. (2008). Plane-wave theory of three-dimensional magnonic crystals, *Phys. Rev. B*, **77**(5), doi:10.1103/ physrevb.77.054437, URL https://doi.org/10.1103/physrevb.77.0544 37.

38. Kostylev, M., Stashkevich, A. A., Roussigné, Y., Grigoryeva, N. A., Mistonov, A. A., Menzel, D., Sapoletova, N. A., Napolskii, K. S.,

Eliseev, A. A., Lukashin, A. V., Grigoriev, S. V. and Samarin, S. N. (2012). Microwave properties of ni-based ferromagnetic inverse opals, *Phys. Rev. B*, **86**(18), doi:10.1103/physrevb.86.184431, URL https://doi.org/10.1103/physrevb.86.184431.

39. Davies, C. S., Francis, A., Sadovnikov, A. V., Chertopalov, S. V., Bryan, M. T., Grishin, S. V., Allwood, D. A., Sharaevskii, Y. P., Nikitov, S. A. and Kruglyak, V. V. (2015). Towards graded-index magnonics: steering spin waves in magnonic networks, *Phys. Rev. B*, **92**(2), doi:10.1103/physrevb.92.020408, URL https://doi.org/10.1103/physrevb.92.020408.
40. Okuda, M., Schwarze, T., Eloi, J.-C., Jones, S. E. W., Heard, P. J., Sarua, A., Ahmad, E., Kruglyak, V. V., Grundler, D. and Schwarzacher, W. (2017). Top-down design of magnonic crystals from bottom-up magnetic nanoparticles through protein arrays, *Nanotechnology*, **28**(15), 155301, doi:10.1088/1361-6528/aa62f3, URL https://doi.org/10.1088/1361-6528/aa62f3.
41. Nipan, G. D., Stognij, A. I. and Ketsko, V. A. (2012). Oxide ferromagnetic semiconductors: coatings and films, *Russ. Chem. Rev.*, **81**(5), 458–475, doi:10.1070/rc2012v081n05abeh004251, URL https://doi.org/10.1070/rc2012v081n05abeh004251.
42. Beginin, E. N., Sadovnikov, A. V., Sharaevskaya, A. Y., Stognij, A. I. and Nikitov, S. A. (2018). Spin wave steering in threedimensional magnonic networks, *Appl. Phys. Lett.*, **112**(12), 122404, doi:10.1063/1.5023138, URL https://doi.org/10.1063/1.5023138.
43. Popov, P., Sharaevskaya, A., Beginin, E., Sadovnikov, A., Stognij, A., Kalyabin, D. and Nikitov, S. (2019). Spin wave propagation in three-dimensional magnonic crystals and coupled structures, *J. Magn. Magn. Mater.*, **476**, 423–427, doi: 10.1016/j.jmmm.2018.12.008, URL https://doi.org/10.1016/j.jmmm.2018.12.008.
44. Popov, P. A., Sharaevskaya, A. Y., Kalyabin, D. V., Stognii, A. I., Beginin, E. N., Sadovnikov, A. V. and Nikitov, S. A. (2018). Volume magnetostatic spin waves in 3D ferromagnetic structures, *J. Commun. Technol. Electron.*, **63**(12), 1431–1438, doi:10.1134/s1064226918120161, URL https://doi.org/10.1134/s1064226918120161.
45. Vansteenkiste, A., Leliaert, J., Dvornik, M., Helsen, M., Garcia-Sanchez, F. and Van Waeyenberge, B. (2014). The design and verification of mumax3, *AIP Adv.*, **4**(10), 107133, doi:10.1063/1.4899186, URL https://doi.org/10.1063/1.4899186.
46. Kalinikos, B. A. and Slavin, A. N. (1986). Theory of dipole-exchange spin wave spectrum for ferromagnetic films with mixed exchange boundary

conditions, *J. Phys. C: Solid State Phys.*, **19**(35), 7013–7033, doi:10. 1088/0022-3719/19/35/014, URL https://doi.org/10.1088/0022-3719/19/35/014.

47. Schlömann, E. (1964). Generation of spin waves in nonuniform magnetic fields. I. Conversion of electromagnetic power into spin-wave power and vice versa, *J. Appl. Phys.*, **35**(1), 159–166, doi:10.1063/1.1713058, URL https://doi.org/10.1063/1.1713058.
48. Jorzick, J., Demokritov, S. O., Hillebrands, B., Bailleul, M., Fermon, C., Guslienko, K. Y., Slavin, A. N., Berkov, D. V. and Gorn, N. L. (2002). Spin wave wells in nonellipsoidal micrometer size magnetic elements, *Phys. Rev. Lett.*, **88**(4), doi: 10.1103/physrevlett.88.047204, URL https://doi.org/10.1103/ physrevlett.88.047204.
49. Bayer, C., Jorzick, J., Demokritov, S. O., Slavin, A. N., Guslienko, K. Y., Berkov, D. V., Gorn, N. L., Kostylev, M. P. and Hillebrands, B. (2006). Spin-wave excitations in finite rectangular elements, in *Topics in Applied Physics* (Springer Berlin Heidelberg), pp. 57–103, doi:10.1007/10938171 2, URL https://doi.org/10.1007/10938171_2.
50. Gubbiotti, G., Conti, M., Carlotti, G., Candeloro, P., Fabrizio, E. D., Guslienko, K. Y., Andre, A., Bayer, C. and Slavin, A. N. (2004). Magnetic field dependence of quantized and localized spin wave modes in thin rectangular magnetic dots, *J. Phys.: Condens. Matter*, **16**(43), 7709–7721, doi:10. 1088/0953-8984/16/43/011, URL https://doi.org/10. 1088/0953-8984/16/43/011.
51. Damon, R. and Eshbach, J. (1961). Magnetostatic modes of a ferromagnet slab, *J. Phys. Chem. Solids*, **19**(3), 308–320, doi:http://dx.doi. org/10.1016/0022-3697(61)90041-5, URL http://wwwsciencedirect. com/science/ article/ pii/ 0022369761900415.
52. Ueda, T., Ueda, Y., Shimasaki, H. and Tsutsumi, M. (2003). Numerical analysis of nonlinear magnetostatic wave propagation by finiteelement method, *IEEE Trans. Magn.*, **39**(5), 3157–3159.
53. Rojdestvenski, I. V., Cottam, M. G. and Slavin, A. N. (1996). Dipole-exchange spin wave spectra of exchangecoupled magnetic multilayers calculated by transfer matrix formalism, *J. Appl. Phys.*, **79**(8), 5724–5726, doi:10.1063/1.362231, URL https://aip.scitation.org/doi/ abs/10.1063/1.362231.
54. Stancil, D. D. And Prabhakar, A. (2009). *Spin Waves: Theory and Applications* (Springer).
55. Kabos, P. And Stalmachov, V. S. (1994). *Magnetostatic Waves and Their Application* (Springer).

56. Sokolovski, M. and Krawczyk, M. (2011). The magnetostatic modes in planar one-dimensional magnonic crystals with nanoscale sizes, *J. Nanopart. Res.*, **13**, 6085.
57. Lisenkov, I., Kalyabin, D., Osokin, S., Kłos, J. W., Krawczyk, M. and Nikitov, S. (2015). Nonreciprocity of edge modes in 1D magnonic crystal, *J. Magn. Magn. Mater.*, **378**, 313–319.
58. Koshiba, M., Hayata, K. and Suzuki, M. (1986). Finite-element solution of anisotropic waveguides with arbitrary tensor permittivity, *J. Lightwave Technol.*, **4**(2), 121–126, doi: 10.1109/jlt.1986.1074687, URL https://doi.org/10.1109/jlt.1986.1074687.
59. Kosaka, H., Kawashima, T., Tomita, A., Notomi, M., Tamamura, T., Sato, T. and Kawakami, S. (1999). Self-collimating phenomena in photonic crystals, *Appl. Phys. Lett.*, **74**(9), 1212–1214, doi:10.1063/1.123502, URL https://doi.org/10.1063/1.123502.
60. Sasaki, H. and Mikoshiba, N. (1979). Directional coupling of magnetostatic surface waves in layered magnetic thin films, *Electron. Lett.*, **15**(6), 172, doi:10.1049/el:19790121, URL https://doi.org/10.1049/el:19790121.
61. Morozova, M. A., Sharaevskaya, A. Y., Sadovnikov, A. V., Grishin, S. V., Romanenko, D. V., Beginin, E. N., Sharaevskii, Y. P. and Nikitov, S. A. (2016). Band gap formation and control in coupled periodic ferromagnetic structures, *J. Appl. Phys.*, **120**(22), 223901, doi:10.1063/1.4971410, URL https://doi.org/10.1063/1.4971410.
62. Osokin, S., Safin, A., Barabanenkov, Y. and Nikitov, S. (2018). Spin waves in finite chain of dipolarly coupled ferromagnetic pillars, *J. Magn. Magn. Mater.*, **465**, 519–523, doi:10.1016/j.jmmm.2018.06.041, URL https://doi.org/10.1016/j.jmmm.2018.06.041.

Chapter 4

Spin Waves in Magnetic Metal-Insulator Hybrid Nanostructures

Chuanpu Liu, Jilei Chen, and Haiming Yu
Fert Beijing Institute, BDBC, School of Microelectronics, Beihang University
Xueyuan Road 37, Beijing, 100191, China
haiming.yu@buaa.edu.cn

4.1 Introduction

Spin-wave (SW) propagation in magnetic thin films and thin-film-based magnetic structures has drawn tremendous attention recently due to its low-energy-consumption property and potential compatibility to the next generation of circuits beyond complementary metal-oxide semiconductors (CMOS) [1–5]. Periodically patterned nanostructures based on magnetic materials with the dimensions down to micrometer size or even smaller are referred as magnonic crystals (MCs) [6–10]. Yttrium iron garnet (YIG)-based MCs are widely investigated due to its low damping and long SW decay length [11–14].

In this chapter, we present recent experimental results based on hybrid MCs composited of YIG thin-film and ferromagnetic (FM)

Three-Dimensional Magnonics: Layered, Micro- and Nanostructures
Edited by Gianluca Gubbiotti

ISBN 978-981-4800-73-0 (Hardcover), 978-0-429-29915-5 (eBook)
www.jennystanford.com

nanowire (NW) arrays [15, 16]. Short-wavelength SWs are observed in the MCs. Considering the SW dispersion relations, such short-wavelength SWs already enter an exchange-dominated regime. The interlayer magnon–magnon coupling is investigated experimentally in such MCs, where a large anticrossing gap up to 1.58 GHz is observed. By further experimental investigation, we find that the magnon–magnon coupling strength can be tuned by different magnetic configurations of a Co/YIG bilayer.

The chapter is organized as follows: in Section 4.2, we present the fabrication procedures of YIG-based MCs and the excitation of short-wavelength SWs. In Section 4.3, we discuss the interlayer magnon–magnon coupling observed in MCs.

4.2 Long-Distance Propagation of Short-Wavelength Spin Waves

4.2.1 Sample Fabrication and All-Electrical Spin-Wave Measurement Technique

A YIG thin film is prepared by sputtering with a thickness of 20 nm [17]. Electron beam lithography (EBL) and ion beam etching (IBE) are used to fabricate an SW waveguide on the YIG thin film. Metallic FM NWs are patterned on top of the YIG waveguide (Fig. 4.1). The detailed fabrication procedure is as follows:

(1) Exposure of cross markers on the YIG thin film for alignment and the lift-off process. A 9% PMMA photoresist is used in the exposure procedure. Then the gold is evaporated on the markers by electron beam evaporation.
(2) Exposure of the YIG waveguide. The size of the waveguide is 250 μm × 90 μm. The waveguide is tapered in the edges to avoid SW reflection and interference.
(3) Etching of the YIG film with Ar ions by IBE. After etching, only the YIG waveguide is left.
(4) Exposure of periodical NWs on top of the YIG waveguide. The length of the NWs is 90 μm, and the width varies from 90 nm

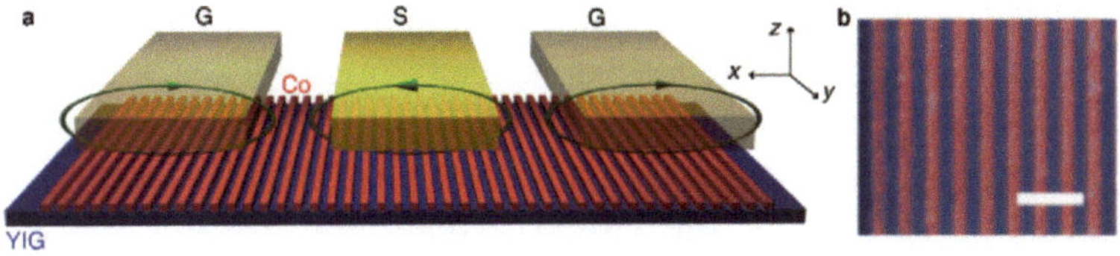

Figure 4.1 Co/YIG-based hybrid nanostructures. (a) Sketch of an YIG (20 nm)/Ti(1 nm)/Co(25 nm) heterostructure. Co NWs are patterned on top of the YIG thin film. (b) SEM image of the heterostructure. The red NWs are Co and the blue part underneath the Co is the YIG thin film. The scale bar is 500 nm. Reproduced from Ref. [15] (https://creativecommons.org/licenses/by/4.0/).

to 300 nm. The period of the NW array varies from 180 nm to 600 nm.

(5) Evaporation FM materials by electron beam evaporation. Magnetic materials such as cobalt and nickel are chosen due to their relatively larger saturation magnetization than YIG. The thickness of the FM NWs is 25 nm. Then 1 nm titanium is evaporated under the FM NWs as an adhesive layer.
(6) Exposure of the antenna on top of the FM NWs. A conventional coplanar waveguide (CPW) is fabricated with the 50 ohm impedance matched. The signal and ground line widths are 2 μm and the gap in between is 1.6 μm. Two identical CPWs are exposed, with the distance varying from 12 μm to 60 μm. The final step of the sample fabrication is the evaporation of the gold on the exposed CPWs.

The vector network analyzer (VNA)-based all-electrical SW spectroscopy technique is utilized to investigate short-wavelength SWs and the magnon–magnon coupling. The frequency range of the VNA (Rohde & Schwarz ZVA 40) is from 10 MHz to 40 GHz. The intermediate frequency (IF) bandwidth can reach 1 Hz. The two ports of a VNA are connected to two CPWs separately with microwave cables and microwave probes. The VNA is first calibrated using the calibration kit provided for the microwave probes. TOSM (open, short, match, and through) calibration is used to correct imperfections for the VNA, microwave cables, and microwave probes.

A microwave signal can be injected into the CPWs by the VNA, and an oscillating magnetic field is then generated, which leads to the excitation of SWs in the FM materials. The magnetic flux changes caused by SWs also apply an inducted microwave current, which could be detected by the VNA. The reflection (S_{11} and S_{22}) and transmission (S_{12} and S_{21}) spectra can be extracted from the VNA measurement.

4.2.2 Different Magnetic Configurations on the Co/YIG Hybrid Nanostructures

In this section, the different magnetic configurations measured on the Co/YIG-based nanostructures are shown. The period of the Co NW array is 200 nm and the center-to-center distance between two CPWs is 30 μm in such hybrid nanostructures. The reflection spectra S_{11} is shown in Fig. 4.2. The external magnetic field is first applied at −3000 Oe and then swept from −400 Oe to 400 Oe.

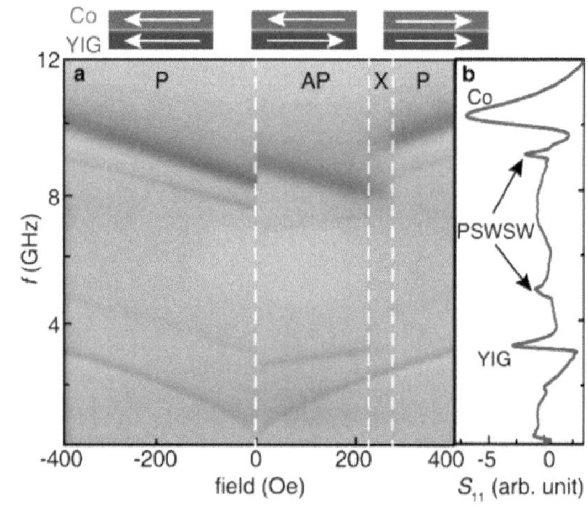

Figure 4.2 Different magnetic configurations of the Co/YIG-based heterostructure. (a) The grayscale reflection spectra S_{11} extracted from the VNA measurement. The applied field is parallel to the NWs and first set to −3000 Oe to saturate the magnetic heterostructure. Then the field is swept from −400 Oe to 400 Oe. (b) A line plot extract from (a) at 400 Oe. Reproduced from Ref. [15] (https://creativecommons.org/licenses/by/4.0/).

The magnetization of both YIG and Co NWs first oriented in the $-y$ direction. In Fig. 4.2a, one can see that four different regions are marked and different magnetic configurations are presented as follows:

(1) Parallel (P) configuration: The magnetization of the YIG film and Co NWs is both orienting in the same direction by applying a large external magnetic field. When sweeping the external magnetic field from -400 Oe to 0 Oe, the P configuration remains unchanged. In the figure, the highest signal branch is the ferromagnetic resonance (FMR) mode of Co NWs, while the lowest branch is the CPW excited mode of the YIG thin film. The frequency of the FMR mode of Co NWs does not reach 0 without an external field due to the shape anisotropy of the NWs.
(2) Antiparallel (AP) configuration: When sweeping the external magnetic field from negative to positive, the magnetization orientation of the YIG film follows the field due to the small coercivity. One can observe that the slope of the FMR of YIG mode changes the sign. However, the slope of the FMR mode of Co NWs remains the same as in the P configuration, which indicates that the magnetization of Co does not switch due to the large shape anisotropy. In this configuration, the magnetization of YIG and Co is AP. The AP configuration remains unchanged until around 230 Oe.
(3) Mixed (X) configuration: This configuration is a hybrid configuration between P and AP configurations. When sweeping a larger external magnetic field over 230 Oe, some of the Co NWs switch the magnetization orientation, while some remain the previous orientations. A multidomain state is formed in the Co NWs. The X state covers a frequency range of about 60 Oe.

4.2.3 Excitation of Short-Wavelength Spin Waves

In this section, we will discuss the excitation of short-wavelength SWs based on the Co/YIG nanostructures. One can observe that in Fig. 4.2a, besides the FMR modes of Co NWs and the YIG film, there are two resonance modes in between. We attribute these modes to high-order SW modes with a large wave vector and a

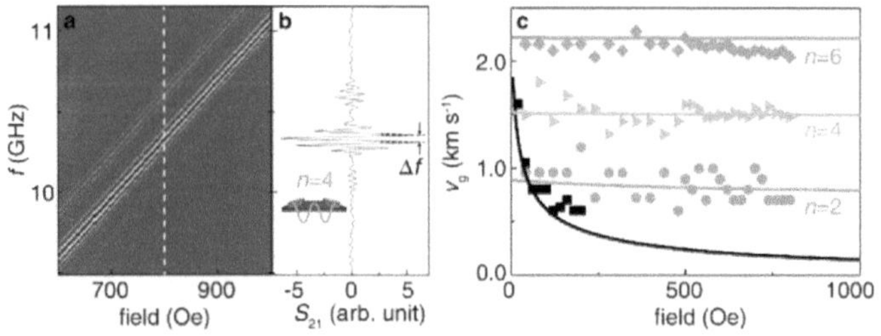

Figure 4.3 Propagating short-wavelength SWs and their group velocities. (a) The grayscale transmission spectra S_{21} measured on the Co/YIG-based heterostructure. The applied field first set to −3000 Oe to saturate the heterostructure and then swept from 600 Oe to 1000 Oe. The field step is 2.5 Oe. (b) A line plot extract from (a) at 800 Oe. (c) Group velocities of different SWs. Black squares present the group velocities of CPW-excited SWs in the YIG thin film extracted from the experimental results on a Co/YIG-based heterostructure. Circles, triangles, and diamonds show the group velocities of high-order SW modes with mode number $n = 2$, $n = 4$, and $n = 6$, respectively. The solid lines are the calculated group velocities based on the derivative of the SW dispersion relation in the YIG film. Reproduced from Ref. [15] (https://creativecommons.org/licenses/by/4.0/).

small wavelength. Compared to the dimension of CPWs, the size of the Co NWs is small and the alternating magnetic fields generated by CPWs can be considered uniformly distributed. The neighboring Co NWs precess in phase in this condition. Through the interlayer magnetic coupling, the precession of magnetization in YIG is driven by the precession of magnetization in the Co NWs. Due to the periodical boundary conditions, only the even-mode-number SWs can be formed, for example, $n = 2, 4, 6, \ldots$, which is opposite to the conventional perpendicular-standing spin waves (PSSWs), which only have odd mode numbers.

Figure 4.3a shows the transmission spectra S_{21} of the Co/YIG hybrid nanostructure taken from the imaginary part. A typical SW signal with the phase information is observed in the spectrum. The high-order SWs with the mode number $n = 4$ are presented, which possess the wavelength down to 100 nm. The frequency separation Δf indicates that the phase of the SW changes 2π, which is accumulated by the SW propagation between two CPWs. The group velocity of the SWs can be calculated using the following equation [18]:

$$v_g = \frac{d\omega}{dk} \approx \frac{2\pi \Delta f}{\frac{2\pi}{s}} = \Delta f \cdot s \qquad (4.1)$$

where $s = 30$ μm is the propagation distance of SWs. Using Eq. 4.1, the group velocities of different high-order SWs with mode number $n = 2, 4, 6, \ldots$ are shown in Fig. 4.3b. The SW that is excited directly by CPWs is also shown in the figure. One can observe that the group velocity of the $n = 6$ mode SW is around 2.2 km/s, which is even higher than the CPW-excited SW in all field range.

4.2.4 Dispersion of Dipole Exchange Spin Waves

In this section, we will discuss the dispersion relation of the observed high-order SW modes. The SW wavelength λ is determined by the periodicity of the Co NWs a with $\lambda = 2a/n$. The corresponding wave vector $\mathbf{k}$ then equals $n\pi/a$. We find the experimentally observed high-order SW frequencies follow well with the equation [19]:

$$\omega = \left[\left(\omega_0 + \omega_M \lambda_{ex} \mathbf{k}^2\right)\left(\omega_0 + \omega_M\left(\lambda_{ex}\mathbf{k}^2 + 1\right)\right)\right]^{1/2}, \qquad (4.2)$$

where $\omega = 2\pi f$, $\omega_0 = \gamma H$, $\omega_M = \gamma \mu_0 M_S$, and $\lambda_{ex} = 3 \times 10^{-12}$ J m^{-1} is the exchange constant. γ stands for the gyromagnetic ratio. The high-order SWs possess a relatively large wave vector than the conventional dipolar SWs and the exchange term $\lambda_{ex}\mathbf{k}^2$ dominates. Therefore, such dispersion relation follows the $f \propto \mathbf{k}^2$ dependance.

The highest-order SW observed in the experiment is the $n = 8$ mode with the wave vector $\mathbf{k} = 125.6$ μm^{-1}, which has a wavelength of 50 nm. The group velocity of such SWs reaches 2.6 km/s, which is calculated by Eq. 4.1. Compared to the other potential spintronics information carriers such as skyrmion and domain wall, the exchange SWs generated by Co NWs/YIG film-based nanostructures transfer information with higher velocity [19, 21]. Note that the short-wavelength SWs with a wave vector over about 20 μm^{-1} cannot be detected by the Brillouin light scattering (BLS) due to its limited wave vector resolution [22].

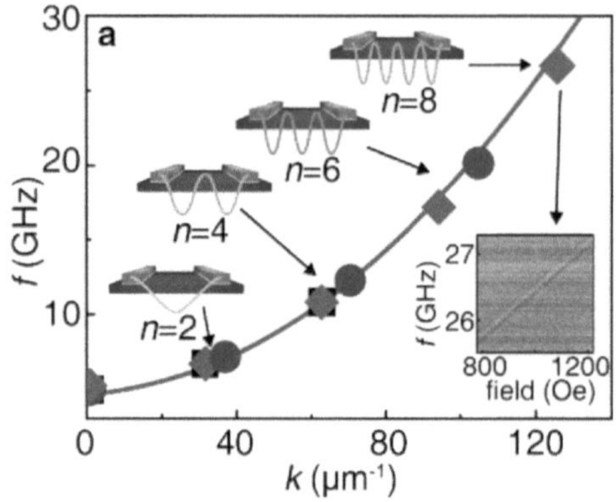

Figure 4.4 SW dispersion relations of the Co/YIG-based heterostructures. Data points shows the frequencies and the wavenumbers of different high-order SWs. The curve is the calculation of the SW dispersion in a YIG film. The inset shows the grayscale transmission spectra S_{21} of the high-order SW modes $n = 8$. Reproduced from Ref. [15] (https://creativecommons.org/licenses/by/4.0/).

To analyze the effective interlayer magnetic coupling that contributes to the formation of the high-order SWs, micromagnetic simulations are also performed. The program OOMMF (http://math.nist.gov/oommf) is utilized. The simulated structure is set exactly the same as the experiment. Two-dimensional fast Fourier transformation is performed and the SW dispersion can be obtained.

First, the interlayer exchange coupling is eliminated and only the interlayer dipolar coupling exists. The simulated dispersion relation shows an agreement to the experiment. The even-mode-number SWs, for example, $n = 2$, 4, 6, and 8, are excited. When adding a relatively small FM interlayer exchange coupling (4.875×10^{-13} J m^{-1}) between Co NWs and the YIG film, the dispersion moves from its former position. In the P state, the dispersion shifts to higher frequencies, while it shifts to lower frequencies in the AP state.

The micromagnetic simulation indicates that the interlayer dipolar interaction is the dominate contribution to the high-order SWs. However, the interlayer exchange coupling will lead to a frequency shift of high-order SW modes.

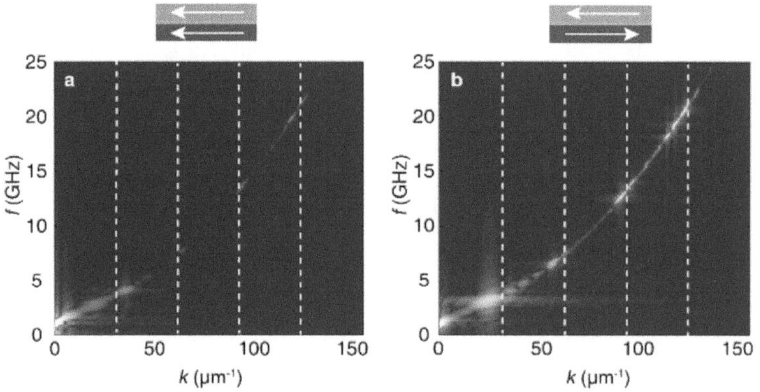

Figure 4.5 SW dispersion relation calculated from the micromagnetic simulation considering interlayer exchange coupling. The simulation structure consists of the Co NW array with the period 200 nm on top of the YIG film. An exchange coupling between the YIG film and the Co NWs is added in the P (a) and AP (b) configurations. Reproduced from Ref. [15] (https://creativecommons.org/licenses/by/4.0/).

4.3 Interlayer Magnon–Magnon Coupling in Ni/YIG- and Co/YIG-Based Magnonic Crystals

Strong coupling systems lies at the heart of many physical processes, including both classical and quantum information processing, due to their characteristic allowing coherent information exchange between different physical systems [23–27]. Magnons, as a quanta of SWs, are collectively spin excitations in magnetic materials that show the potential to couple with other systems.

Recently, magnon-photon coupling between microwave cavity photon and FMR of magnetic materials has widely been investigated [28–32]. The coupling strength of the hybrid coupled system obeys the Dicke model $g \propto \sqrt{N}$, where N is the total spin number [33]. However, when the magnetic materials are in the nanosized scale, which is compatible to the nanotechnology-based modern CMOS technique, the Dicke model indicates weak coupling.

In this section, we will discuss the strong magnon–magnon coupling observed in the Ni or Co NWs and YIG-based MCs [16]. The coupled magnon–magnon system consists of the FMR mode of the

FM NWs and the high-order SWs in the YIG film. An anticrossing gap as large as 1.58 GHz at a frequency of about 7.5 GHz is observed in the experiment, indicating that such hybrid system reaches the ultrastrong coupling (USC) regime.

4.3.1 Magnon–Magnon Coupling in a Ni/YIG-Based Hybrid Structure

In the experiment, a YIG-based MC structure was fabricated. The thickness of the YIG thin film was 20 nm. Periodical nickel NW arrays were patterned on top of the YIG film with periodicity 600 nm. The thickness of the Ni film was 20 nm. A conventional CPW was fabricated on top of the NW arrays and connected to a VNA via a microwave probe. An external magnetic field was applied parallel to the NW length (easy magnetization direction). The NW arrays acted as Bragg scattering gratings to generate high-order SWs. Analogous with the conventional PSSW, the high-order SWs, which possess a short wavelength, were named in-plane standing spin waves (iSSWs).

The reflection spectra S_{11} were extracted from the VNA measurement shown in Fig. 4.6. Two main branches of resonance

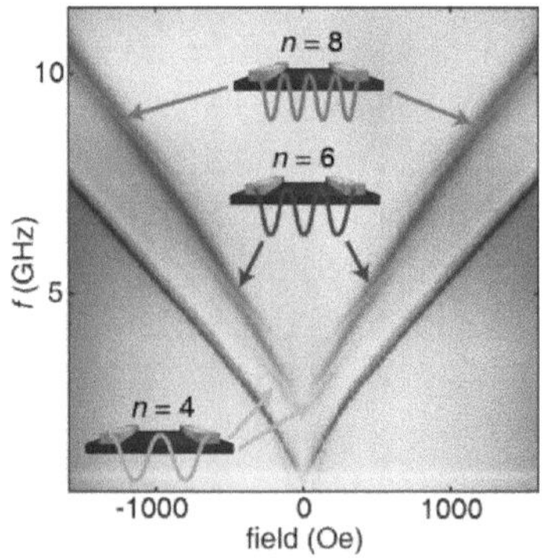

Figure 4.6 Grayscale reflection spectra S_{11} measured on the Ni/YIG heterostructure. The arrows highlight the anticrossings between the FMR mode of Ni NWs and the resonance modes of high-order iSSWs. Reprinted (figure) with permission from Ref. [16]. Copyright (2018) by the American Physical Society.

modes are observed in the spectrum. The lower mode is the CPW excited mode of YIG and the higher mode is the FMR mode of Ni NWs. The neighboring Ni NWs are precessing in phase in the resonance frequency. Through the interlayer magnetic coupling, the precession of magnetization in YIG is driven by the Ni NWs. Due to the periodical boundary conditions, only the even-mode-number SWs can be formed, for example, $n = 2, 4, 6, \ldots$ However, an interesting anticrossing gap is found when the resonances of the iSSWs approach the resonance of the Ni NWs. In Fig. 4.6, three distinct anticrossings are observed in both the negative and positive magnetic fields, which are attribute to the iSSW mode with mode number $n = 4$, $n = 6$, and $n = 8$.

4.3.2 Coupling Strength and Magnon–Magnon Cooperativity

The coupling strength is defined as the half value of the minimal peak-to-peak frequency spacing in the anticrossing. Figure 4.7 shows the anticrossing behavior of the mode $n = 8$ iSSW. The coupling strength g can be extracted from the transmission spectra and the coupling strength of *the* $n = 8$ mode is 120 MHz. For the perpendicular-standing SWs, which have confined wavelength

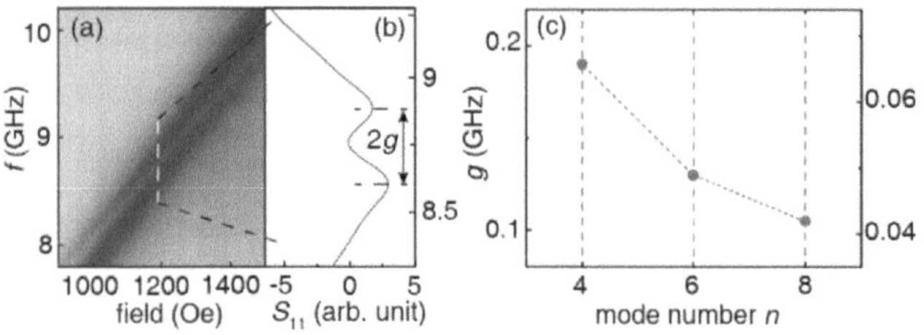

Figure 4.7 Anticrossing behavior and magnon–magnon coupling strength. (a) The grayscale reflection spectra S_{11} measured on the Ni/YIG-based heterostructure. The applied field first set to −3000 Oe to saturate the heterostructure and then swept from 900 Oe to 1500 Oe. The field step is 2.5 Oe. (b) A line plot extracted from (a) at 1200 Oe. (c) Magnon–magnon coupling strength dependence on the iSSW mode numbers. Reprinted (figure) with permission from Ref. [16]. Copyright (2018) by the American Physical Society.

by the film boundary, the coupling strength decreases when the mode number increases by the rule $g = \sqrt{d}/n$ where d is the thickness. We also observe a similar relation between the coupling strength and the wavenumber in our experiment, which is shown in Fig. 4.7c.

We can extract the damping rate for both the Ni mode and the YIG mode where damping rates represent the full width at half maximum (FWHM) of the corresponding modes. By extracting the FWHMs, we have $\kappa_{\mathrm{m}}^{\mathrm{Ni}} = 0.63$ GHz and $\kappa_{\mathrm{m}}^{\mathrm{YIG}} = 0.06$ GHz, where $\kappa_{\mathrm{m}}^{\mathrm{Ni}} > g > \kappa_{\mathrm{m}}^{\mathrm{YIG}}$. This condition indicates the magnetically induced transparency (MIT) for magnon transmission [31]. The magnon–magnon cooperativity can be also calculated by $C = g^2/(\kappa_{\mathrm{m}}^{\mathrm{Ni}} \times \kappa_{\mathrm{m}}^{\mathrm{YIG}})$, where $C = 0.38$ in our case.

4.3.3 Tuning the Magnon–Magnon Coupling by Different Magnetic Configurations

In this section, we will present the magnon–magnon coupling observed in the Co/YIG-based heterostructure. The period of the Co NWs shrinks to 180 nm. Compared to Ni, Co has larger saturation magnetization. The external magnetic field is applied parallel to the NWs from -1000 Oe to 1000 Oe. The reflection spectra S_{11} are shown in Fig. 4.8a. Three different magnetic configurations, P, AP, and X, are presented, which have been explained in Section 4.2. The high-order iSSWs are also observed in the VNA spectra. In the P configuration, we observe similar anticrossing behavior as shown in the Ni/YIG based MC (Fig. 4.8c). The anticrossing behavior is observed when the $n = 4$ iSSW mode crosses with the FMR mode of Co NWs. The anticrossing gap in the P configuration is 284 MHz and the system is still in the MIT regime.

However, in the AP configuration, a relatively large anticrossing gap can be observed with the gap width 1.58 GHz, which is shown in Fig. 4.8g. We have $\kappa_{\mathrm{m}}^{\mathrm{Co}} = 0.50$ GHz and $\kappa_{\mathrm{m}}^{\mathrm{YIG}} = 0.06$ GHz, where $g > \kappa_{\mathrm{m}}^{\mathrm{YIG}}$ and $g > \kappa_{\mathrm{m}}^{\mathrm{Co}}$. This condition indicates strong coupling where the magnon–magnon cooperativity can be calculated as $C = 21$ in the AP configuration. In this case, the anticrossing frequency is at about $\omega_{\mathrm{a}} = 7.5$ GHz, and the ratio $g/\omega_{\mathrm{a}} = 10.5\%$ indicates the coupled system reaches the USC regime. We also measured the Co/YIG

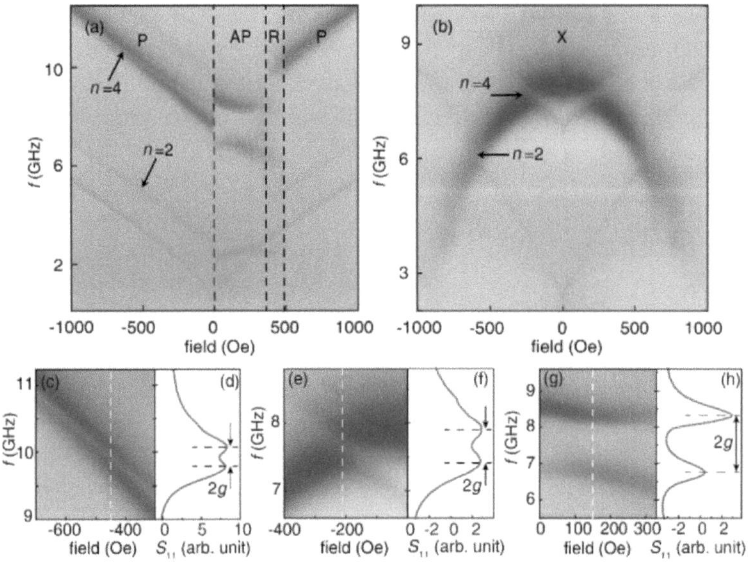

Figure 4.8 Reflection spectra of the Co/YIG heterostructure. (a and b) Full-range spectra with the external field parallel (a) and perpendicular (b) to the NWs. The external field is first applied at −3000 Oe and then swept from −1000 Oe to 1000 Oe. (c, e, and (g) Zoom-in spectra of the anticrossing range in P, X, and AP configurations, respectively. (d, f, and h) Single line plots extracted from the white dashed lines of each spectrum indicating the coupling strength.

heterostructure in the noncollinear state. In this configuration, the external magnetic field is applied perpendicular to the Co NWs. The spectra of the noncollinear state are shown in Fig. 4.8c,e. In this case, we observe an anticrossing gap of 0.47 GHz, which indicates such state is the intermediate of P and AP states.

The observed USC may be attribute to the enhanced interlayer exchange coupling due to the exchange spring effect. Note that there is a 1-nm-thick Ti interlayer between Co NWs and the YIG film as an adhesive layer. Such Ti layer is thin enough to support the FM interlayer exchange coupling. In the P configuration, there are no spin textures in the hybrid nanostructure. As we rotate the magnetization of one of the two magnetic layers, while the magnetization of the double layers remains collinear at the interface due to the FM exchange coupling, we create an exchange spring near the interface that penetrates into the YIG layer over the exchange length that can be large for soft magnetic materials. In addition, the existence of the interlayer dipolar coupling could also contribute to the observed large anticrossing, as we demonstrated in Section 4.2.

References

1. Chumak, A. V., Vasyuchka, V. I., Serga, A. A. and Hillebrands, B. (2015). Magnon spintronics, *Nat. Phys.*, **11**, 453–461.
2. Kruglyak, V. V., Demokritov, S. O. and Grundler, D. (2010). Magnonics, *J. Phys. D: Appl. Phys.*, **43**, 264001.
3. Demidov, V. E., Urazhdin, S., de Loubens, G., Klein, O., Cros, V., Anane, A. and Demokritov, S. O. (2017). Magnetization oscillations and waves driven by pure spin currents, *Phys. Rep.*, **673**, 1–23.
4. Yu, H., Xiao, J. and Pirro, P. (2018). Magnon spintronics, *J. Magn. Magn. Mater.*, **450**, 1–2.
5. Neusser, S. and Grundler D. (2009). Magnonics: spin waves on the nanoscale, *Adv. Mater.*, **21**, 2927–2932.
6. Neusser, S., Duerr, G., Bauer, H. G., Tacchi, S., Madami, M., Woltersdorf, G., Gubbiotti, G., Back, C. H. and Grundler, D. (2010). Anisotropic propagation and damping of spin waves in a nanopatterned antidot lattice, *Phys. Rev. Lett.*, **105**, 067208.
7. Ding, J., Kostylev, M. P. and Adeyeye, A. O. (2012). Magnonic crystal as a medium with tunable disorder on a periodical lattice, *Phys. Rev. Lett.*, **107**, 047205.
8. Gubbiotti, G., Tacchi, S., Madami, M., Carlotti, G., Jain, S., Adeyeye, A. O. and Kostylev, M. P. (2012). Collective spin waves in a bicomponent two-dimensional magnonic crystal, *Appl. Phys. Lett.*, **100**, 162407.
9. Chumak, A. V., Serga, A. A., Hillebrands, B. and Kostylev, M. P. (2008). Scattering of backward spin waves in a one-dimensional magnonic crystal, *Appl. Phys. Lett.*, **93**, 022508.
10. Yu, H., Duerr, G., Huber, R., Bahr, M., Schwarze, T., Brandl, F. and Grundler, D. (2013). Omnidirectional spin-wave nanograting coupler, *Nat. Commun.*, **4**, 2702.
11. Serga, A. A., Chumak, A. V. and Hillebrands, B. (2010). YIG magnonics, *J. Phys. D: Appl. Phys.*, **43**, 264002.

12. Chumak, A. V., Serga, A. A. and Hillebrands, B. (2014). Magnon transistor for all-magnon data processing, *Nat. Commun.*, **5**, 4700.

13. Yu, H., d'Allivy Kelly, O., Cros, V., Bernard, R., Bortolotti, P., Anane, A., Brandl, F., Heimbach, F. and Grundler, D. (2016). Approaching soft X-ray wavelengths in nanomagnet-based microwave technology, *Nat. Commun.*, **7**, 11255.

14. Papp, A., Porod, W. and Csaba, G. (2015). Hybrid yttrium iron garnet-ferromagnet structures for spin-wave devices, *J. Appl. Phys.*, **117**, 17E101.

15. Liu, C., Chen, J., Liu, T., Heimbach, F., Yu, H., Xiao, Y., Hu, J., Liu, M., Chang, H., Stueckler, T., Tu, S., Zhang, Y. G., Zhang, Y., Gao, P., Liao, Z., Yu, D., Xia, K., Lei, N., Zhao, W. and Wu, M. (2018). Long-distance propagation of short-wavelength spin waves, *Nat. Commun.*, **9**, 738.

16. Chen, J., Liu, C., Liu, T., Xiao, Y., Xia, K., Bauer, G. E. W., Wu, M. and Yu, H. (2018). Strong interlayer magnon-magnon coupling in magnetic metal-insulator hybrid nanostructures, *Phys. Rev. Lett.*, **120**, 217202.

17. Chang, H., Li, P., Zhang, W., Liu, T., Hoffmann, A., Deng, L. and Wu, M. (2014). Nanometer-thick yttrium iron garnet films with extremely low damping, *IEEE Magn. Lett.*, **5**, 6700104.

18. Yu, H., d' Allivy Kelly, O., Cros, V., Bernard, R., Bortolotti, P., Anane, A., Brandl, F., Huber, R., Stasinopoulos, I. and Grundler, D. (2014). Magnetic thin-film insulator with ultra-low spin wave damping for coherent nanomagnonics, *Sci. Rep.*, **4**, 6848.

19. Stancil, D. D. and Prabhakar, A. (2009). *Spin Waves: Theory and Applications* (Springer, New York).

20. Kim, K. J., Kim, S. K., Hirata, Y., Oh, S., Tono, T., Kim, D., Okuno, T., Ham, W. S., Kim, S., Go, G., Tserkovnyak, Y., Tsukamoto, A., Moriyama, T., Lee, K. and Ono, T. (2017). Fast domain wall motion in the vicinity of the angular momentum compensation temperature of ferrimagnets, *Nat. Mater.*, **16**, 1187.

21. Woo, S., Litzius, K., Krüger, B., Im, M.-Y., Caretta, L., Richter, K., Mann, M., Krone, A., Reeve, R. M., Weigand, M., Agrawal, P., Lemesh, I., Mawass, M-A., Fischer, P., Kläui, M. and Beach, G. S. D. (2016). Observation of room-temperature magnetic skyrmions and their current-driven dynamics in ultrathin metallic ferromagnets, *Nat. Nanotechnol.*, **15**, 501–506.

22. Madami, M., Bonetti, S., Consolo, G., Tacchi, S., Carlotti, G., Gubbiotti, G., Mancoff, F. B., Yar, M. A. and Åkerman, J. (2011). Direct observation

of a propagating spin wave induced by spin-transfer torque, *Nat. Nanotechnol.*, **6**, 635–638.

23. Tabuchi, Y., Ishino, S., Noguchi, A., Ishikawa, T., Yamazaki, R., Usami, K. and Nakamura, Y. (2015). Coherent coupling between a ferromagnetic magnon and a superconducting qubit, *Science*, **349**, 405.
24. Zhu, X., Shiro, S., Kemp, A., Kakuyanagi, K., Karimoto, S.-I., Nakano, H., Munro, W. J., Tokura, Y., Everitt, M. S., Nemoto, K., Kasu, M., Mizuochi, N. and Semba, K. (2011). Coherent coupling of a superconducting flux qubit to an electron spin ensemble in diamond, *Nature*, **478**, 221.
25. Majer, J., Chow, J. M., Gambetta, J. M., Koch, J., Johnson, B. R., Schreier, J. A., Frunzio, L., Schuster, D. I., Houck, A. A., Wallraff, A., Blais, A., Devoret, M. H., Girvin, S. M. and Schoelkopf, R. J. (2007). Coupling superconducting qubits via a cavity bus, *Nature*, **449**, 443.
26. Imamoglu, A. (2009). Cavity QED based on collective magnetic dipole coupling: spin ensembles as hybrid two-level systems, *Phys. Rev. Lett.*, **102**, 083602.
27. Wesenberg, J. H., Ardavan, A., Briggs, G. A. D., Morton, J. J. L., Schoelkopf, R. J., Schuster, D. I. and Molmer, K. (2009). Quantum computing with an electron spin ensemble, *Phys. Rev. Lett.*, **103**, 070502.
28. Bai, L., Harder, M., Chen, Y. P., Fan, X., Xiao, J. Q. and Hu, C. M. (2015). Spin pumping in electrodynamically coupled magnon-photon systems, *Phys. Rev. Lett.*, **114**, 227201.
29. Huebl, H., Zollitsch, C. W., Lotze, J., Hocke, F., Greifenstein, M., Marx, A., Gross, R. and Goennenwein, S. T. B. (2013). High cooperativity in coupled microwave resonator ferrimagnetic insulator hybrids, *Phys. Rev. Lett.*, **111**, 127003.
30. Cao, Y., Yan, P., Huebl, H., Goennenwein, S. T. B. and Bauer, G. E.W. (2015). Exchange magnon-polaritons in microwave cavities, *Phys. Rev. B*, **91**, 094423.
31. Zhang, X., Zou, C.-L., Jiang, L. and Tang, H. X. (2014). Strongly coupled magnons and cavity microwave photons, *Phys. Rev. Lett.*, **113**, 156401.
32. Haigh, J. A., Lambert, N. J., Doherty, A. C. and Ferguson, A. J. (2015). Dispersive readout of ferromagnetic resonance for strongly coupled magnons and microwave photons, *Phys. Rev. B*, **91**, 104410.
33. Dicke, R. H. (1954). Coherence in spontaneous radiation processes, *Phys. Rev.*, **93**, 99.

Chapter 5

Spin Waves in Thin Films and Magnonic Crystals with Dzyaloshinskii–Moriya Interactions

Rodolfo A. Gallardo,[a,b] **David Cortés-Ortuño,**[c]
Roberto E. Troncoso,[d] **and Pedro Landeros**[a,b]

[a]*Departamento de Física, Universidad Técnica Federico Santa María, Avenida España 1680, Valparaíso, Chile*
[b]*Center for the Development of Nanoscience and Nanotechnology (CEDENNA), 917-0124 Santiago, Chile*
[c]*Faculty of Engineering and Physical Sciences, University of Southampton, Southampton SO17 1BJ, United Kingdom*
[d]*Center for Quantum Spintronics, Department of Physics, Norwegian University of Science and Technology, NO-7491 Trondheim, Norway*
pedro.landeros@usm.cl

5.1 Introduction

In the late fifties of the past century, Dzyaloshinskii proposed in a seminal paper a phenomenological theory of antisymmetric exchange coupling between spins to explain the phenomenon of weak ferromagnetism in antiferromagnetic compounds [1]. Two years later, Moriya derived this interaction as a spin–orbit

Three-Dimensional Magnonics: Layered, Micro- and Nanostructures
Edited by Gianluca Gubbiotti

ISBN 978-981-4800-73-0 (Hardcover), 978-0-429-29915-5 (eBook)
www.jennystanford.com

coupling between electrons within the framework of superexchange theory [2–4]. It was then shown that this anisotropic exchange interaction arises in materials that lack inversion symmetry and where strong spin–orbit coupling effects are present. Nowadays, this antisymmetric exchange coupling is known as the Dzyaloshinskii–Moriya interaction (DMI) and has been a key ingredient for the description of the magnetic properties of a variety of compounds with broken symmetry [5–10]. These include noncentrosymmetric bulk ferromagnets, multiferroics, perovskites, cuprates, and ferromagnetic thin films [11–16], among others. The study of materials with DMIs has been pursued with high interest because it has been thoroughly established, both by theory and experiment, that DMIs induce chiral, topological, and nonreciprocal features. A primary consequence is the occurrence of chiral spin textures such as magnetic helices, skyrmions, skyrmion lattices, and chiral domain walls (DWs) in ferromagnetic materials [17–42].

Several theories have been developed to understand the origin of interfacial DMI in ultrathin ferromagnetic films in contact with a heavy metal. Originally, Levy and Fert [43–46] developed a theory for disordered magnetic alloys with heavy metal impurities that involves an additional contribution to the Ruderman–Kittel–Kasuya–Yosida (RKKY) interaction [47, 48]. This extra term is of DMI type and arises from the spin–orbit scattering of the conduction electron gas with the heavy metal impurities [45]. Recently, it has been proposed that the origin of the interfacial DMI is closely related to the proximity-induced magnetic moment in the heavy metal [49], while ab initio calculations reveal that the proximity-induced magnetic moment in the heavy metal has no direct correlation with the DMI in Co/Pt interfaces [50]. Using a density functional approach, Belabbes et al. [51] demonstrated that the sign and magnitude of the DMI are related to the degree of 3d–5d orbital hybridization around the Fermi level. The temperature dependence of the DMI and its interplay with the anisotropy of the orbital magnetic moment and the magnetic dipole moment in Pt/Co/MgO trilayers were studied experimentally and theoretically by Kim et al. [52], where the interfacial DMI originates from the asymmetric charge distribution caused by the breaking of inversion symmetry. According to these studies, the underlying physical mechanism behind the DMI in ferromagnetic ultrathin films is not completely understood and is still a topic under discussion.

Among the different kinds of elementary excitations in condensed matter physics, spin waves play a significant role because of the shorter wavelengths that can be attained at GHz frequencies and also due to the reduced losses by heating. Spin waves have been observed from millimeter down to micrometer [53] and, more recently, to submicrometer length scales [54–62]. These spin excitations can be generated using microwaves in resonant cavities, alternating Oersted fields produced by radiofrequency currents, inelastic neutron scattering, spin-transfer torques from electric currents, visible light as in Brillouin light scattering (BLS), and the time-resolved magneto-optical Kerr effect [63–67]. A more sophisticated experimental prove of spin waves requires a synchrotron and is based on time-resolved scanning transmission X-ray microscopy [60–62].

The influence of the DMI on the spin-wave spectra has been widely studied in the last years. The first observation of nonreciprocal spin waves [16, 68, 69] caused by interfacial DMI was achieved by Zakeri et al. [70, 71] using the spin-polarized electron energy loss (SPEEL) experimental technique. A few years later, nonreciprocal spin waves were observed by means of BLS spectroscopy in several ferromagnet/heavy metal interfaces [72–87]. In agreement with the results of Zakeri et al. [70, 71], these numerous experiments have shown that interfacial DMI creates a noticeable frequency asymmetry in the spin-wave dispersion of counterpropagating Damon–Eshbach (DE) spin waves. The DMI-induced magnon asymmetry has been theoretically established by several groups [88–101]. From these calculations, one can see that the strength of the DMI can be estimated experimentally by measuring the frequency difference of counterpropagating waves, as proposed in Ref. [96]. Furthermore, it was also predicted that in systems with bulk DMI a frequency difference could be observed in the backward-volume geometry [91, 96], that is, with the magnetization and wave vector parallel to each other and along the plane of the film. This effect has been measured in a few chiral lattice magnets [69, 102–105]. Beyond the study of spin waves in field-polarized states, spin excitations in non-uniform magnetic textures, such as helical, conical, and skyrmion phases, have also been reported [106–111].

The dynamics of spin waves under the influence of bulk and interfacial DMI, both in thin films and magnonic crystals (MCs), is reviewed in this chapter. Nonreciprocal properties induced by different classes of DMI in thin-film magnets are discussed in Sections 5.2 and 5.3. A detailed analysis from the point of view of micromagnetic simulations is given in Section 5.4. The band structure of a chiral MC defined with a periodic DMI is discussed in Section 5.5, where both the plane-wave method and micromagnetic simulations are used to theoretically analyze the system. Conclusions are summarized in Section 5.6.

5.2 Spin Waves in Thin Films with Bulk DMI

In this section, a theoretical background is given for the spin-wave dynamics in thin films with bulk DMI. It is known that two counterpropagating spin waves show a frequency difference in magnetic films with bulk DMI but in the backward-volume geometry [91, 96], that is, with the magnetization and wave vector both parallel to each other and along the film's plane. This behavior has been measured in some chiral lattice magnets [102–105, 112] and also in the noncentrosymmetric antiferromagnet $\alpha Cu_2V_2O_7$ [113].

The dynamics of the magnetization is described theoretically by following Refs. [96, 114, 115], where the magnetization is composed by a static and a dynamic part: $\mathbf{M}(\mathbf{r};t) = M_Z\widehat{Z} + \mathbf{m}(\mathbf{r};t)$. The static part is directed toward the Z axis, which is chosen along the equilibrium magnetization orientation (see Fig. 5.1). The dynamic magnetization can be written as $\mathbf{m}(\mathbf{r};t) = \mathbf{m}(\mathbf{r})e^{i\omega t}$ so that the spatial term is $\mathbf{m}(\mathbf{r}) = m_X(\mathbf{r})\widehat{X} + m_Y(\mathbf{r})\widehat{Y}$. For small deviations of the magnetization from the equilibrium position, it can be assumed that the dynamic components satisfy $m_{X,Y} \ll M_s$ and then $M_Z \approx M_s - (m_X^2 + m_Y^2)/(2M_s)$ [114]. The dynamic magnetization components are represented in Fourier form in terms of the wave vector $\mathbf{k}$ as

$$m_{\alpha,\beta}(\mathbf{r}) = \frac{1}{\sqrt{L^2 d}}\sum_{\mathbf{k}} m_{\alpha,\beta}(\mathbf{k})\, e^{i\mathbf{k}\cdot\mathbf{r}}, \tag{5.1}$$

where $\mathbf{r}$ and $\mathbf{k}$ lie in the plane of the ferromagnetic film of area L^2 and thickness d, and α, $\beta = X, Y$. The spin wave Hamiltonian can be expressed as [114, 115]

$$\mathcal{H} = \frac{1}{2M_s}\sum\nolimits_{\mathbf{k}}\sum\nolimits_{\alpha,\beta}\mathcal{W}_{\alpha,\beta}(\mathbf{k})m_\alpha^*(\mathbf{k})m_\beta(\mathbf{k}), \tag{5.2}$$

where $\mathcal{W}_{\alpha,\beta}(\mathbf{k})$ are the stiffness fields, which contain contributions from the dipolar energy, Zeeman energy, exchange interaction, anisotropies, and the DMI. These energies were calculated in Refs. [114, 115], while the role of DMI was studied in Refs. [96, 110] for several Lifshitz invariants representing different crystal symmetries [12, 17–19, 116].

In the micromagnetic limit, the DMI Hamiltonian can be written as

$$\mathcal{H}_{\mathrm{DM}}(\mathbf{r}) = \int_V w_{\mathrm{DM}} dV, \tag{5.3}$$

where the energy density w_{DM} involves combinations of first-order spatial derivatives of the magnetization known as *Lifshitz invariants* [12, 17–19], which are defined as

$$\mathcal{L}_{ij}^{(k)} = M_i\frac{\partial M_j}{\partial x_k} - M_j\frac{\partial M_i}{\partial x_k}. \tag{5.4}$$

Depending on the symmetry of the noncentrosymmetric crystal, w_{DM} assumes different forms. One commonly studied sum of invariants has the structure (see page 18 in Ref. [116])

$$\mathcal{L} = \mathcal{L}_{yx}^{(z)} + \mathcal{L}_{xz}^{(y)} + \mathcal{L}_{zy}^{(x)} = \mathbf{M}\cdot(\nabla\times\mathbf{M}), \tag{5.5}$$

that is characteristic of noncentrosymmetric cubic crystals, for instance, MnSi, FeGe, MnGe, or Cu_2OSeO_3 [110, 116]. The DM energy density is given by $w_{\mathrm{DM}} = \mu_0\lambda_{\mathrm{DM}}\mathcal{L}$, where $\lambda_{\mathrm{DM}} = D/\mu_0 M_s^2$ has units of length, and is proportional to the usual DMI strength D in units of Joule/meter^2. For cubic crystals, the DM energy reads

$$\mathcal{H}_{\mathrm{DM}}(\mathbf{r}) = \frac{D}{M_s^2}\int_V \mathbf{M}\cdot(\nabla\times\mathbf{M})dV. \tag{5.6}$$

For different crystal symmetries, one can write the DMI in terms of other combinations of Lifshitz invariants, as shown in Table 5.1, which was adapted from Ref. [96], where cgs units were used with the strength of the DMI represented by the length λ_{DM}. In Table 5.1, as in the rest of the chapter, SI units are used, while the strength of

Table 5.1 DM energy density and corresponding asymmetry in the spin-wave dispersion relation for different symmetry classes. The case of interfacial DMI corresponds to the C_{nv} symmetry group [96].

Symmetry class	Energy density w_{DM}	Frequency asymmetry Δf
$\boldsymbol{T}$	$\left(D/M_s^2\right)\left(\mathcal{L}_{yx}^{(z)} + \mathcal{L}_{xz}^{(y)} + \mathcal{L}_{zy}^{(x)}\right)$	$2\gamma D\,(\pi M_s)^{-1}\,k\cos\phi_{\mathbf{k}}\cos\phi_{\mathrm{m}}$
$\boldsymbol{C}_{\mathrm{nv}}$	$\left(D/M_s^2\right)\left(\mathcal{L}_{zy}^{(z)} + \mathcal{L}_{xy}^{(x)}\right)$	$-2\gamma D\,(\pi M_s)^{-1}\,k\sin\phi_{\mathbf{k}}\cos\phi_{\mathrm{m}}$
$\boldsymbol{D}_{\mathrm{2d}}$	$\left(D/M_s^2\right)\left(\mathcal{L}_{xy}^{(z)} + \mathcal{L}_{zy}^{(x)}\right)$	$-2\gamma D\,(\pi M_s)^{-1}\,k\cos\phi_{\mathbf{k}}\cos\phi_{\mathrm{m}}$
$\boldsymbol{D}_{\mathrm{n}}$	$\left(D^{(1)}/M_s^2\right)\left(\mathcal{L}_{zy}^{(x)} - \mathcal{L}_{xy}^{(z)}\right)$ $+\left(D^{(2)}/M_s^2\right)\left(\mathcal{L}_{zx}^{(y)}\right)$	$-2\gamma D^{(1)}\,(\pi M_s)^{-1}\,k\cos\phi_{\mathbf{k}}\cos\phi_{\mathrm{m}}$
$\boldsymbol{C}_{\mathrm{n}}$	$\left(D^{(1)}/M_s^2\right)\left(\mathcal{L}_{zy}^{(z)} + \mathcal{L}_{xy}^{(x)}\right)$ $+\left(D^{(2)}/M_s^2\right)\left(\mathcal{L}_{zy}^{(x)} - \mathcal{L}_{xy}^{(z)}\right)$	$-2\gamma\,(\pi M_s)^{-1}\,k[D^{(1)}\sin\phi_{\mathbf{k}}$ $-D^{(2)}\cos\phi_{\mathbf{k}}]\cos\phi_{\mathrm{m}}$

the DMI is represented by the parameter D [97, 110]. Following Refs. [96, 115], the magnetization components in the fixed coordinate system xyz (see Fig. 5.1) are specified as

$$M_x = m_X(\mathbf{r}) \tag{5.7a}$$

$$M_y = M_Z \sin\phi_{\mathrm{m}} + m_Y(\mathbf{r})\cos\phi_{\mathrm{m}} \tag{5.7b}$$

$$M_z = M_Z \cos\phi_{\mathrm{m}} - m_Y(\mathbf{r})\sin\phi_{\mathrm{m}}, \tag{5.7c}$$

where ϕ_{m} is the angle between the equilibrium magnetization and the plane of the film.

By expanding the dynamic components of the magnetization up to second order, the sum of invariants (see Eq. 5.5) are expressed as

$$\mathcal{L} = M_s \frac{\partial m_Y}{\partial x} + M_s \sin\phi_{\mathrm{m}} \frac{\partial m_X}{\partial z} + m_Y \cos\phi_{\mathrm{m}} \frac{\partial m_X}{\partial z} - m_X \cos\phi_{\mathrm{m}} \frac{\partial m_Y}{\partial z}. \tag{5.8}$$

In the thin-film limit, the dynamic magnetization components do not depend on the normal coordinate y, so $m_{X,Y}(\mathbf{r}) = m_{X,Y}(x, z)$, and then it is possible to integrate along the thickness to obtain

$$\mathcal{H}_{\mathrm{DM}} = \frac{Dd\,\cos\phi_{\mathrm{m}}}{M_s^2} \int \left[m_Y \frac{\partial m_X}{\partial z} - m_X \frac{\partial m_Y}{\partial z} \right] dx\,dz, \tag{5.9}$$

where the linear terms have been disregarded, since they cancel out due to the equilibrium condition. Using the Fourier representation

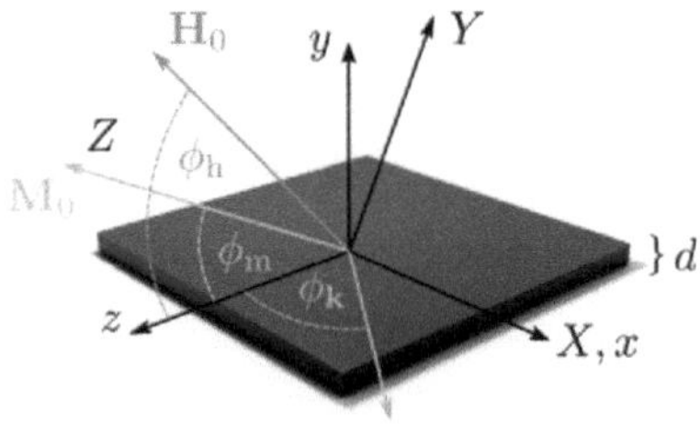

Figure 5.1 (a) Geometry and notation. Vectors $\mathbf{M_0}$ and $\mathbf{H_0}$ correspond, respectively, to the equilibrium magnetization and applied field, while d denotes the thickness of the ferromagnetic film. The coordinate system (x, y, z) is fixed in the film, while the coordinates (X, Y, Z) are defined according to the equilibrium magnetization, such that Z is always parallel to $\mathbf{M_0}$, which makes an angle ϕ_{m} with the film. Spin waves propagate in the plane at an angle $\phi_{\mathbf{k}}$ with the z axis.

(see Eq. 5.1) the Hamiltonian associated to the bulk DMI is given by

$$\mathcal{H}_{\mathrm{DM}} = \frac{1}{2M_{\mathrm{s}}}\sum_{\mathbf{k}} [\mathcal{W}_{XY}^{\mathrm{DM}}(\mathbf{k})\, m_X^*(\mathbf{k})\, m_Y(\mathbf{k}) + \mathcal{W}_{YX}^{\mathrm{DM}}(\mathbf{k})\, m_Y^*(\mathbf{k}) m_X(\mathbf{k})], \tag{5.10}$$

in which the DMI stiffness fields are given by

$$\mathcal{W}_{XY}^{\mathrm{DM}}(\mathbf{k}) = -\frac{2D}{M_{\mathrm{s}}} ik\, \cos\phi_{\mathbf{k}}\, \cos\phi_{\mathrm{m}}, \tag{5.11}$$

and $\mathcal{W}_{XY}^{\mathrm{DM}}(\mathbf{k}) = -\mathcal{W}_{YX}^{\mathrm{DM}}(\mathbf{k})$. Here, k is the magnitude of the wave vector and $\phi_{\mathbf{k}}$ is the angle between the wave vector and the projection of the equilibrium magnetization onto the plane, as shown in Fig. 5.1. Using the other energetic contributions (see Refs. [96, 115] for details) that add to the total Hamiltonian (Eq. 5.2), which include the DM term (Eq. 5.10), the stiffness fields are calculated as

$$\mathcal{W}_{XX}(\mathbf{k}) = \mathcal{W}_{XX}(0) + \mu_0 M_{\mathrm{s}} F(kd) \sin^2\phi_{\mathbf{k}} + D_{\mathrm{ex}} k^2, \tag{5.12a}$$

$$\mathcal{W}_{YY}(\mathbf{k}) = \mathcal{W}_{YY}(0) - \mu_0 M_{\mathrm{s}} F(kd)[\cos 2\phi_{\mathrm{m}} + \sin^2\phi_{\mathbf{k}} \sin^2\phi_{\mathrm{m}}] + D_{\mathrm{ex}} k^2, \tag{5.12b}$$

$$\mathcal{W}_{XY}(\mathbf{k}) = -\mu_0 M_{\mathrm{s}} F(kd) \sin\phi_{\mathbf{k}} \cos\phi_{\mathbf{k}}\, \sin\phi_{\mathrm{m}} - \frac{2D}{M_{\mathrm{s}}} ik\, \cos\phi_{\mathbf{k}}\, \cos\phi_{\mathrm{m}}, \tag{5.12c}$$

$$\mathcal{W}_{YX}(\mathbf{k}) = -\mu_0 M_{\mathrm{s}} F(kd) \sin\phi_{\mathbf{k}} \cos\phi_{\mathbf{k}}\, \sin\phi_{\mathrm{m}} + \frac{2D}{M_{\mathrm{s}}} ik\, \cos\phi_{\mathbf{k}}\, \cos\phi_{\mathrm{m}}, \tag{5.12d}$$

where $D_{\mathrm{ex}} = 2A/M_{\mathrm{s}}$ is the exchange stiffness and $F(x) = 1 - (1 - e^{-x})/x$. Additionally,

$$\mathcal{W}_{XX}(0) = \mu_0 H_0 \cos(\phi_{\mathrm{h}} - \phi_{\mathrm{m}}) - \mu_0 M_{\mathrm{eff}} \sin^2\phi_{\mathrm{m}}, \tag{5.13a}$$

$$\mathcal{W}_{YY}(0) = \mu_0 H_0 \cos(\phi_{\mathrm{h}} - \phi_{\mathrm{m}}) + \mu_0 M_{\mathrm{eff}} \cos 2\phi_{\mathrm{m}}. \tag{5.13b}$$

In Eqs. 5.13, H_0 is a static magnetic field applied at an angle ϕ_{h} respect to the plane, and $M_{\mathrm{eff}} = M_{\mathrm{s}} - H_{\mathrm{s}}$ is the effective magnetization, with H_{s} the perpendicular surface anisotropy. The angle ϕ_{m} can be calculated from a simple equilibrium condition [96, 115].

To calculate the spin-wave dispersion relation, the dynamic magnetization components are promoted to operators, with well-known commutation rules [114, 117]. The temporal evolution of the dynamic magnetization is given by

$$i\hbar\dot{m}_{X,Y}(\mathbf{k};t) = [m_{X,Y}(\mathbf{k};t), \mathcal{H}]. \tag{5.14}$$

Then, using $m_{X,Y}(-\mathbf{k};t) = m^*_{X,Y}(\mathbf{k};t)$, the Hamiltonian given by Eqs. 5.2 and 5.12, the commutation rules from Refs. [114, 117] and the definition of the absolute value of the gyromagnetic ratio $\gamma = -\mu\hbar^{-1}$, with μ as the magnetic moment of a magnetic ion in the film, it is possible to obtain a pair of first-order differential equations that can be written in matrix form,

$$\begin{pmatrix} \dot{m}_X \\ \dot{m}_Y \end{pmatrix} = \gamma \begin{pmatrix} -\mathcal{W}_{YX} & -\mathcal{W}_{YY} \\ \mathcal{W}_{XX} & \mathcal{W}_{XY} \end{pmatrix} \begin{pmatrix} m_X \\ m_Y \end{pmatrix}. \tag{5.15}$$

The dispersion relation of the spin waves can now be computed by assuming harmonic solutions for the dynamic magnetization components and reads

$$\omega_{\pm}(\mathbf{k}) = \gamma\left\{-\mathcal{I}[\mathcal{W}_{XY}] \pm \sqrt{\mathcal{W}_{XX}\mathcal{W}_{YY} - \mathcal{R}[\mathcal{W}_{XY}]^2}\right\}. \tag{5.16}$$

Here, $\mathcal{I}[\mathcal{W}_{XY}]$ and $\mathcal{R}[\mathcal{W}_{XY}]$ denote the imaginary and the real part of the function $\mathcal{W}_{XY}$. Notice that the above formula is valid for any symmetry class, provided that $\mathcal{W}_{XY}(\mathbf{k}) = \mathcal{W}^*_{YX}(\mathbf{k})$ [96]. For a crystal with symmetry class T, the dispersion relation becomes

$$\omega(\mathbf{k}) = \frac{2\gamma D}{M_{\mathrm{s}}} k \cos\phi_{\mathbf{k}} \cos\phi_{\mathrm{m}} + \gamma\left[\mathcal{W}_{XX}(\mathbf{k})\mathcal{W}_{YY}(\mathbf{k}) - \left(\frac{\mu_0 M_{\mathrm{s}}}{2} kd \sin\phi_{\mathbf{k}} \cos\phi_{\mathbf{k}} \sin\phi_{\mathrm{m}}\right)^2\right]^{\frac{1}{2}}. \tag{5.17}$$

The DMI appears only in the linear term proportional to D. For the Lifshitz invariants describing chiral lattice ferromagnets (symmetry class T in Table 5.1), the influence of the DMI on the spectrum is strongest when the magnetization lies in the plane ($\phi_m = 0$) and when wave vectors are parallel to the equilibrium magnetization ($\phi_k = 0$), which are known as backward-volume magnetostatic spin waves. Under wave vector inversion $\mathbf{k} \rightarrow -\mathbf{k}$, where $\mathbf{k}$ rotates π radians, one must change ϕ_k by $\phi_k \pm \pi$. In this case, the DM term in the dispersion relation changes its sign, since $\cos(\phi_k \pm \pi) = -\cos\phi_k$, and therefore, because of this asymmetry, the relation is nonreciprocal [91, 96, 102, 104, 110]. According to this, the frequency $f(\mathbf{k}) = \omega(\mathbf{k})/(2\pi)$ can be used to define a frequency asymmetry given by

$$\Delta f = f(\mathbf{k}) - f(-\mathbf{k}). \tag{5.18}$$

In the case of the symmetry class T, the nonreciprocity in the spin-wave dispersion relation is linear with respect to the wave vector and reads

$$\Delta f = \frac{2\gamma D}{\pi M_s} k \cos\phi_k \cos\phi_m. \tag{5.19}$$

For this particular crystallographic class, and in the backward-volume configuration $\phi_k = 0$, nonreciprocal spin waves have been observed recently in different magnetic compounds. These include the bulk noncentrosymmetric ferromagnets $LiFe_5O_8$ [102], Cu_2OSeO_3 [104], MnSi [103, 105, 112], and the noncentrosymmetric antiferromagnet $\alpha Cu_2V_2O_7$ [113].

Similar results are also obtained for the other allowed combinations of Lifshitz invariants, which are governed by the symmetry of the crystallographic classes. The Δf frequency expressions of the different invariants are summarized in Table 5.1. They follow a common structure because in the thin-film limit, there is no dependence of the dynamic magnetization on the y coordinate and the second-order terms share the same structure. Furthermore, as Eq. 5.19 dictates, all the asymmetries vanish if the system is saturated perpendicularly to the film plane ($\phi_m = \pi/2$), and Δf is largest when $\mathbf{M}$ is parallel to the film plane. The differences in the frequency asymmetries between different symmetry classes are the dependence on the angle $\phi_{\mathbf{k}}$ between **M** and **k** and the type of DMI constants involved.

The full set of response functions for microwave excitations was also calculated (see Eqs. 34–37 in Ref. [96]), which allows us to estimate the nonreciprocal absorption spectrum in systems with bulk and interfacial DMIs.

5.3 Spin Waves in Thin Films with Interfacial DMI

A different mechanism for the presence of DMI in ferromagnets is the spin–orbit coupling arising at the interface of a thin-film ferromagnet adjacent to a heavy metal layer. Under the presence of interfacial DMI, spin waves exhibit similar effects than the ones in bulk DMI systems. These effects have been theoretically studied by several groups [86, 88–101]. Furthermore, in these film samples, of particular interest is the dependence of the interfacial DMI strength [86] on the heavy metal thickness, which requires the evaluation of the DMI vector. The thickness dependence of the DMI vector and the frequency nonreciprocity in interfacial DMI films are discussed in this section.

Starting with the analysis of the heavy metal thickness, the theoretical approach used here is based on the microscopic DMI Hamiltonian, which reads

$$\mathcal{H}_{\mathrm{DM}} = \sum\nolimits_{\langle i,j\rangle} \mathbf{D}_{ij} \cdot \left(\mathbf{S}_i \times \mathbf{S}_j\right). \tag{5.20}$$

The DMI couples any pair of neighboring atomic spins $\mathbf{S}_i$ and $\mathbf{S}_j$ in the interfacial layer of the ferromagnetic film through a third heavy metal site. The indexes i and j label a pair of interacting ferromagnetic spins at the interface. The DM vector $\mathbf{D}_{ij}$ links ferromagnetic spins at sites i and j with a third heavy metal site, labeled by a lattice vector $\mathbf{l}$ in the heavy metal, that is perpendicular to the triangle described by the three sites [11, 12]. This theory follows the formalism developed by Fert and Levy for disordered magnetic alloys with heavy metal impurities [45, 46]. The existence of a DMI at the ferromagnet/heavy metal interface is a well-established phenomenon, in which the DMI strength decreases with the thickness d of the ferromagnetic layer, due to the interface nature of the DMI [75], and increases up to a saturation value with the thickness of the heavy metal layer [86, 87]. This behavior suggests that there is a significant number of heavy metal atoms contributing to the strength of the interfacial DMI. Therefore, to calculate the DM vector, the thickness and lattice structure of the heavy metal need to be considered. The DM vector that links a pair of ferromagnetic spins includes contributions from a group of neighboring heavy metal atoms and can be evaluated by summing over several heavy metal lattice vectors **l** (see Fig. 1 in Ref. [86]), that is,

$$\mathbf{D}_{ij} = \sum_{\mathbf{l}} V_1 \frac{\hat{\mathbf{R}}_i^{\mathbf{l}} \cdot \hat{\mathbf{R}}_j^{\mathbf{l}}}{R_i^{\mathbf{l}} R_j^{\mathbf{l}} R_{ij}} \hat{\mathbf{R}}_i^{\mathbf{l}} \times \hat{\mathbf{R}}_j^{\mathbf{l}} \sin\left[k_{\mathrm{F}}\left(R_i^{\mathbf{l}} + R_j^{\mathbf{l}} + R_{ij}\right) + \frac{\pi Z_d}{10}\right], \tag{5.21}$$

where $V_1 = [(135\pi\lambda_d\Gamma^2)/(32E_{\mathrm{F}}^2k_{\mathrm{F}}^3)]\sin(\pi Z_d/10)$ has units of Joule m^3 and depends on the spin–orbit coupling constant λ_d, the Fermi energy E_{F} and Fermi wave vector $\mathbf{k}_{\mathrm{F}}$ of the conduction electrons, the coupling constant Γ between ferromagnetic spins, and the number of incomplete subshell electrons Z_d (see Eqs. 5 and 6 in Ref. [45] and Eq. 26 in Ref. [46] for details). The position vectors $\mathbf{R}_i^{\mathbf{l}}$ and $\mathbf{R}_j^{\mathbf{l}}$, join the ferromagnetic sites $\mathbf{S}_i$ and $\mathbf{S}_j$ to the heavy metal sites. The DM vector between spins at sites i and $i + x$ can be written as $\mathbf{D}_{i,j=i+x} = -D\hat{\mathbf{z}}$, while the corresponding DM vector between spins at i and $i + z$ reads $\mathbf{D}_{i,j=i+z} = D\hat{\mathbf{x}}$ [86]. The y component of $\mathbf{D}_{ij}$ vanishes due to the assumed perfectly flat interface.

The calculation of the spin-wave dispersion relation and the frequency nonreciprocity follows the same approach of the previous section but with a different combination of Lifshitz invariants, namely $\mathcal{L} = \mathcal{L}_{zy}^{(z)} + \mathcal{L}_{xy}^{(x)}$. After replacing into $\mathcal{L}$ the magnetization components M_x, M_y, and M_z (see Eq. 5.7) up to the second order, we obtain

$$\mathcal{L} \approx M_{\mathrm{s}}\frac{\partial m_Y}{\partial z} - M_{\mathrm{s}}\sin\phi_{\mathrm{m}}\frac{\partial m_X}{\partial x} - m_Y\cos\phi_{\mathrm{m}}\frac{\partial m_X}{\partial x} + m_X\cos\phi_{\mathrm{m}}\frac{\partial m_Y}{\partial x} \tag{5.22}$$

In the limit of a thin film, the DMI Hamiltonian becomes

$$\mathcal{H}_{\mathrm{DM}} = \frac{Dd\cos\phi_{\mathrm{m}}}{M_{\mathrm{s}}^2}\int\left[m_X\frac{\partial m_Y}{\partial x} - m_Y\frac{\partial m_X}{\partial x}\right]dx\,dz. \tag{5.23}$$

Within a Fourier representation of the dynamic magnetization, the frequency dispersion of the spin waves is separated into two contributions, $f(\mathbf{k}) = f_r(\mathbf{k}) + f_{DM}(\mathbf{k})$, a reciprocal part and a nonreciprocal one associated to the DMI, which is given by

$$f_{DM}(\mathbf{k}) = \frac{\gamma D(d_{HM})}{\pi M_s} k \sin\phi_{\mathbf{k}} \cos\phi_m. \tag{5.24}$$

In this expression $D(d_{HM}) \equiv S^2/(na_0 d)]|\mathbf{D}_{ij}|$ is the volume-averaged DMI strength, γ is the gyromagnetic ratio, a_0 is the separation between nearest-neighbor ferromagnetic spins, and n is an integer introduced to consider first, second, or even third neighbors [86]. The reciprocal part $f_r(\mathbf{k})$ includes exchange, dipolar, anisotropy, and Zeeman contributions [96]. The frequency difference of oppositely propagating spin waves, $\Delta f = f(\mathbf{k}) - f(-\mathbf{k})$, is then calculated as

$$\Delta f(\mathbf{k}, d_{HM}) = \frac{2\gamma D(d_{HM})}{\pi M_s} k \sin\phi_{\mathbf{k}} \cos\phi_m. \tag{5.25}$$

Most of the parameters involved in the frequency difference can be controlled by the experimental setup. For example, ϕ_m is controlled with the orientation of the applied field, $\phi_{\mathbf{k}}$ and k are controlled with the excitation mechanism, while γ and M_s can be measured by conventional magnetometry or ferromagnetic resonance (FMR). Then, by measuring Δf, the strength of the interfacial DMI can be directly measured, as was theoretically proposed in Ref. [96], and then measured by several groups, mainly using BLS [72–87], as summarized in Tables 5.2 and 5.3. It is worth to mention that the estimation of the DMI strength through measurement of the frequency difference between counterpropagating spin waves, in the field-polarized case, should be the most reliable method, since there are no assumptions about the possible magnetic texture. Moreover, there is no need to neglect or approximate the dipolar interactions, as in the case of DW-based methods. As can be seen from Tables 5.2 and 5.3, some groups have measured the DMI strength with both BLS and DW-based methods, where different values of D are found [81, 127].

The theoretical model developed previously has been applied to analyze experimental observations of the strength of the DM constant in Pt(d_{HM})/CoFeB films. Specifically, Tacchi et al. [86] studied the influence of the heavy metal thickness (d_{HM}) on the induced

Table 5.2 Reported magnitudes for DM couplings in various heavy metal/ferromagnet interfacial systems. Experimental measurements include BLS, electrical and domain wall (DW)-based methods, micromagnetic simulations, and microscopy techniques. The thickness of each layer comprising the structure is measured in nanometers (nm).

Material	Reference	$\|D\|$ (mJ/m^2)	Method
Pt(3)/CoFe(0.6)/MgO	[37]	0.5	DW
Pt(5)/Co(0.7)/Ir(0.46)/Pt(3)	[118]	0.12	DW
Pt(4)/Co(1.6)/Ni	[72, 73]	0.44	BLS
Pt(2)/CoFeB(0.8)/MgO	[74]	1.0	BLS
Pt(4)/CoFeB(1–1.6)/AlO_x	[75]	0.8–0.4	BLS
Pt(4)/Co(1–2)/AlO_x	[75]	1.2–0.9	BLS
Pt(6)/NiFe(1.3–10)/SiN	[76]	0.15–0.025	BLS
Pt(6)/NiFe(4)/Si	[77]	1.2	BLS
Pt(3)/Co(0.6–1.2)/AlO_x	[78]	2.7–1.57	BLS
Ta/Pt(4)/Co(1.35)/AlO_x	[79]	1.65	BLS
Pd/Fe/Ir	[119]	3.9	SPSTM/DW Simulation
$(Ir(1)/Co(0.6)/Pt(1))_{10}$	[41]	1.6–1.9	STXM/DW Simulation
Ir(4)/Co(1.2–3)/AlO_x	[80]	0.3–0.7	BLS
W(2-3)/CoFeB(1)/MgO	[81]	0.25–0.27	BLS
W(2-3)/CoFeB(1)/MgO	[81]	0.23–0.12	DW
TaN(1)/CoFeB(1)/MgO	[81]	0.31	BLS
TaN(1)/CoFeB(1)/MgO	[81]	0.05	DW
Hf(1)/CoFeB(1)/MgO	[81]	0.15	BLS
Hf(1)/CoFeB(1)/MgO	[81]	0.01	DW
Pt(3)/Co(16)/MgO	[120]	0.25	Electrical
Ta(5)/CoFeB(0.8)/MgO(2)	[121]	0.057	Kerr/DW

DMI by means of BLS measurements of ultrathin CoFeB films. As a result, it was found that the strength of the interfacial DMI increases with heavy metal thickness, reaching a saturation value for d_{HM} larger than a few nanometers. In the BLS measurements reported in Ref. [86], the Stokes and anti-Stokes peaks, corresponding to spin waves propagating in opposite directions, are simultaneously observed with comparable intensity. These peaks are characterized

Table 5.3 Continuation of Table 5.2. The magnitudes reported in the section at the bottom of this table refer to DMI values from atomistic parameters or spatial units.

Material	References	$|D|$ (mJ/m^2)	Method
Pt(3)/CoFe(1–1.6)/MgO	[82]	1.4–2.3	BLS
Pt(4)/CoFe(0.8–1.6)/MgO	[82]	0.9–1.5	BLS
W/CoFeB(1)/SiO_2	[83]	0.25	BLS
Ta(5)/CoFeB(1)/Pt(0.15)/MgO	[84]	0.056	BLS
Pt(0.4–6)/CoFeB(2)/Cu	[86]	0–0.45	BLS
Co(1.4)/Pt	[122]	>0.5	DW
Pt(5.4)/Co(2.5)/ Au,Ir(0–2.5)/Pt(2.6)	[123]	0.6	BLS
Ta/CoFeB/TaO_x	[87]	0.22	BLS
Ir/Co_2FeAl(0.9)/Si	[124]	0.34	BLS
MgO(5)/Fe(3)/Pt(4)	[125]	0.349	BLS
SiO_2/Fe(3)/Pt(4)	[125]	0.225	BLS
Pt(6)/MgO(367)/Fe(3)/Pt(4)	[125]	0.335	BLS
Pt(6)/MgO(367)/Fe(3)/ Al(5)/SiO_2(5)	[125]	0.1025	BLS
Pt(6)/Co(3)	[126]	0.33–0.43	BLS
Pt/Co(0.8)/Ir(0–2)/Ta	[127]	0.75–1.68	BLS
Pt/Co(0.8)/Ir(0–2)/Ta	[127]	0.48–1.25	DW
Graphene/$Ni_{80}Fe_{20}$(3)/Ta(2)	[128]	0.067	BLS
FeFe/W	[129]	7.5 meV/nm	SPSTM simulation
FeFe/W	[70]	0.5–0.9 meV	SPEELS
$(Co/Ni)_n$/Ir/Pt	[35]	0.12 meV/atom	BLS
$(Ni/Co)_n$/Ir	[130]	0.36 meV/atom	BLS
Graphene/Co/Ru	[131]	0.11 meV/atom	SPLEEM/DW

by a frequency shift that increases with heavy metal thickness. The frequency of both peaks interchanges when reversing the direction of the applied magnetic field, due to the reversal of the propagation direction of the spin waves. It was also found that the frequency asymmetry exhibits a linear dependence as a function of k, and it becomes more pronounced when increasing the thickness of the heavy metal. The in-plane angular dependence of the frequency shift was also measured with BLS [73, 75, 86, 87], which evidences a clear sinusoidal dependence with $\phi_{\mathbf{k}}$, in agreement with Eq. 5.25, as predicted theoretically [96].

5.4 Micromagnetic Simulations of Spin Waves with Interfacial DMI

Micromagnetic simulations are a powerful numerical method for the analysis of spin dynamics in magnetic samples of a few microns. In particular, this numerical technique can be used to calculate the spectrum of spin waves, and when a DMI is present, it reveals different properties characteristic of these systems, such as the spin-wave asymmetry. In this context, simulations have successfully supported theoretical formulations [80, 96, 97, 99, 132–137] and experimental measurements [72–87, 138] on thin-film systems with a homogeneous DMI.

A micromagnetic simulation is based on numerically discretizing the continuum description of the magnetic system into a mesh of magnetic moments whose arrangement depends on the discretization method. The two most commonly used methods are finite differences and finite elements. For the former, the sample is divided into a regular grid of cuboids, which each represents a magnetic moment (see Fig. 5.2a), and the expressions of the magnetic interactions are approximated according to this numerical method. The spacing between cuboids is usually chosen with a magnitude smaller than the exchange length. To simulate the dynamics of the magnetization, the Landau–Lifshitz–Gilbert equation of motion is numerically integrated for every magnetic moment. Three publicly available finite difference micromagnetic simulation codes are OOMMF [139], MuMax3 [140], and Fidimag [141].

The method for the simulation of spin waves propagating in a single direction has been discussed in detail in Refs. [142, 143] and in Ref. [144] for the case of a thin film with DMI. Following the techniques of the aforementioned studies, a similar system has been simulated (see Fig. 5.2) using the parameters shown in Table 5.4. To simulate the infinite film, it is customary to specify a sufficiently

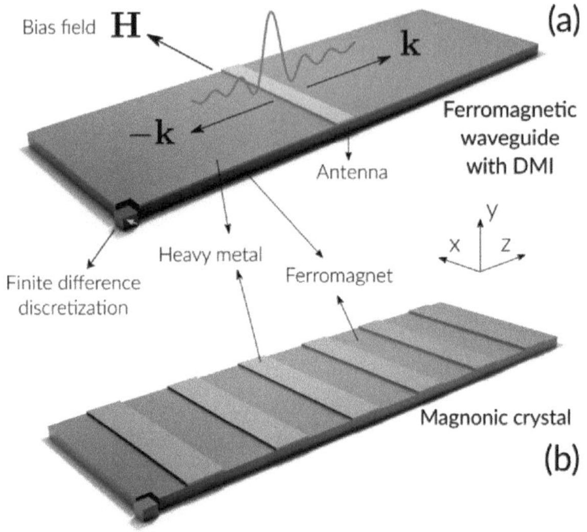

Figure 5.2 (a) Thin-film waveguide system with interfacial DMI, where a ferromagnetic film is in contact with a heavy metal surface and the sample is saturated with a bias field. An antenna at the center of the sample can be used to excite counterpropagating spin waves (characterized by a spin-wave vector **k**) across the waveguide. Micromagnetic simulations are performed using a regular grid of cuboids representing magnetic moments. (b) Magnonic crystal specified by a regular pattern of heavy metal stripes.

long waveguide along the spin-wave propagation direction, in this case the $x$direction. Consequently, the sample is saturated with a bias field **H**. To excite the spin waves, a weak periodic magnetic field $\mathbf{h}_{\mathrm{exc}}$ is applied in a region at the center of the waveguide in order to generate waves propagating in the two opposite directions, and it is applied perpendicular to the bias field. Experimentally, the spin-wave pulses can be generated using an antenna [145] (see Fig. 5.2a). The amplitude of the simulated periodic field is significantly smaller than the bias field. It is suggested to use a sinc function dependent on time for the field, delayed by time t_0, in order to excite all possible wave modes within a specific range of frequencies given by a cut-off frequency f_0 in the sinc function argument. A similar method can be applied to the spatial coordinates to restrict the range of wave vector magnitudes [142]. When exciting the system, the magnetization components are saved for every magnetic moment in

Table 5.4 Parameters for the simulation of spin waves propagating in a thin-film waveguide with DMI and with periodicity along the y direction. Dimensions are $2000 \times 128 \times 1$ nm^3.

Symbol	Magnetic parameter	Value
A	Exchange	11.1 pJ/m
D	Dzyaloshinskii–Moriya strength	3.0 mJ/m^2
M_s	Saturation magnetization	658 kA/m
$\mu_0\mathbf{H}$	External field	(0, 0.25, 0) T
$\mu_0\mathbf{h}_{exc}$	Excitation field	(0.025, 0, 0) T
f_0	Cut-off frequency	60 GHz
t_0	Excitation delay	50 ps
γ	Gyromagnetic ratio	1.76×10^{11} Hz/T
α	Gilbert damping parameter	0.01
τ	Excitation time	4 ns
$\Delta\tau$	Saving time step	0.5 ps

regularly spaced time steps $\Delta\tau$. Finally, after a sufficiently long time τ, a 2D Fourier transform in space and time variables is performed in order to obtain the spin-wave spectrum. To avoid noise in the spectrum caused by the reflection of the spin waves at the sample edges, it has been shown that using a strong damping towards the sample edges minimizes this effect [146]. Alternatively, it is also possible to use periodic boundary conditions.

The result of the simulated spin-wave spectrum in a thin film of permalloy, specified in Table 5.4, is depicted in Fig. 5.3, where the spin waves propagate perpendicular to the saturation field. The excited modes in this case are known as DE modes. This simulation was obtained with the MuMax3 code using periodic boundary conditions along the waveguide width. The spin-wave spectrum is shown as the intensity of the 2D Fourier transform of the x component of the magnetization, which is scaled logarithmically, and is directly compared with the theory [96, 97]. A test for the accuracy of the simulation is the estimation of the DMI from the spin-wave asymmetry. Accordingly, a second-order polynomial is fitted using the data from $k = -0.0785$ nm^{-1} up to $k = 0.2985$ nm^{-1}, and the asymmetry predicts a DMI magnitude of 2.89 mJ/m^2, in good agreement with the original value of 3 mJ/m^2. It is possible to extend the lower limit of the data for the curve fit; however, the polynomial

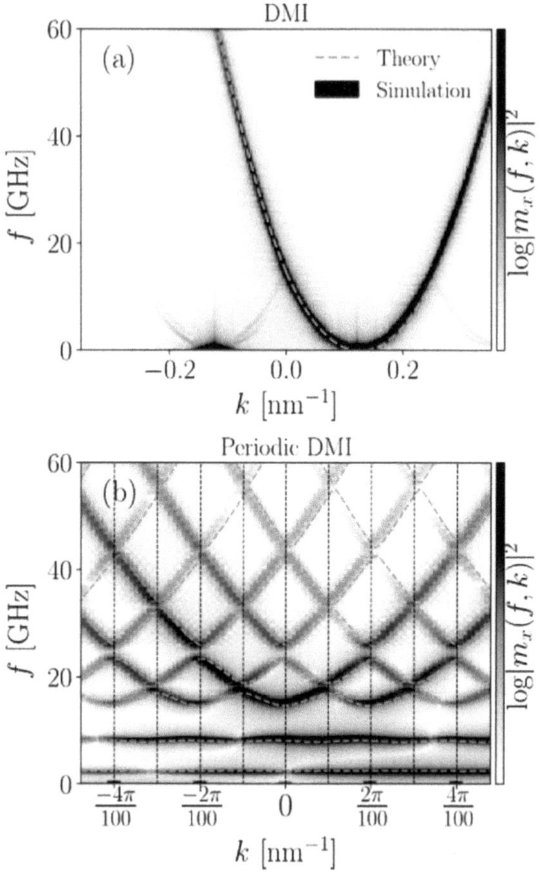

Figure 5.3 (a) Micromagnetic simulations of spin-wave propagation in permalloy thin films with (a) homogeneous DMI from a heavy metal film on top of the ferromagnet and (b) periodic DMI from regularly patterned heavy metal stripes. Simulations were obtained with the MuMax3 code. Details of simulation (a) are specified in Table 5.4. Simulation (b) is similar, but the DMI is set in regions of 50 nm width with a periodicity of 100 nm.

becomes less accurate in reproducing the intensity peaks away from the minimum, which can be improved by setting a larger simulation time. Moreover, since a cut-off frequency was set to 60 GHz, the spectrum is approximately valid down to $k \approx -0.1$ nm^{-1}.

From the result discussed previously, it can be concluded that micromagnetic simulations are highly accurate to reproduce the theoretical predictions. It is now straightforward to extend the simulations to more complex scenarios. For instance, in Fig. 5.3b the same permalloy system is simulated by setting a periodic DMI along the long axis of the waveguide. This can be obtained by patterning the top or bottom surface of the ferromagnetic system with regularly spaced heavy metal wires [147]. The simulated system of Fig. 5.3b was specified by alternating regions of material with and without a DMI of 50 nm width. In this case, the periodic DMI induces flat bands and indirect bandgaps, which are not characteristic of a magnonic system [147]. These effects are thoroughly discussed in Section 5.5.

5.5 Spin Waves in Thin Films with Periodic DMI

Spin waves propagating in a periodic magnetic media, such as an MC [148]—the magnetic counterpart of a photonic crystal—exhibit magnonic bandgaps induced by Bragg scattering, which means there are frequencies where propagation is forbidden. These bandgaps appear at the boundaries of the Brillouin zones (BZs) and are induced when waves with wave vector $\mathbf{k} = \pm n\pi/a$, with a being the lattice parameter and n an integer that satisfies the Bragg condition. On the other hand, the Bragg condition is modified by the breaking of symmetry induced by DMIs, since two counterpropagating waves at the same frequency have different wavelengths. Therefore, neither magnonic bandgaps nor the standing waves appear at the BZ edges anymore. According to this, MCs with a periodic chiral interaction exhibit indirect bandgaps, where band edges lie outside the borders of the BZs. Different cases of MCs with chiral properties have been studied recently [100, 101], reporting the observation of indirect bandgaps. Topological properties were also found in a theoretical study of periodic arrays of magnetic nanoislands on top of a heavy metal layer [149]. In addition, it has been shown, by means of micromagnetic simulations, how the DMI induces a nontrivial temporal evolution of standing spin waves in a geometrically confined structure [150]. Recently, it has been shown how a periodic DMI can produce all these effects in a magnonic thin-film system [147], namely (i) indirect bandgaps, (ii) low-frequency flat bands, and (iii) an unconventional temporal evolution of the standing waves at the bandgap edges. Following this latter work, this section is dedicated to the theoretical background of the role of a periodic DMI.

An overview of a chiral MC is shown in Fig. 5.2b, where a periodic array of heavy metal stripes is put in contact with a ferromagnetic layer. Owing to the high spin–orbit coupling of the heavy metal and the ultrathin thickness of the ferromagnetic layer, a periodic DMI is expected to occur in the system. In the continuum limit, effective fields associated with the different interactions are required to study the dynamic of the MC by means of both the Landau–Lifshitz equation and the plane-wave method. In the latter approach, the periodic quantities are expanded into Fourier series so that the equation of motion is treated as an eigenvalue problem [147]. While the effective fields associated with anisotropies, dipolar interactions, and exchange interactions are very well known, the effective field due to the DMI ($\mathbf{h}^{\mathrm{DM}}$) becomes nontrivial, because the DMI strength (D) is now a space-dependent periodic parameter. To derive the DM effective field, it is necessary to start from the atomic Hamiltonian $\mathcal{H}_{\mathrm{DM}} = \sum_{\langle i,j \rangle} \mathbf{D}_{ij} \cdot (\mathbf{S}_i \times \mathbf{S}_j)$. Since the DMI is short ranged, one can see that only the terms $\mathbf{D}_{i-1,i} \cdot (\mathbf{S}_{i-1} \times \mathbf{S}_i) + \mathbf{D}_{i,i+1} \cdot (\mathbf{S}_i \times \mathbf{S}_{i+1}) = \mathbf{S}_i \cdot (\mathbf{D}_{i-1,i} \times \mathbf{S}_{i-1} - \mathbf{D}_{i,i+1} \times \mathbf{S}_{i+1})$ are involved in $\mathcal{H}_{\mathrm{DM}}$, thus

$$\mathcal{H}_{\mathrm{DM}} = \sum_i \mathbf{S}_i \cdot (\mathbf{D}_{i-1,i} \times \mathbf{S}_{i-1} - \mathbf{D}_{i,i+1} \times \mathbf{S}_{i+1}). \tag{5.26}$$

Here, $\mathbf{S}_{i\pm1}$ represent the neighbor spins of $\mathbf{S}_i$, and the following effective field for the spin at site i is obtained:

$$\mathbf{h}_i^{\mathrm{DM}} = \mathbf{D}_{i-1,i} \times \mathbf{S}_{i-1} - \mathbf{D}_{i,i+1} \times \mathbf{S}_{i+1}. \tag{5.27}$$

In a very extended array of spins, the variation of the spin orientation between neighboring sites is very smooth; therefore, in the continuum approximation, the elements located around site i can be written as

$$\mathbf{S}_{i\pm1} \simeq \mathbf{S}_i \pm \frac{\partial \mathbf{S}_i}{\partial z} \delta z, \tag{5.28}$$

$$\mathbf{D}_{i,i+1} \simeq \mathbf{D}_{i-1,i} + \frac{\partial \mathbf{D}_{i-1,i}}{\partial z} \delta z. \tag{5.29}$$

Notice that it has been assumed that the DM coupling only varies along the z axis (see Fig. 5.2b); thus, the effective field in Eq. 5.27

becomes

$$\mathbf{h}_i^{\mathrm{DM}} \simeq \left(-2\mathbf{D}_{i-1,i} \times \tfrac{\partial \mathbf{S}_i}{\partial z} - \tfrac{\partial \mathbf{D}_{i-1,i}}{\partial z} \times \mathbf{S}_i\right)\delta z. \tag{5.30}$$

Taking into account that $\mathbf{D}_{i-1,i} \perp (\mathbf{r}_i - \mathbf{r}_{i\pm1})$, the effective DMI field in the continuum limit is

$$\mathbf{h}^{\mathrm{DM}} \simeq \frac{1}{\mu_0 M_s^2}\left\{\left[2D(\mathbf{r})\frac{\partial}{\partial z}m_z(\mathbf{r}) + m_z(\mathbf{r})\frac{\partial}{\partial z}D(\mathbf{r})\right]\hat{y}\right.$$
$$\left. - \left[2D(\mathbf{r})\frac{\partial}{\partial z}m_y(\mathbf{r}) + m_y(\mathbf{r})\frac{\partial}{\partial z}D(\mathbf{r})\right]\hat{z}\right\}, \tag{5.31}$$

where $D(\mathbf{r})$ is the DM strength that is a periodic function of z and can be expanded as $D(\mathbf{r}) = \sum_{\mathbf{G}} D(\mathbf{G})e^{i\mathbf{G}\cdot\mathbf{r}}$, with $\mathbf{G} = (2\pi n/a)\hat{z}$ a reciprocal lattice vector. By including all the energy contributions into the Landau–Lifshitz equation [151], the following eigenvalue problem is obtained:

$$\tilde{\mathbf{T}}\,\mathbf{m}_{\mathbf{G}} = i\frac{\omega}{\mu_0\gamma}\mathbf{m}_{\mathbf{G}}, \tag{5.32}$$

where $\mathbf{m}_{\mathbf{G}}^{\mathrm{T}} = [m_z(G_1), \ldots, m_z(G_N), m_y(G_1), \ldots, m_y(G_N)]$ is the eigenvector and $\tilde{\mathbf{T}}$ is given by

$$\tilde{\mathbf{T}} = \begin{pmatrix} \tilde{\mathbf{T}}^{zz} & \tilde{\mathbf{T}}^{zy} \\ \tilde{\mathbf{T}}^{yz} & \tilde{\mathbf{T}}^{yy} \end{pmatrix}. \tag{5.33}$$

The matrix elements of the submatrices are defined as follows:

$$\mathbf{T}^{zz}_{\mathbf{G},\mathbf{G}'} = \mathbf{T}^{yy}_{\mathbf{G},\mathbf{G}'} = -i\frac{2D(\mathbf{G}-\mathbf{G}')}{\mu_0 M_s}\left(\frac{\mathbf{G}+\mathbf{G}'}{2} + \mathbf{k}\right)\cdot\hat{z}, \tag{5.34a}$$

$$\mathbf{T}^{zy}_{\mathbf{G},\mathbf{G}'} = -\left[M_s(\mathbf{G}+\mathbf{k})^2\lambda_{\mathrm{ex}}^2 + H + M_s F_{\mathbf{G},\mathbf{k}}\right]\delta_{\mathbf{G},\mathbf{G}'}, \tag{5.34b}$$

$$\mathbf{T}^{yz}_{\mathbf{G},\mathbf{G}'} = \left[M_s(\mathbf{G}+\mathbf{k})^2\lambda_{\mathrm{ex}}^2 + H + M_s(1 - F_{\mathbf{G},\mathbf{k}})\right]\delta_{\mathbf{G},\mathbf{G}'}. \tag{5.34c}$$

In Eq. 5.34, $F_{\mathbf{G},\mathbf{k}} = e^{-|\mathbf{G}+\mathbf{k}|d/2}$, λ_{ex} is the exchange length and d is the thickness of the ferromagnet.

To understand the dynamic properties of the system shown in Fig. 5.2b, where the equilibrium magnetization is pointing along x and the spin wave propagation is along z, typical magnetic parameters of permalloy are used (see Table 5.4). In the zero-DMI case, the dispersion is completely reciprocal, as shown in Fig. 5.4a.

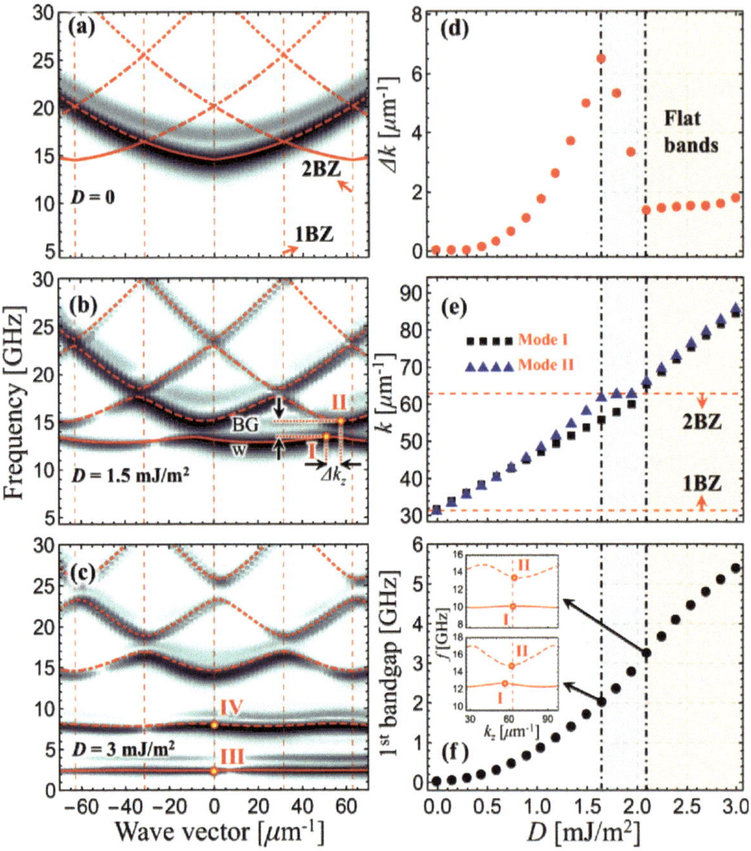

Figure 5.4 (a–c) Dispersion relation for a Py film covered with an array of heavy metal wires, where a = 100 nm, D = 0, 15 and 3 mJ/m^2. The lines correspond to the theory and the color code to simulations with the OOMMF code, where darker (lighter) color represents an intensity maximum (minimum). Panel (d) depicts the wave vector shift $\Delta\mathbf{k}$, while (e) illustrates the modes I and II [see (b)] as a function of D. In (f) the first bandgap as a function of the Dzyaloshinskii–Moriya strength D is shown. The shaded areas correspond to three ranges of D-values where different behaviors are predicted. Insets in (f) show the first bandgap behavior during the transition between the shaded areas [147].

When a periodic DMI is present, indirect bandgaps emerge, which are generated by the nonreciprocal character of the DMI. Interestingly, when the DMI strength is increased the low-frequency branches decrease in frequency and become flat (see Fig. 5.4c,d). One particular feature of the spin-wave dynamics with a periodic DMI is that the transition from reciprocal dispersion (see Fig. 5.4a) up to the emergence of low-frequency flat bands is not monotonic, as illustrated in Fig. 5.4d,f. When the parameter D is increased, the indirect character of the spin waves also increases. Then, the mode II (minimum of the bandgap edge shown in Fig. 5.4b) reaches the second Brillouin zone (2BZ), and it remains pinned under the increasing of D, as shown by the triangles in Fig. 5.4e. At this point, mode I is still shifting to high k values, so when mode I also reaches the 2BZ, the flat region is reached and the mode shifts to lower frequencies. If D further increases, the upper-frequency branches also become flat. The emergence of these flat bands is enhanced if the lattice parameter increases (not shown), and it can be manipulated by changing the geometrical properties of the heavy metal stripes [147]. These flat bands possibly have important consequences, such as the Bose–Einstein condensation [152] predicted in bosonic systems hosting flat bands [153], and frustrated quantum magnets [154, 155]. Topological properties of higher-dimensional periodic DMI structures are still to be explored.

It is important to remark that the influence of a continuous heavy metal film in contact with a ferromagnetic layer is not observed in the FMR response [96]. The inclusion of a periodic DMI modifies the band structure in a way that additional modes can be now observed in FMR experiments [101, 147, 150]. A difficulty arises, however, when trying to understand the temporal evolution of these waves, because the standing waves around the gaps originate outside the BZs. Indeed, while a typical standing behavior is observed in the zones without DMI, the spin waves have a nonzero phase velocity in the areas in contact with the heavy metal stripes. This time dependence can be understood by taking into account the boundary conditions induced by DMI at the interfaces, since the DMI produces canted states of the dynamic magnetization, inducing, thus, a nontrivial temporal evolution of the spin waves along the whole system [147]. Furthermore, it can be shown that the inclusion of Gilbert damping does not significantly modify the band structure obtained in Fig. 5.4. However, it is well known that if a ferromagnet is coupled to a heavy metal, the damping increases, decreasing then the propagation length of the waves, which is not favorable for applications. Hence, other kinds of nonreciprocal sources could be of interest, as, for instance, (i) an electric current using the adiabatic spin transfer torque [156, 157]; (ii) the flexoelectric interaction [158]; (iii) metallized MCs [159, 160]; (iv) a bicomponent MC composed of a normal ferromagnet and a bulk periodic DMI material; (v) periodic ferromagnetic bilayers, where nonreciprocity is induced by dynamic dipolar interaction [62, 68, 161–165]; or even (vi) films with graded magnetization along the thickness [166].

5.6 Conclusions

The influence of DMI on the behavior of spin waves in chiral lattice magnets, ultrathin ferromagnetic films capped with a heavy metal layer, and chiral MCs has been reviewed. Such anisotropic exchange coupling produces nonreciprocal features on the spin-wave spectrum of a magnetic system, a phenomenon that occurs for both bulk and interfacial DMI. For bulk DMI systems, such as MnSi, FeGe, and Cu_2OSeO_3, it was theoretically shown that spin-wave nonreciprocity should appear mostly in backward-volume configuration. On the other hand, the SW nonreciprocity induced in thin films with interfacial DMI, as, for instance, Pt/Co/Ni or Ir/Co/AlO_x, is maximum in the DE geometry. Interestingly, for both types of DMI, the frequency nonreciprocity is proportional to the wave vector, the strength of the DMI, and the in-plane component of the static magnetization, vanishing for perpendicularly magnetized thin films. In the case of interfacial DMI, it has been also shown that the frequency shift decreases with the thickness of the ferromagnetic film, as a consequence of the interfacial nature of the coupling, while the nonreciprocity increases with the thickness of the heavy metal layer and saturates in a length scale of a few nanometers. This behavior is a result of the spin–orbit coupling between the sea of conduction electrons and the heavy metal atoms around the ferromagnetic sites.

The concept of a chiral MC has been introduced theoretically, where the interfacial DMI is considered as periodic along a single direction. The effect of this periodicity includes additional features in the magnon spectrum such as flat bands, indirect gaps, and an unusual spin-wave evolution, with standing waves showing finite phase velocities in the zones where the DMI is nonzero. These results have been obtained with micromagnetic simulations and using a theoretical approach based on the plane-wave method. These chiral MCs with periodic DMI, which may be attained, for instance, by covering a thin ferromagnetic film with an array of heavy metal wires, host interesting physical properties, encouraging future experimental studies to prove and evidence these phenomena.

Acknowledgments

We acknowledge fruitful discussions with R. Arias, D. L. Mills, T. Schneider, J. Lindner, K. Lenz, A. Roldán-Molina, G. Gubbiotti, S. Tacchi, M. Ahlberg, H. Fangohr, and F. Ma. This work was supported by FONDECYT grants 1161403, 11170736, and 3150372 and Centers of Excellence with Basal/CONICYT financing, grant FB0807, CEDENNA. D. C.-O. acknowledges support from the EPSRC Programme grant on Skyrmionics (EP/N032128/1).

References

1. Dzyaloshinsky, I. (1958). A thermodynamic theory of weak ferromagnetism of antiferromagnetics, *J. Phys. Chem. Solids*, **4**, 241.
2. Anderson, P. W. (1959). New approach to the theory of superexchange interactions, *Phys. Rev.*, **115**, 2.
3. Moriya, T. (1960). New mechanism of anisotropic superexchange interaction, *Phys. Rev. Lett.*, **4**, 228.
4. Moriya, T. (1960). Anisotropic superexchange interaction and weak ferromagnetism, *Phys. Rev.*, **120**, 91.

5. Pincus, P. (1960). Theory of magnetic resonance in αFe_2O_3, *Phys. Rev. Lett.*, **5**, 13.
6. Cheong, S.-W., Thompson, J. D. and Fisk, Z. (1989). Metamagnetism in La_2CuO_4, *Phys. Rev. B*, **39**, 4395.
7. Coffey, D., Bedell, K. S. and Trugman, S. A. (1990). Effective spin Hamiltonian for the CuO planes in La_2CuO_4 and metamagnetism, *Phys. Rev. B*, **42**, 6509.
8. Coffey, D., Rice, T. M. and Zhang, F. C. (1991). Dzyaloshinskii-Moriya interaction in the cuprates, *Phys. Rev. B*, **44**, 10112.
9. Uchida, M., Onose, Y., Matsui, Y. and Tokura, Y. (2006). Real-space observation of helical spin order, *Science*, **311**, 359.
10. Nagaosa, N. and Tokura, Y. (2013). Topological properties and dynamics of magnetic skyrmions, *Nat. Nanotechnol.*, **8**, 899.
11. Fert, A. (1991). Magnetic and transport properties of metallic multilayers, in *Metallic Multilayers*, Vol. 59 of Materials Science Forum (Trans Tech Publications), pp. 439–480.
12. Crépieux, A. and Lacroix, C. (1998). Dzyaloshinsky-Moriya interactions induced by symmetry breaking at a surface, *J. Magn. Magn. Mater.*, **182**, 341.
13. Bode, M., Heide, M., von Bergmann, K., Ferriani, P., Heinze, S., Bihlmayer, G., Kubetzka, A., Pietzsch, O., Blügel, S. and Wiesendanger, R. (2007). Chiral magnetic order at surfaces driven by inversion asymmetry, *Nature*, **447**, 190.
14. De Luca, G. M., Ghiringhelli, G., Moretti Sala, M., Di Matteo, S., Haverkort, M. W., Berger, H., Bisogni, V., Cezar, J. C., Brookes, N. B. and Salluzzo, M. (2010). Weak magnetism in insulating and superconducting cuprates, *Phys. Rev. B*, **82**, 214504.
15. Wiesendanger, R. (2016). Nanoscale magnetic skyrmions in metallic films and multilayers: a new twist for spintronics, *Nat. Rev. Mater.*, **1**, 16044.
16. Cheong, S.-W., Talbayev, D., Kiryukhin, V. and Saxena, A. (2018). Broken symmetries, non-reciprocity, and multiferroicity, *npj Quantum Mater.*, **3**, 19.
17. Dzyaloshinskii, I. E. (1964). Theory of helicoidal structures in antiferromagnets. I. Nonmetals, *Sov. Phys. JETP*, **19**, 960.
18. Bogdanov, A. N. and Yablonskii, D. A. (1989). Thermodynamically stable vortices in magnetically ordered crystals. The mixed state of magnets, *Sov. Phys. JETP*, **68**, 101.

19. Bogdanov, A. N. and Rößler, U. K. (2001). Chiral symmetry breaking in magnetic thin films and multilayers, *Phys. Rev. Lett.*, **87**, 037203.
20. Rößler, U. K., Bogdanov, A. N. and Pfleiderer, C. (2006). Spontaneous skyrmion ground states in magnetic metals, *Nature*, **442**, 797.
21. von Bergmann, K., Heinze, S., Bode, M., Vedmedenko, E. Y., Bihlmayer, G., Blügel, S. and Wiesendanger, R. (2006). Observation of a complex nanoscale magnetic structure in a hexagonal Fe monolayer, *Phys. Rev. Lett.*, **96**, 167203.
22. Vedmedenko, E. Y., Udvardi, L., Weinberger, P. and Wiesendanger, R. (2007). Chiral magnetic ordering in two-dimensional ferromagnets with competing Dzyaloshinsky-Moriya interactions, *Phys. Rev. B*, **75**, 104431.
23. Ferriani, P., von Bergmann, K., Vedmedenko, E. Y., Heinze, S., Bode, M., Heide, M., Bihlmayer, G., Blügel, S. and Wiesendanger, R. (2008). Atomic-scale spin spiral with a unique rotational sense: Mn monolayer on W(001), *Phys. Rev. Lett.*, **101**, 027201.
24. Heide, M., Bihlmayer, G. and Blügel, S. (2008). Dzyaloshinskii-Moriya interaction accounting for the orientation of magnetic domains in ultrathin films: Fe/W(110), *Phys. Rev. B*, **78**, 140403.
25. Moore, T. A., Miron, I. M., Gaudin, G., Serret, G., Auffret, S., Rodmacq, B., Schuhl, A., Pizzini, S., Vogel, J. and Bonfim, M. (2008). High domain wall velocities induced by current in ultrathin Pt/Co/AlO_x wires with perpendicular magnetic anisotropy, *Appl. Phys. Lett.*, **93**, 262504.
26. Mühlbauer, S., Binz, B., Jonietz, F., Pfleiderer, C., Rosch, A., Neubauer, A., Georgii, R. and Böni, P. (2009). Skyrmion lattice in a chiral magnet, *Science*, **323**, 915.
27. Münzer, W., Neubauer, A., Adams, T., Mühlbauer, S., Franz, C., Jonietz, F., Georgii, R., Böni, P., Pedersen, B., Schmidt, M., Rosch, A. and Pfleiderer, C. (2010). Skyrmion lattice in the doped semiconductor $Fe_{1-x}Co_xSi$, *Phys. Rev. B*, **81**, 041203.
28. Yu, X. Z., Onose, Y., Kanazawa, N., Park, J. H., Han, J. H., Matsui, Y., Nagaosa, N. and Tokura, Y. (2010). Real-space observation of a two-dimensional skyrmion crystal, *Nature*, **465**, 901.
29. Yu, X. Z., Kanazawa, N., Onose, Y., Kimoto, K., Zhang, W. Z., Ishiwata, S., Matsui, Y. and Tokura, Y. (2011). Near room-temperature formation of a skyrmion crystal in thin-films of the helimagnet FeGe, *Nat. Mater.*, **10**, 106.
30. Heinze, S., von Bergmann, K., Menzel, M., Brede, J., Kubetzka, A., Wiesendanger, R., Bihlmayer, G. and Blügel, S. (2011). Spontaneous

atomic-scale magnetic skyrmion lattice in two dimensions, *Nat. Phys.*, **7**, 713.

31. Onose, Y., Okamura, Y., Seki, S., Ishiwata, S. and Tokura, Y. (2012). Observation of magnetic excitations of skyrmion crystal in a helimagnetic insulator Cu_2OSeO_3, *Phys. Rev. Lett.*, **109**, 037603.
32. Seki, S., Yu, X. Z., Ishiwata, S. and Tokura, Y. (2012). Observation of skyrmions in a multiferroic material, *Science*, **336**, 198.
33. Thiaville, A., Rohart, S., Jué, É., Cros, V. and Fert, A. (2012). Dynamics of Dzyaloshinskii domain walls in ultrathin magnetic films, *Europhys. Lett.*, **100**, 57002.
34. Rohart, S. and Thiaville, A. (2013). Skyrmion confinement in ultrathin film nanostructures in the presence of Dzyaloshinskii-Moriya interaction, *Phys. Rev. B*, **88**, 184422.
35. Chen, G., Ma, T., N'Diaye, A. T., Kwon, H., Won, C., Wu, Y. and Schmid, A. K. (2013). Tailoring the chirality of magnetic domain walls by interface engineering, *Nat. Commun.*, **4**, 2671.
36. Je, S.-G., Kim, D.-H., Yoo, S.-C., Min, B.-C., Lee, K.-J. and Choe, S.-B. (2013). Asymmetric magnetic domain-wall motion by the Dzyaloshinskii-Moriya interaction, *Phys. Rev. B*, **88**, 214401.
37. Emori, S., Bauer, U., Ahn, S.-M., Martinez, E. and Beach, G. S. D. (2013). Current-driven dynamics of chiral ferromagnetic domain walls, *Nat. Mater.*, **12**, 611.
38. Fert, A., Cros, V. and Sampaio, J. (2013). Skyrmions on the track, *Nat. Nanotechnol.*, **8**, 152.
39. Pizzini, S., Vogel, J., Rohart, S., Buda-Prejbeanu, L. D., Jué, É., Boulle, O., Miron, I. M., Safeer, C. K., Auffret, S., Gaudin, G. and Thiaville, A. (2014). Chirality-induced asymmetric magnetic nucleation in Pt/Co/AlO_x ultrathin microstructures, *Phys. Rev. Lett.*, **113**, 047203.
40. Torrejon, J., Kim, J., Sinha, J., Mitani, S., Hayashi, M., Yamanouchi, M. and Ohno, H. (2014). Interface control of the magnetic chirality in CoFeB/MgO heterostructures with heavy-metal underlayers, *Nat. Commun.*, **5**, 4655.
41. Moreau-Luchaire, C., Moutafis, C., Reyren, N., Sampaio, J., Vaz, C. A. F., Horne, N. V., Bouzehouane, K., Garcia, K., Deranlot, C., Warnicke, P., Wohlhter, P., George, J.-M., Weigand, M., Raabe, J., Cros, V. and Fert, A. (2016). Additive interfacial chiral interaction in multilayers for stabilization of small individual skyrmions at room temperature, *Nat. Nanotechnol.*, **11**, 444.

42. Fert, A., Reyren, N. and Cros, V. (2017). Magnetic skyrmions: advances in physics and potential applications, *Nat. Rev. Mater.*, **2**, 17031.
43. Levy, P. M. (1969). Indirect exchange via spin-orbit coupled states, *Solid State Commun.*, **7**, 1813.
44. Smith, D. (1976). New mechanisms for magnetic anisotropy in localised S-state moment materials, *J. Magn. Magn. Mater.*, **1**, 214.
45. Fert, A. and Levy, P. M. (1980). Role of anisotropic exchange interactions in determining the properties of spin-glasses, *Phys. Rev. Lett.*, **44**, 1538
46. Levy, P. M. and Fert, A. (1981). Anisotropy induced by nonmagnetic impurities in CuMn spin-glass alloys, *Phys. Rev. B*, **23**, 4667.
47. Ruderman, M. A. and Kittel, C. (1954). Indirect exchange coupling of nuclear magnetic moments by conduction electrons, *Phys. Rev.*, **96**, 99.
48. Yosida, K. (1957). Magnetic properties of Cu-Mn alloys, *Phys. Rev.*, **106**, 893.
49. Ryu, K.-S., Yang, S.-H., Thomas, L. and Parkin, S. S. P. (2014). Chiral spin torque arising from proximity-induced magnetization, *Nat. Commun.*, **5**, 3910.
50. Yang, H., Thiaville, A., Rohart, S., Fert, A. and Chshiev, M. (2015). Anatomy of Dzyaloshinskii-Moriya interaction at Co/Pt interfaces, *Phys. Rev. Lett.*, **115**, 267210.
51. Belabbes, A., Bihlmayer, G., Bechstedt, F., Blügel, S. and Manchon, A. (2016). Hund's rule-driven Dzyaloshinskii-Moriya interaction at 3D—5D interfaces, *Phys. Rev. Lett.*, **117**, 247202.
52. Kim, S., Ueda, K., Go, G., Jang, P.-H., Lee, K.-J., Belabbes, A., Manchon, A., Suzuki, M., Kotani, Y., Nakamura, T., Nakamura, K., Koyama, T., Chiba, D., Yamada, K. T., Kim, D.-H., Moriyama, T., Kim, K.-J. and Ono, T. (2018). Correlation of the Dzyaloshinskii–Moriya interaction with Heisenberg exchange and orbital asphericity, *Nat. Commun.*, **9**, 1648.
53. Serga, A. A., Chumak, A. V. and Hillebrands, B. (2010). YIG magnonics, *J. Phys. D: Appl. Phys.*, **43**, 264002.
54. Demidov, V. E., Kostylev, M. P., Rott, K., Münchenberger, J., Reiss, G. and Demokritov, S. O. (2011). Excitation of short-wavelength spin waves in magnonic waveguides, *Appl. Phys. Lett.*, **99**, 082507.
55. Yu, H., Duerr, G., Huber, R. Bahr, M., Schwarze, T., Brandl, F. and Grundler, D. (2013). Omnidirectional spin-wave nanograting coupler, *Nat. Commun.*, **4**, 2702.

56. Chumak, A. V., Vasyuchka, V. I., Serga, A. A. and Hillebrands, B. (2015). Magnon spintronics, *Nat. Phys.*, **11**, 453.
57. Yu, H., d'Allivy Kelly, O., Cros, V., Bernard, R., Bortolotti, P., Anane, A., Brandl, F., Heimbach, F. and Grundler, D. (2016). Approaching soft X-ray wavelengths in nanomagnet-based microwave technology, *Nat. Commun.*, **7**, 11255.
58. Ciubotaru, F., Devolder, T., Manfrini, M., Adelmann, C. and Radu, I. P. (2016). All electrical propagating spin wave spectroscopy with broadband wavevector capability, *Appl. Phys. Lett.*, **109**, 012403.
59. Liu, C., Chen, J., Liu, T., Heimbach, F., Yu, H., Xiao, Y., Hu, J., Liu, M., Chang, H., Stueckler, T., Tu, S., Zhang, Y., Zhang, Y., Gao, P., Liao, Z., Yu, D., Xia, K., Lei, N., Zhao, W. and Wu, M. (2018). Long-distance propagation of short-wavelength spin waves, *Nat. Commun.*, **9**, 738.
60. Stoll, H., Noske, M., Weigand, M., Richter, K., Krüger, B., Reeve, R. M., Hänze, M., Adolff, C. F., Stein, F.-U., Meier, G., Kläui, M. and Schütz, G. (2015). Imaging spin dynamics on the nanoscale using X-ray microscopy, *Front. Phys.*, **3**, 26.
61. Wintz, S., Tiberkevich, V., Weigand, M., Raabe, J., Lindner, J., Erbe, A., Slavin, A. and Fassbender, J. (2016). Magnetic vortex cores as tunable spin-wave emitters, *Nat. Nanotechnol.*, **11**, 948.
62. Sluka, V., Schneider, T., Gallardo, R. A., Kákay, A., Weigand, M., Warnatz, T., Mattheis, R., Roldán-Molina, A., Landeros, P., Tiberkevich, V., Slavin, A., Schütz, G., Erbe, A., Deac, A., Lindner, J., Raabe, J., Fassbender, J. and Wintz S. (2019). Emission and propagation of 1D and 2D spin waves with nanoscale wavelengths in anisotropic spin textures, *Nat. Nanotechnol.*, **14**, 328.
63. Farle, M. (1998). Ferromagnetic resonance of ultra thin metallic layers, *Rep. Prog. Phys.*, **61**, 755.
64. Hillebrands, B. (2000). *Brillouin Light Scattering from Layered Magnetic Structures* (Springer, Berlin, Heidelberg), pp. 174–289.
65. Gubbiotti, G., Tacchi, S., Madami, M., Carlotti, G., Adeyeye, A. O. and Kostylev, M. (2010). Brillouin light scattering studies of planar metallic magnonic crystals, *J. Phys. D: Appl. Phys.*, **43**, 264003.
66. Lenk, B., Ulrichs, H., Garbs, F. and Münzenberg, M. (2011). The building blocks of magnonics, *Phys. Rep.*, **507**, 107.
67. Farle, M., Silva, T. and Woltersdorf, G. (2013). *Spin Dynamics in the Time and Frequency Domain* (Springer, Berlin, Heidelberg), pp. 37–83.
68. Camley, R. (1987). Non-reciprocal surface waves, *Surf. Sci. Rep.*, **7**, 103.

69. Tokura, Y. and Nagaosa, N. (2018). Non-reciprocal responses from non-centrosymmetric quantum materials, *Nat. Commun.*, **9**, 3740.
70. Zakeri, K., Zhang, Y., Prokop, J., Chuang, T.-H., Sakr, N., Tang, W. X. and Kirschner, J. (2010). Asymmetric spin-wave dispersion on Fe(110): direct evidence of the Dzyaloshinskii-Moriya interaction, *Phys. Rev. Lett.*, **104**, 137203.
71. Zakeri, K., Zhang, Y., Chuang, T.-H. and Kirschner, J. (2012). Magnon life times on the Fe(110) surface: the role of spin-orbit coupling, *Phys. Rev. Lett.*, **108**, 197205.
72. Di, K., Zhang, V. L., Lim, H. S., Ng, S. C., Kuok, M. H., Yu, J., Yoon, J., Qiu, X. and Yang, H. (2015). Direct observation of the Dzyaloshinskii-Moriya interaction in a Pt/Co/Ni film, *Phys. Rev. Lett.*, **114**, 047201.
73. Zhang, V. L., Di, K., Lim, H. S., Ng, S. C., Kuok, M. H., Yu, J., Yoon, J., Qiu, X. and Yang, H. (2015). In-plane angular dependence of the spin-wave nonreciprocity of an ultrathin film with Dzyaloshinskii-Moriya interaction, *Appl. Phys. Lett.*, **107**, 022402.
74. Di, K., Zhang, V. L., Lim, H. S., Ng, S. C., Kuok, M. H., Qiu, X. and Yang, H. (2015). Asymmetric spin-wave dispersion due to Dzyaloshinskii-Moriya interaction in an ultrathin Pt/CoFeB film, *Appl. Phys. Lett.*, **106**, 052403.
75. Cho, J., Kim, N.-H., Lee, S., Kim, J.-S., Lavrijsen, R., Solignac, A., Yin, Y., Han, D.-S., van Hoof, N. J. J., Swagten, H. J. M., Koopmans, B. and You, C.-Y. (2015). Thickness dependence of the interfacial Dzyaloshinskii–Moriya interaction in inversion symmetry broken systems, *Nat. Commun.*, **6**, 7635.
76. Nembach, H. T., Shaw, J. M., Weiler, M., Jué, E. and Silva, T. J. (2015). Linear relation between Heisenberg exchange and interfacial Dzyaloshinskii-Moriya interaction in metal films, *Nat. Phys.*, **11**, 825.
77. Stashkevich, A. A., Belmeguenai, M., Roussigné, Y., Cherif, S. M., Kostylev, M., Gabor, M., Lacour, D., Tiusan, C. and Hehn, M. (2015). Experimental study of spin-wave dispersion in Py/Pt film structures in the presence of an interface Dzyaloshinskii-Moriya interaction, *Phys. Rev. B*, **91**, 214409.
78. Belmeguenai, M., Adam, J.-P., Roussigné, Y., Eimer, S., Devolder, T., Kim, J.-V., Cherif, S. M., Stashkevich, A. and Thiaville, A. (2015). Interfacial Dzyaloshinskii-Moriya interaction in perpendicularly magnetized Pt/Co/AlO_x ultrathin films measured by Brillouin light spectroscopy, *Phys. Rev. B*, **91**, 180405.

79. Kim, N.-H., Han, D.-S., Jung, J., Cho, J., Kim, J.-S., Swagten, H. J. M. and You, C.-Y. (2015). Improvement of the interfacial Dzyaloshinskii-Moriya interaction by introducing a Ta buffer layer, *Appl. Phys. Lett.*, **107**, 142408.
80. Kim, N.-H., Jung, J., Cho, J., Han, D.-S., Yin, Y., Kim, J.-S., Swagten, H. J. M. and You, C.-Y. (2016). Interfacial Dzyaloshinskii-Moriya interaction, surface anisotropy energy, and spin pumping at spin orbit coupled Ir/Co interface, *Appl. Phys. Lett.*, **108**, 142406.
81. Soucaille, R., Belmeguenai, M., Torrejon, J., Kim, J.-V., Devolder, T., Roussigné, Y., Chérif, S.-M., Stashkevich, A. A., Hayashi, M. and Adam, J.-P. (2016). Probing the Dzyaloshinskii-Moriya interaction in CoFeB ultrathin films using domain wall creep and Brillouin light spectroscopy, *Phys. Rev. B*, **94**, 104431.
82. Belmeguenai, M., Gabor, M. S., Roussigné, Y., Stashkevich, A., Chérif, S. M., Zighem, F. and Tiusan, C. (2016). Brillouin light scattering investigation of the thickness dependence of Dzyaloshinskii-Moriya interaction in $Co_{0.5}Fe_{0.5}$ ultrathin films, *Phys. Rev. B*, **93**, 174407.
83. Chaurasiya, A. K., Banerjee, C., Pan, S., Sahoo, S., Choudhury, S., Sinha, J. and Barman, A. (2016). Direct observation of interfacial Dzyaloshinskii-Moriya interaction from asymmetric spin-wave propagation in W/CoFeB/SiO_2 heterostructures down to sub-nanometer CoFeB thickness, *Sci. Rep.*, **6**, 32592.
84. Ma, X., Yu, G., Li, X., Wang, T., Wu, D., Olsson, K. S., Chu, Z., An, K., Xiao, J. Q., Wang, K. L. and Li, X. (2016). Interfacial control of Dzyaloshinskii-Moriya interaction in heavy metal/ferromagnetic metal thin film heterostructures, *Phys. Rev. B*, **94**, 180408.
85. Kwon, J. H., Yoon, J., Deorani, P., Lee, J. M., Sinha, J., Lee, K.-J., Hayashi, M. and Yang, H. (2016). Giant non-reciprocal emission of spin waves in Ta/Py bilayers, *Sci. Adv.*, **2**, e1501892.
86. Tacchi, S., Troncoso, R. E., Ahlberg, M., Gubbiotti, G., Madami, M., Åkerman, J. and Landeros, P. (2017). Interfacial Dzyaloshinskii-Moriya interaction in Pt/CoFeB films: effect of the heavy-metal thickness, *Phys. Rev. Lett.*, **118**, 147201.
87. Chaurasiya, A. K., Choudhury, S., Sinha, J. and Barman, A. (2018). Dependence of interfacial Dzyaloshinskii-Moriya interaction on layer thicknesses in Ta/CoFeB/TaO_x heterostructures from Brillouin light scattering, *Phys. Rev. Appl.*, **9**, 014008.
88. Melcher, R. L. (1973). Linear contribution to spatial dispersion in the spin-wave spectrum of ferromagnets, *Phys. Rev. Lett.*, **30**, 125.

89. Bak, P. and Jensen, M. H. (1980). Theory of helical magnetic structures and phase transitions in MnSi and FeGe, *J. Phys. C: Solid State Phys.*, **13**, L881.
90. Nakanishi, O., Yanase, A., Hasegawa, A. and Kataoka, M. (1980). The origin of the helical spin density wave in MnSi, *Solid State Commun.*, **35**, 995.
91. Kataoka, M. (1987). Spin waves in systems with long period helical spin density waves due to the antisymmetric and symmetric exchange interactions, *J. Phys. Soc. Jpn.*, **56**, 3635.
92. Belitz, D., Kirkpatrick, T. R. and Rosch, A. (2006). Theory of helimagnons in itinerant quantum systems, *Phys. Rev. B*, **73**, 054431.
93. Maleyev, S. V. (2006). Cubic magnets with Dzyaloshinskii-Moriya interaction at low temperature, *Phys. Rev. B*, **73**, 174402.
94. Udvardi, L. and Szunyogh, L. (2009). Chiral asymmetry of the spin-wave spectra in ultrathin magnetic films, *Phys. Rev. Lett.*, **102**, 207204.
95. Costa, A. T., Muniz, R. B., Lounis, S., Klautau, A. B. and Mills, D. L. (2010). Spin-orbit coupling and spin waves in ultrathin ferromagnets: the spin-wave Rashba effect, *Phys. Rev. B*, **82**, 014428.
96. Cortés-Ortuño, D. and Landeros, P. (2013). Influence of the Dzyaloshinskii–Moriya interaction on the spin-wave spectra of thin films, *J. Phys.: Condens. Matter*, **25**, 156001.
97. Moon, J.-H., Seo, S.-M., Lee, K.-J., Kim, K.-W., Ryu, J., Lee, H.-W., McMichael, R. D. and Stiles, M. D. (2013). Spin-wave propagation in the presence of interfacial Dzyaloshinskii-Moriya interaction, *Phys. Rev. B*, **88**, 184404.
98. Kostylev, M. (2014). Interface boundary conditions for dynamic magnetization and spin wave dynamics in a ferromagnetic layer with the interface Dzyaloshinskii-Moriya interaction, *J. Appl. Phys.*, **115**, 233902.
99. Garcia-Sanchez, F., Borys, P., Vansteenkiste, A., Kim, J.-V. and Stamps, R. L. (2014). Non-reciprocal spin-wave channeling along textures driven by the Dzyaloshinskii-Moriya interaction, *Phys. Rev. B*, **89**, 224408.
100. Ma, F. and Zhou, Y. (2014). Interfacial Dzialoshinskii-Moriya interaction induced nonreciprocity of spin waves in magnonic waveguides, *RSC Adv.*, **4**, 46454.
101. Mruczkiewicz, M. and Krawczyk, M. (2016). Influence of the Dzyaloshinskii-Moriya interaction on the FMR spectrum of magnonic crystals and confined structures, *Phys. Rev. B*, **94**, 024434.

102. Iguchi, Y., Uemura, S., Ueno, K. and Onose, Y. (2015). Non-reciprocal magnon propagation in a noncentrosymmetric ferromagnet $LiFe_5O_8$, *Phys. Rev. B*, **92**, 184419.
103. Grigoriev, S. V., Sukhanov, A. S., Altynbaev, E. V., Siegfried, S.-A., Heinemann, A., Kizhe, P. and Maleyev, S. V. (2015). Spin waves in full-polarized state of Dzyaloshinskii-Moriya helimagnets: small-angle neutron scattering study, *Phys. Rev. B*, **92**, 220415.
104. Seki, S., Okamura, Y., Kondou, K., Shibata, K., Kubota, M., Takagi, R., Kagawa, F., Kawasaki, M., Tatara, G., Otani, Y. and Tokura, Y. (2016). Magnetochiral nonreciprocicity of volume spin wave propagation in chiral-lattice ferromagnets, *Phys. Rev. B*, **93**, 235131.
105. Sato, T. J., Okuyama, D., Hong, T., Kikkawa, A., Taguchi, Y., Arima, T. and Tokura, Y. (2016). Magnon dispersion shift in the induced ferromagnetic phase of noncentrosymmetric MnSi, *Phys. Rev. B*, **94**, 144420.
106. Nicklow, R. M., Wakabayashi, N., Wilkinson, M. K. and Reed, R. E. (1971). Spin-wave dispersion relation for Er metal at 4.5°K, *Phys. Rev. Lett.*, **27**, 334.
107. Janoschek, M., Bernlochner, F., Dunsiger, S., Pfleiderer, C., Böni, P., Roessli, B., Link, P. and Rosch, A. (2010). Helimagnon bands as universal excitations of chiral magnets, *Phys. Rev. B*, **81**, 214436.
108. Schwarze, T., Waizner, J., Garst, M., Bauer, A., Stasinopoulos, I., Berger, H., Pfleiderer, C. and Grundler, D. (2015). Universal helimagnon and skyrmion excitations in metallic, semiconducting and insulating chiral magnets, *Nat. Mater.*, **14**, 478.
109. Takagi, R., Morikawa, D., Karube, K., Kanazawa, N., Shibata, K., Tatara, G., Tokunaga, Y., Arima, T., Taguchi, Y., Tokura, Y. and Seki, S. (2017). Spin-wave spectroscopy of the Dzyaloshinskii-Moriya interaction in room-temperature chiral magnets hosting skyrmions, *Phys. Rev. B*, **95**, 220406.
110. Garst, M., Waizner, J. and Grundler, D. (2017). Collective spin excitations of helices and magnetic skyrmions: review and perspectives of magnonics in non-centrosymmetric magnets, *J. Phys. D: Appl. Phys.*, **50**, 293002.
111. Weber, T., Waizner, J., Tucker, G. S., Georgii, R., Kugler, M., Bauer, A., Pfleiderer, C., Garst, M. and Böni, P. (2018). Field dependence of non-reciprocal magnons in chiral MnSi, *Phys. Rev. B*, **97**, 224403.
112. Shirane, G., Cowley, R., Majkrzak, C., Sokoloff, J. B., Pagonis, B., Perry, C. H. and Ishikawa, Y. (1983). Spiral magnetic correlation in cubic MnSi, *Phys. Rev. B*, **28**, 6251.

113. Gitgeatpong, G., Zhao, Y., Piyawongwatthana, P., Qiu, Y., Harriger, L. W., Butch, N. P., Sato, T. J. and Matan, K. (2017). Non-reciprocal magnons and symmetry-breaking in the $\alpha Cu_2V_2O_7$ non-centrosymmetric antiferromagnet, *Phys. Rev. Lett.*, **119**, 047201.

114. Arias, R. and Mills, D. L. (1999). Extrinsic contributions to the ferromagnetic resonance response of ultrathin films, *Phys. Rev. B*, **60**, 7395.

115. Landeros, P., Arias, R. E. and Mills, D. L. (2008). Two magnon scattering in ultrathin ferromagnets: the case where the magnetization is out of plane, *Phys. Rev. B*, **77**, 214405.

116. Leonov, A. (2012). Twisted, localized, and modulated states described in the phenomenological theory of chiral and nanoscale ferromagnets, PhD thesis, IFW Dresden.

117. Kittel, C. (1963). *Quantum Theory of Solids* (John Wiley and Sons, New York).

118. Hrabec, A., Porter, N. A., Wells, A., Benitez, M. J., Burnell, G., McVitie, S., McGrouther, D., Moore, T. A. and Marrows, C. H. (2014). Measuring and tailoring the Dzyaloshinskii-Moriya interaction in perpendicularly magnetized thin films, *Phys. Rev. B*, **90**, 020402.

119. Romming, N., Kubetzka, A., Hanneken, C., von Bergmann, K. and Wiesendanger, R. (2015). Field-dependent size and shape of single magnetic skyrmions, *Phys. Rev. Lett.*, **114**, 177203.

120. Lee, J. M., Jang, C., Min, B.-C., Lee, S.-W., Lee, K.-J. and Chang, J. (2016). All-electrical measurement of interfacial Dzyaloshinskii-Moriya interaction using collective spin-wave dynamics, *Nano Lett.*, **16**, 62.

121. Khan, R. A., Shepley, P. M., Hrabec, A., Wells, A. W. J., Ocker, B., Marrows, C. H. and Moore, T. A. (2016). Effect of annealing on the interfacial Dzyaloshinskii-Moriya interaction in Ta/CoFeB/MgO trilayers, *Appl. Phys. Lett.*, **109**, 132404.

122. Corredor, E. C., Kuhrau, S., Kloodt-Twesten, F., Frömter, R. and Oepen, H. P. (2017). SEMPA investigation of the Dzyaloshinskii-Moriya interaction in the single, ideally grown Co/Pt(111) interface, *Phys. Rev. B*, **96**, 060410.

123. Rowan-Robinson, R. M., Stashkevich, A. A., Roussigné, Y., Belmeguenai, M., Chérif, S.-M., Thiaville, A., Hase, T. P. A., Hindmarch, A. T. and Atkinson, D. (2017). The interfacial nature of proximity-induced magnetism and the Dzyaloshinskii-Moriya interaction at the Pt/Co interface, *Sci. Rep.*, **7**, 16835.

124. Belmeguenai, M., Gabor, M. S., Roussigné, Y., Petrisor, T., Mos, R. B., Stashkevich, A., Chérif, S. M. and Tiusan, C. (2018). Interfacial Dzyaloshinskii-Moriya interaction sign in Ir/Co_2FeAl systems investigated by Brillouin light scattering, *Phys. Rev. B*, **97**, 054425.

125. Zhang, W., Zhong, H., Zang, R., Zhang, Y., Yu, S., Han, G., Liu, G. L., Yan, S. S., Kang, S. and Mei, L. M. (2018). Electrical field enhanced interfacial Dzyaloshinskii-Moriya interaction in MgO/Fe/Pt system, *Appl. Phys. Lett.*, **113**, 122406.

126. Bouloussa, H., Yu, J., Roussigné, Y., Belmeguenai, M., Stashkevitch, A., Yang, H. and Chérif, S. M. (2018). Brillouin light scattering investigation of interfacial Dzyaloshinskii–Moriya interaction in ultrathin Co/Pt nanostripe arrays, *J. Phys. D: Appl. Phys.*, **51**, 225005.

127. Shahbazi, K., Kim, J.-V., Nembach, H. T., Shaw, J. M., Bischof, A., Rossell, M. D., Jeudy, V., Moore, T. A. and Marrows, C. H. (2019). Domain wall motion and interfacial Dzyaloshinskii-Moriya interactions in Pt/Co/Ir(tIr)/Ta multilayers, *Phys. Rev. B*, **99**, 094409.

128. Chaurasiya, A. K., Kumar, A., Gupta, R., Chaudhary, S., Muduli, P. K. and Barman, A. (2019). Direct observation of unusual interfacial Dzyaloshinskii-Moriya interaction in graphene/NiFe/Ta heterostructures, *Phys. Rev. B*, **99**, 035402.

129. Meckler, S., Mikuszeit, N., Preßler, A., Vedmedenko, E. Y., Pietzsch, O. and Wiesendanger, R. (2009). Real-space observation of a right-rotating inhomogeneous cycloidal spin spiral by spin-polarized scanning tunneling microscopy in a triple axes vector magnet, *Phys. Rev. Lett.*, **103**, 157201.

130. Chen, G., N'Diaye, A. T., Wu, Y. and Schmid, A. K. (2015). Ternary superlattice boosting interface-stabilized magnetic chirality, *Appl. Phys. Lett.*, **106**, 062402.

131. Yang, H., Chen, G., Cotta, A. A. C., N'Diaye, A. T., Nikolaev, S. A., Soares, E. A., Macedo, W. A. A., Liu, K., Schmid, A. K., Fert, A. and Chshiev, M. (2018). Significant Dzyaloshinskii-Moriya interaction at graphene-ferromagnet interfaces due to the Rashba effect, *Nat. Mater.*, **17**, 605.

132. Manchon, A., Ndiaye, P. B., Moon, J.-H., Lee, H.-W. and Lee, K.-J. (2014). Magnon-mediated Dzyaloshinskii-Moriya torque in homogeneous ferromagnets, *Phys. Rev. B*, **90**, 224403.

133. Garcia-Sanchez, F., Sampaio, J., Reyren, N., Cros, V. and Kim, J.-V. (2016). A skyrmion-based spin-torque nano-oscillator, *New J. Phys.*, **18**, 075011.

134. Lee, S.-J., Moon, J.-H., Lee, H.-W. and Lee, K.-J. (2017). Spin-wave propagation in the presence of inhomogeneous Dzyaloshinskii-Moriya interactions, *Phys. Rev. B*, **96**, 184433.

135. Wang, Z., Zhang, B., Cao, Y. and Yan, P. (2018). Probing the Dzyaloshinskii-Moriya interaction via the propagation of spin waves in ferromagnetic thin films, *Phys. Rev. Appl.*, **10**, 054018.

136. Zivieri, R., Giordano, A., Verba, R., Azzerboni, B., Carpentieri, M., Slavin, A. N. and Finocchio, G. (2018). Theory of non-reciprocal spin-wave excitations in spin Hall oscillators with Dzyaloshinskii-Moriya interaction, *Phys. Rev. B*, **97**, 134416.

137. Mulkers, J., Van Waeyenberge, B. and Milošević, M. V. (2018). Tunable Snell's law for spin waves in heterochiral magnetic films, *Phys. Rev. B*, **97**, 104422.

138. Brandão, J., Azzawi, S., Hindmarch, A. T. and Atkinson, D. (2017). Understanding the role of damping and Dzyaloshinskii-Moriya interaction on dynamic domain wall behaviour in platinum-ferromagnet nanowires, *Sci. Rep.*, **7**, 4569.

139. Donahue, M. J. and Porter, D. G. (1999). OOMMF user's guide, version 1.0. National Institute of Standards and Technology, Gaithersburg, MD.

140. Vansteenkiste, A., Leliaert, J., Dvornik, M., Helsen, M., Garcia-Sanchez, F. and Van Waeyenberge, B. (2014). The design and verification of MuMax3, *AIP Adv.*, **4**, 107133.

141. Bisotti, M.-A., Cortés-Ortuño, D., Pepper, R., Wang, W., Beg, M., Kluyver, T. and Fangohr, H. (2018). Fidimag: a finite difference atomistic and micromagnetic simulation package, *J. Open Res. Software*, **6**, 22.

142. Venkat, G., Kumar, D., Franchin, M., Dmytriiev, O., Mruczkiewicz, M., Fangohr, H., Barman, A., Krawczyk, M. and Prabhakar, A. (2013). Proposal for a standard micromagnetic problem: spin wave dispersion in a magnonic waveguide, *IEEE Trans. Magn.*, **49**, 524.

143. Kumar, D., Dmytriiev, O., Ponraj, S. and Barman, A. (2012). Numerical calculation of spin wave dispersions in magnetic nanostructures, *J. Phys. D: Appl. Phys.*, **45**, 015001.

144. Cortés-Ortuño, D., Beg, M., Nehruji, V., Breth, L., Pepper, R., Kluyver, T., Downing, G., Hesjedal, T., Hatton, P., Lancaster, T., Hertel, R., Hovorka, O. and Fangohr, H. (2018). Proposal for a micromagnetic standard problem for materials with Dzyaloshinskii-Moriya interaction, *New J. Phys.*, **20**, 113015.

145. Sebastian, T., Schultheiss, K., Obry, B., Hillebrands, B. and Schultheiss, H. (2015). Micro-focused Brillouin light scattering: imaging spin waves at the nanoscale, *Front. Phys.*, **3**, 35.
146. Venkat, G., Fangohr, H. and Prabhakar, A. (2018). Absorbing boundary layers for spin wave micromagnetics, *J. Magn. Magn. Mater.*, **450**, 34.
147. Gallardo, R. A., Cortés-Ortuño, D., Schneider, T., Roldán-Molina, A., Ma, F., Troncoso, R. E., Lenz, K., Fangohr, H., Lindner, J. and Landeros, P. (2019). Flat bands, indirect gaps, and unconventional spin-wave behavior induced by a periodic Dzyaloshinskii-Moriya interaction, *Phys. Rev. Lett.*, **122**, 067204.
148. Puszkarski, H. and Krawczyk, M. (2003). Magnonic crystals the magnetic counterpart of photonic crystals, in *Interfacial Effects and Novel Properties of Nanomaterials*, Vol. 94 of Solid State Phenomena (Trans Tech), pp. 125–134.
149. Iacocca, E. and Heinonen, O. (2017). Topologically nontrivial magnon bands in artificial square spin ices with Dzyaloshinskii-Moriya interaction, *Phys. Rev. Appl.*, **8**, 034015.
150. Zingsem, B. V., Farle, M., Stamps, R. L. and Camley, R. E. (2017). The unusual nature of confined modes in a chiral system, arXiv:1609.03417.
151. Vasseur, J. O., Dobrzynski, L., Djafari-Rouhani, B. and Puszkarski, H. (1996). Magnon band structure of periodic composites, *Phys. Rev. B*, **54**, 1043.
152. Demokritov, S. O., Demidov, V. E., Dzyapko, O., Melkov, G. A., Serga, A. A., Hillebrands, B. and Slavin, A. N. (2006). Bose–Einstein condensation of quasi-equilibrium magnons at room temperature under pumping, *Nature*, **443**, 430.
153. Huber S. D. and Altman, E. (2010). Bose condensation in flat bands, *Phys. Rev. B*, **82**, 184502.
154. Gardner, J. S., Gingras, M. J. P. and Greedan, J. E. (2010). Magnetic pyrochlore oxides, *Rev. Mod. Phys.*, **82**, 53.
155. Zapf, V., Jaime, M. and Batista, C. D. (2014). Bose-Einstein condensation in quantum magnets, *Rev. Mod. Phys.*, **86**, 563.
156. Fernández-Rossier, J., Braun, M., Núñez, A. S. and MacDonald, A. H. (2004). Influence of a uniform current on collective magnetization dynamics in a ferromagnetic metal, *Phys. Rev. B*, **69**, 174412.
157. Vlaminck, V. and Bailleul, M. (2008). Current-induced spin-wave doppler shift, *Science*, **322**, 410.

158. Mills, D. L. and Dzyaloshinskii, I. E. (2008). Influence of electric fields on spin waves in simple ferromagnets: role of the flexoelectric interaction, *Phys. Rev. B*, **78**, 184422.

159. Mruczkiewicz, M., Krawczyk, M., Gubbiotti, G., Tacchi, S., Filimonov, Y. A., Kalyabin, D. V., Lisenkov, I. V. and Nikitov, S. A. (2013). Nonreciprocity of spin waves in metallized magnonic crystal, *New J. Phys.*, **15**, 113023.

160. Mruczkiewicz, M., Pavlov, E. S., Vysotsky, S. L., Krawczyk, M., Filimonov, Y. A. and Nikitov, S. A. (2014). Observation of magnonic band gaps in magnonic crystals with non-reciprocal dispersion relation, *Phys. Rev. B*, **90**, 174416.

161. Grünberg, P., Schreiber, R., Pang, Y., Brodsky, M. B. and Sowers, H. (1986). Layered magnetic structures: evidence for antiferromagnetic coupling of Fe layers across Cr interlayers, *Phys. Rev. Lett.*, **57**, 2442.

162. Zhang, P. X. and Zinn, W. (1987). Spin-wave modes in antiparallel magnetized ferromagnetic double layers, *Phys. Rev. B*, **35**, 5219.

163. Henry, Y., Gladii, O. and Bailleul, M. (2016). Propagating spin-wave normal modes: a dynamic matrix approach using plane-wave demagnetizating tensors, arXiv:1611.06153.

164. Gallardo, R. A., Schneider, T., Chaurasiya, A. K., Oelschlägel, A., Arekapudi, S. S. P. K., Roldán-Molina, A., Hübner, R., Lenz, K., Barman, A., Fassbender, J., Lindner, J., Hellwig, O. and Landeros, P. (2019). Reconfigurable spin-wave non-reciprocity induced by dipolar interaction in a ferromagnetic bilayer, submitted.

165. Hillebrands, B. (1990). Spin-wave calculations for multilayered structures, *Phys. Rev. B*, **41**, 530.

166. Gallardo, R. A., Alvarado-Seguel, P., Schneider, T., Gonzalez-Fuentes, C., Roldán-Molina, A., Lenz, K., Lindner, J. and Landeros, P. (2019). Spin-wave non-reciprocity in magnetization-graded ferromagnetic films, *New J. Phys.*, **21**, 033026.

Chapter 6

Emission and Active Manipulation of Spin Waves in Multiferroic Heterostructures

Sampo Hämäläinen, Huajun Qin, and Sebastiaan van Dijken
NanoSpin, Department of Applied Physics, Aalto University School of Science, P.O. Box 15100, FI-00076 Aalto, Finland
sebastiaan.van.dijken@aalto.fi

In this chapter, we review our work on strain-coupled multiferroic heterostructures and discuss their potential as programmable hybrids in magnonics research. Attractive properties of multiferroic heterostructures include the imprinting of magnetic domain patterns with a regular modulation of magnetic anisotropy, the creation of abrupt anisotropy boundaries, strong pinning of straight magnetic domain walls, and electric field control of magnetic anisotropy, magnetic switching, and magnetic domain wall motion. We show that these attributes provide an attractive platform for the emission of short-wavelength spin waves, the confinement of spin-wave modes in structurally uniform films, and the manipulation of spin-wave transmission. Many of the magnonic concepts that we demonstrate in multiferroic heterostructures are extendable to other material systems.

Three-Dimensional Magnonics: Layered, Micro- and Nanostructures
Edited by Gianluca Gubbiotti

ISBN 978-981-4800-73-0 (Hardcover), 978-0-429-29915-5 (eBook)
www.jennystanford.com

6.1 Strain-Coupled Multiferroic Heterostructures

In our studies, the multiferroic heterostructures consist of a ferroelectric $BaTiO_3$ substrate and a strain-coupled ferromagnetic film with a 10–200 nm thickness. At room temperature, $BaTiO_3$ exhibits a tetragonal crystal structure with a tetragonality of 1.1% and the ferroelectric polarization aligns along the unit-cell axis with the largest lattice parameter. Three different domain patterns can form in tetragonal $BaTiO_3$, two of which are ferroelectric *and* ferroelastic and thus allow for electric field control of magnetism via interface strain coupling. The ferroelastic stripe patterns are characterized by 90° rotations of the ferroelectric polarization and corresponding changes of the in-plane lattice structure. In the so-called a_1–a_2 domain pattern (Fig. 6.1a), the polarization is oriented in the substrate plane. Consequently, the unit cell in the $BaTiO_3$ (001) surface plane has a rectangular shape in both domains. In this geometry, both the polarization and the axis of the largest lattice parameter rotate by 90° at the domain boundaries. In a–c stripe domains (Fig. 6.1b), the symmetry of the crystal lattice in the $BaTiO_3$ (001) surface plane alternates between cubic (c domains with perpendicular polarization) and rectangular (a domains with in-plane polarization).

If a ferromagnetic film is grown on top of a $BaTiO_3$ substrate with a_1–a_2 or a–c stripe domains, pattern transfer from the ferroelectric material to the ferromagnetic layer can be attained. Full domain correlations in strain-coupled multiferroic heterostructures were first demonstrated for CoFe films on top of $BaTiO_3$ substrates with an a_1–a_2 domain pattern [1–3], as shown in Fig. 6.1c. In this system, strain-induced magnetoelastic anisotropy dominates over magnetocrystalline anisotropy, which is small because of the atomic composition ($Co_{60}Fe_{40}$ [4]) and polycrystalline structure of the CoFe film. Because of the positive magnetostriction constant of CoFe, the strain-induced uniaxial magnetic anisotropy axis and ferroelectric polarization are collinear. In zero magnetic field, the magnetization of the ferromagnetic film follows the 90° polarization rotations at the a_1–a_2 boundaries. Pattern transfer from $BaTiO_3$ a–c domains to ferromagnetic films has also been

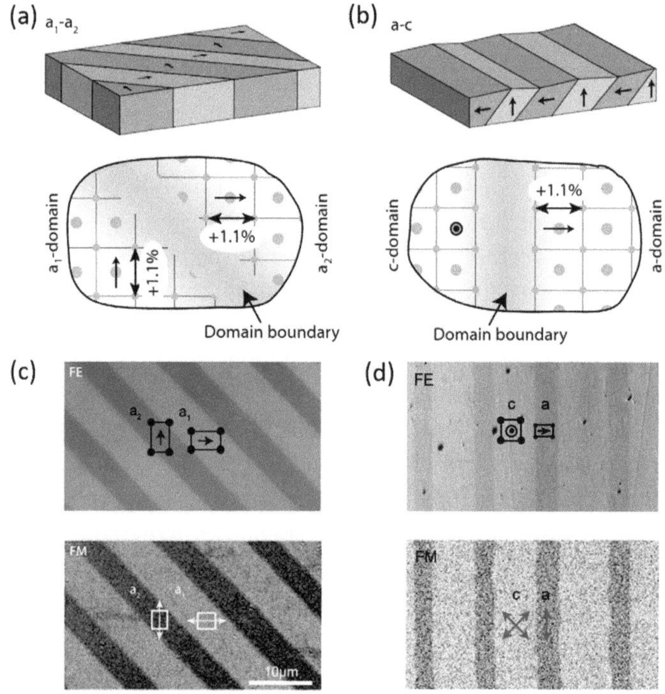

Figure 6.1 (a and b) Schematic of a_1–a_2 and a–c domain patterns in a $BaTiO_3$ (001) substrate. (c and d) Ferroelectric (top) and ferromagnetic (bottom) images of (c) a 15 nm $CoFe/BaTiO_3$ sample with a_1–a_2 stripe domains and (d) a 20 nm $Fe/BaTiO_3$ multiferroic heterostructure with an a–c domain configuration. In both examples, interface strain coupling imprints the ferroelectric stripe domains into the ferromagnetic film.

realized [5, 6]. An example is shown in Fig. 6.1d for epitaxial Fe. Again, the magnetoelastic anisotropy is determined by the structural symmetry of the ferroelectric crystal in the substrate plane. Due to negative magnetostriction of Fe, the uniaxial magnetic easy axis on top of a ferroelectric a domain is oriented parallel to the domain boundary, whereas the biaxial anisotropy axes on top of a c domain are aligned at 45°. Besides CoFe and Fe, domain pattern transfer from ferroelectric $BaTiO_3$ crystals has also been demonstrated for CoFeB [7–9], $CoFe_2O_4$ [10], $(La,Sr)MnO_3$ [11], Ni [12], NiFe [13], FeGa [14], and FeRh [15], indicating effective strain coupling for a variety of ferromagnetic materials.

The imprinting of ferroelectric stripe domains into a ferromagnetic film produces several attractive properties. First, pattern transfer induces a regular modulation of magnetic anisotropy. At the domain boundaries, the anisotropy changes in symmetry, orientation, and/or strength. Since the underlying ferroelectric boundaries are typically 2–5 nm wide [16–18], the strain-induced changes in magnetic anisotropy are nearly abrupt. In zero magnetic field, the alignment of magnetization along one of the easy anisotropy axes produces a nearly uniform magnetization state within the stripe domains. Second, the abrupt anisotropy boundaries strongly pin straight magnetic domain walls. In a magnetic field, the domain walls do, therefore, not move but change their internal spin structure. One particularly interesting example is field-controlled toggling between narrow head-to-tail and broad head-to-head or tail-to-tail domain walls in multiferroic heterostructures with a_1–a_2 stripe domains [19]. This nonvolatile switching effect enables active control of spin-wave (SW) transmission by reprogramming of the domain wall structure [20].

The application of an electric field across the $BaTiO_3$ substrate of a multiferroic heterostructure changes the ferroelectric domain pattern via abrupt switching or lateral domain wall motion. In multiferroic heterostructures, this also alters the magnetic properties of the strain-coupled ferromagnetic film. Several electric-field-controlled magnetic effects have been demonstrated, including the writing and erasure of regular magnetic domain patterns [1–3], local magnetic switching [1–3], and reversible motion of magnetic domain walls [3, 6]. Besides the in-plane magnetized films discussed thus far, we note that imprinting of $BaTiO_3$ domains into Cu/Ni multilayers with perpendicular magnetic anisotropy has also been demonstrated [21, 22]. Electric-field-induced effects in this strain-coupled system include local switching between perpendicular and in-plane magnetization [21] and reversible magnetic domain wall motion [22].

In the following sections, we review the potential of $BaTiO_3$-based multiferroic heterostructures for the emission and active manipulation of SWs. We consider CoFeB films on top of ferroelectric a_1–a_2 stripe domains and Fe films on top of a–c domain patterns. Reasonably low magnetic damping and uniform (CoFeB) or epitaxial (Fe) film growth motivate the selection of these ferromagnets.

6.2 Spin-Wave Emission from a Pinned Domain Wall

SW emission is an essential part of magnonic devices. Microwave antennas are often used for excitation. A drawback of this approach is the impossibility of exciting SWs with wavelengths that are shorter than the antenna width [23]. Motivated by this shortcoming, alternative approaches for the emission of short-wavelength SWs have been proposed. Examples include SW injection by spin-torque nano-oscillators [24–27] or wavelength conversions near microwave antennas using tapered waveguides [28], waveguides coupled to a magnetic antenna [29], grating couplers [30, 31], isolated antidots [32, 33], and nanowire or nanodisk microwave-to-SW transducers [34, 35]. Topological defects in ferromagnetic films such as vortices and domain walls provide another attractive route toward the realization of localized SW sources [36–40]. When brought into oscillation or gyration, the noncollinear spin structure of such defects induces regular magnetic perturbations that propagate through the ferromagnetic layer as SWs.

Pinned magnetic domain walls in multiferroic heterostructures such as the ones depicted in Figs. 6.1c and 6.1d provide a good testbed for SW emission. To gauge the performance of a domain wall SW source, we performed micromagnetic simulations in MuMax3 [41] using experimentally derived values of the strain-induced magnetic anisotropy strength as an input parameter [37]. We considered an a_1–a_2 domain pattern in zero magnetic field and excited the pinned domain wall by a spin-polarized alternating current [42, 43]. Via the spin-transfer torque effect, this produces driven oscillations of a 90° domain wall in the narrow pinning potential of the magnetic anisotropy boundary. For an analytical description and discrete

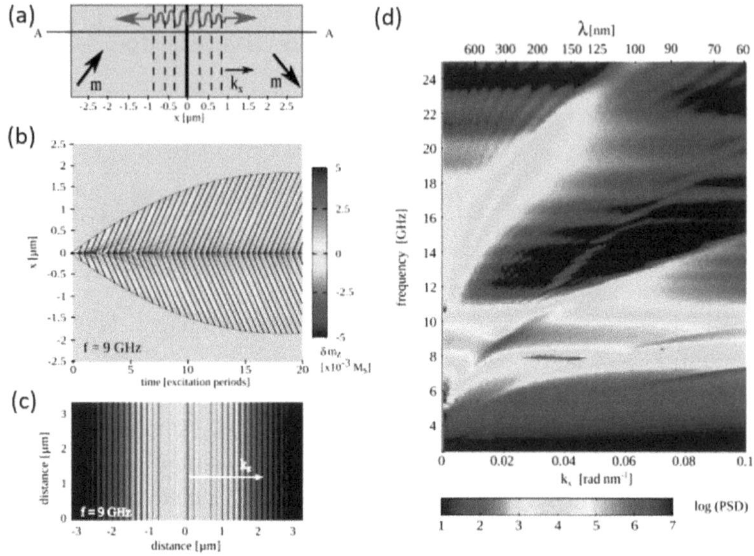

Figure 6.2 (a) Schematic illustration of spin-wave emission from an infinitely long oscillating domain wall in a multiferroic heterostructure with a_1–a_2 domains. (b) Spin-wave profiles along line A in panel (a) as a function of excitation time. Solid black lines indicate the location of spin-wave nodes. Dotted lines denote the location where the spin-wave amplitude drops below 0.015% of the saturation magnetization. The excitation frequency is 9 GHz. (c) Logarithmic plot of the spin-wave profile after 20 excitation cycles. (d) Dispersion graph of the emitted spin waves after 20 excitation periods. All data were obtained for an uniaxial magnetic anisotropy strength of 5×10^4 J/m^3. Reprinted from Ref. [37]. (https://creativecommons.org/licenses/by/4.0/).

Heisenberg model of the dynamic properties of a pinned 90° head-to-tail domain wall, we refer to Ref. [44]. Here, we briefly summarize our micromagnetic findings.

Results for a pinned domain wall and a current density of $j = 1 \times 10^{12}$ A/m^2 are shown in Fig. 6.2. Because the magnetic domain wall is infinitely long in these simulations (use of 2D periodic boundary conditions), the wave fronts of the emitted SWs are parallel to the wall despite the 45° angle between the wave vector and magnetization. Figure 6.2b shows how SWs with a frequency of 9 GHz propagate into the ferromagnetic domains. The solid black lines denote the location of SW nodes. Their slopes correspond to the phase velocity of the SWs. The dotted lines in Fig. 6.2b mark locations where the SW amplitude drops below 0.015% of the saturation magnetization. While this envelope is initially determined by the product of the phase velocity and time, it converges to a constant value after about 15 excitation periods. Beyond this excitation time, the energy that the spin-polarized alternating current pumps into the system is balanced by energy dissipation through SW damping. Figure 6.2c shows the out-of-plane magnetization profile after 20 excitation periods. Spatial 1D Fourier transformation of this profile gives the wave vector spectrum at this excitation frequency. Repetition of this procedure for frequencies ranging from 2.5 to 25 GHz produces the full dispersion diagram shown in Fig. 6.2d. The graph exhibits two main dispersion bands. The broad horizontal and thus nondispersive band around 9.3 GHz corresponds to the fundamental domain wall oscillation mode. The most intense dispersive branch ($f_{\mathrm{d}}(\mathbf{k})$) reflects the properties of the propagating SWs. The oscillating domain wall emits SWs with wavelengths down to 150 nm and frequencies up to 24 GHz. At higher frequencies, the magnetic domain wall can no longer follow the spin-polarized alternating current and SW emission into the ferromagnetic domains shuts down. The weaker dispersive bands seen in Fig. 6.2d correspond to harmonics with $f_{\mathrm{n}} = f_{\mathrm{d}}/n$. For more details on the characterization of SW modes, we refer to Ref. [37].

The properties of the domain wall SW source are tunable by patterning of the ferromagnetic film into a nanowire. In this geometry, Damon–Eshbach-like standing SWs are excited inside the domain wall along the direction of the nanowire width. These confined modes complement the back-and-forth oscillatory motion of the domain wall. Since standing SWs are excited up to much higher frequency, the wavelength of the propagating SWs that they emit can be very short. For example, we simulated the emission of near-monochromatic SWs with a wavelength of 25 nm at an excitation frequency of 50 GHz [37].

6.3 Spin-Wave Confinement and Emission by Uniform Field Excitation

Two main effects occur when multiferroic heterostructures with regular modulations of magnetic anisotropy are excited by a spatially uniform magnetic field. First, standing SWs form within the stripe domains when the confinement condition $\mathbf{k} = n\pi/w$ is met. Here, $\mathbf{k}$ is the wave vector of the SWs at the excitation frequency, n is the order number, and w is the width of the domain. Second, SWs are emitted from the magnetic anisotropy boundaries. To demonstrate both effects, we studied a 50-nm-thick $Co_{40}Fe_{40}B_{20}$ film grown on top of a $BaTiO_3$ substrate with a_1–a_2 stripe domains [45, 46]. The magnetic structure of this sample is similar to the one depicted in Fig. 6.1c, with the alternating a_1 and a_2 domains having a width of 5 ±1.5 μm and 7.5 ± 1.5 μm, respectively. SW excitations in the $Co_{40}Fe_{40}B_{20}$ film were measured using broadband SW spectroscopy. In the experiments, the sample was placed face down onto a coplanar waveguide with a 20-μm-wide signal line, as illustrated in Fig. 6.3a. The coplanar waveguide was connected to a vector network analyzer and produced a long-wavelength microwave field (H_{rf}) with a main excitation at $|\mathbf{k}| = 0.096$ rad/μm. In all measurements, H_{rf} was oriented perpendicular to the stripe domains of the $Co_{40}Fe_{40}B_{20}/BaTiO_3$ sample. A quadrupole electromagnet was used to provide an external bias field (H_{ext}) along different in-plane directions.

Figure 6.3b shows a broadband SW spectrum for H_{ext} oriented along the uniaxial magnetic anisotropy axis of one of the stripe domains. Two prominent branches are measured. These branches correspond to first-order standing SW modes in the two different stripe domains. The branch with the two minima is produced by a SW mode in the stripe domains with the smaller effective magnetic field (anisotropy axis perpendicular to the bias field). SW excitations in domains with the anisotropy axis parallel to the bias field are responsible for the branch at higher frequencies. Figure 6.3c shows the angular dependence of the broadband SW spectrum for an external bias field of 60 mT. We note that this field nearly saturates the magnetization of the $Co_{40}Fe_{40}B_{20}$ film. Consequently,

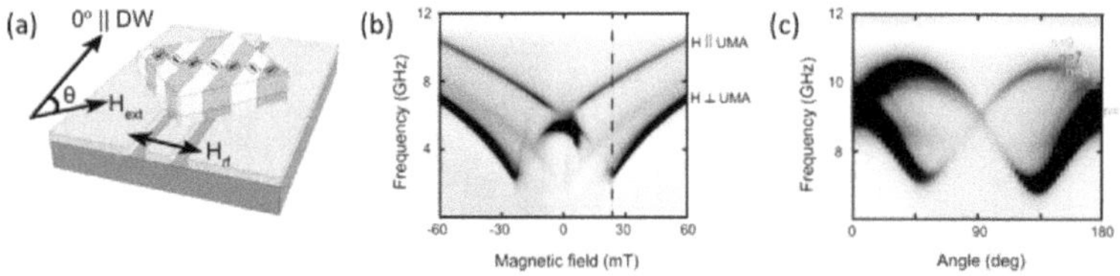

Figure 6.3 (a) Broadband spin-wave spectroscopy measurement geometry. The CPW and sample are not plotted to scale. (b) Grayscale plot of the field-dependent absorption spectrum of a 50 nm $Co_{40}Fe_{40}B_{20}/BaTiO_3$ sample measured for H_{ext} oriented along the uniaxial magnetic anisotropy axis of one of the stripe domains. (c) Angular dependence of spin-wave excitations in the same sample at a constant magnetic bias field of 60 mT. In addition to first-order standing spin waves in both stripe domains (two main branches), higher-order confined modes are also measured. Reprinted (figure) with permission from Ref. [46]. Copyright (2017) by the American Physical Society.

magnetic domain walls are not present in the experiments, which makes the magnetic configuration different from the remanent state discussed in the previous section. In Fig. 6.3c, the frequencies of the two main branches oscillate and become degenerate when H_{ext} is oriented either parallel or perpendicular to the domain boundaries. In addition to the two prominent branches, we also identify a set of SW excitations with smaller intensities (labeled $n = 3$, 5, 7, and 9 in Fig. 6.3c). These modes correspond to higher-order standing SWs in the stripe domains with the smaller effective magnetic field.

To analyze SW excitation in our multiferroic heterostructures in more detail, we performed micromagnetic simulations in MuMax3. Figure 6.4a illustrates the effective magnetic field for a 60 mT bias field perpendicular to the easy anisotropy axis of the 7.5-μm-wide stripe domain. In this geometry, higher-order standing SWs are excited if their wave vector matches the confinement condition (Fig. 6.4b). As an example, the excitation of a fifth-order mode is depicted in Fig. 6.4c. Since no modes exist in the neighboring domains at the excitation frequency, these excitations are perfectly confined. If we rotate the in-plane bias field by 90°, the effective magnetic field of the 7.5-μm-wide stripe domain becomes larger than that of its neighbors. Under this condition, standing SWs are excited

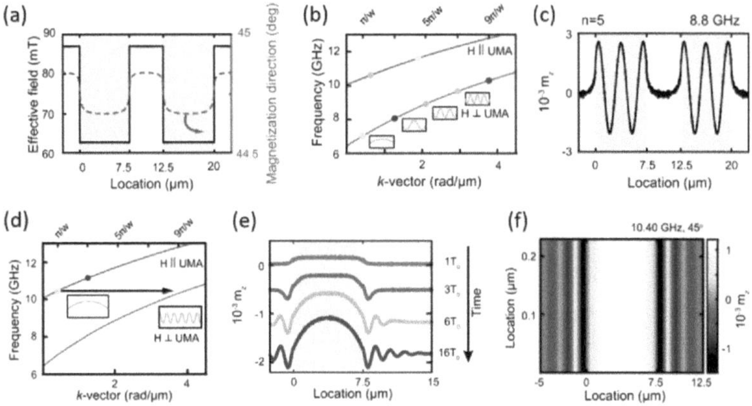

Figure 6.4 (a) Simulated effective magnetic field and magnetization direction of a 50 nm $Co_{40}Fe_{40}B_{20}/BaTiO_3$ sample for a 60 mT bias field perpendicular to the easy anisotropy axis of the 7.5-μm-wide stripe domain. (b) Calculated dispersion relations for both domains. (c) Simulated spin-wave profile of the fifth-order mode formed in the domain with a smaller effective magnetic field. (d) Illustration of wavelength coversion mechanism from a standing spin wave in the domain with a larger effective field to a propagating spin wave in the domain with a smaller effective field. (e) Formation of standing and propagating spin waves as a function of time after the onset of uniform field excitation. (f) Profile of emitted spin waves from a first-order confined mode for a 60 mT bias field parallel to the easy anisotropy axis of the 7.5-μm-wide stripe domain. Reprinted (figure) with permission from Ref. [46]. Copyright (2017) by the American Physical Society.

at higher frequencies. However, since modes with larger wave vectors are now available in neighboring domains, these excitations trigger the emission of short-wavelength propagating SWs. Figure 6.4d illustrates the wavelength conversion mechanism in our anisotropy-modulated system. The buildup of a first-order standing SW mode in the domain with higher effective magnetic field and the propagation of a short-wavelength SW from the boundary into the neighboring domain are demonstrated in Fig. 6.4e. Figure 6.4f depicts the SW profile for the same configuration during steady-state excitation. The wavelength of emitted SWs can be tuned from a few micrometers down to the nanoscale by in-plane rotation of the external bias field (for details see Ref. [46]).

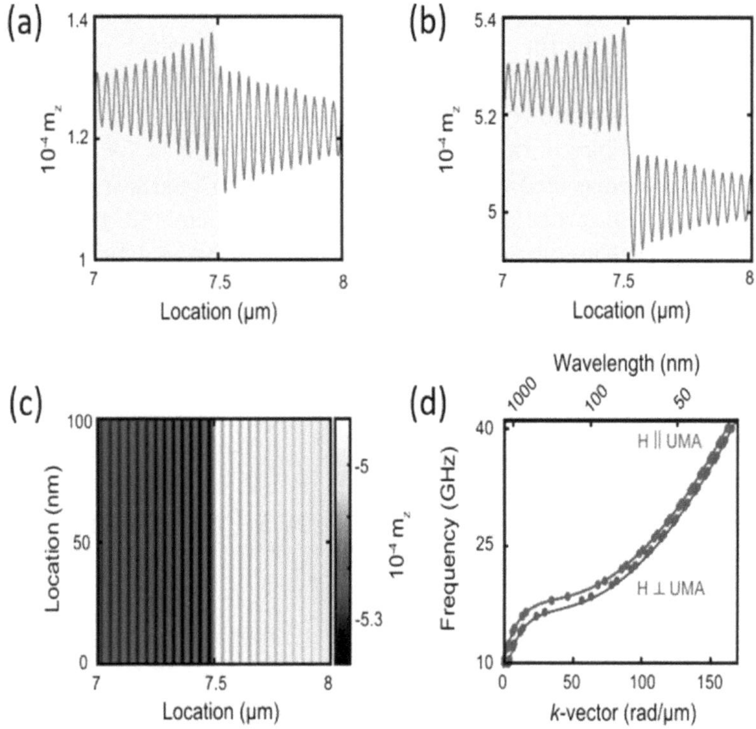

Figure 6.5 (a, b) Simulated profiles of spin waves that are excited by a 40 GHz uniform excitation field when a 60 mT bias field is applied along the easy magnetic anisotropy axis of one of the domains. The snapshots in (a) and (b), which are recorded with a phase difference of about 75° in the microwave excitation signal, illustrate differing forced magnetization precessions in the two domains. (c) Two-dimensional profile of emitted spin waves at 40 GHz. (d) Simulated (symbols) and calculated (lines) dispersion curves for spin waves propagating into the two domains. Reprinted (figure) with permission from Ref. [46]. Copyright (2017) by the American Physical Society.

Besides the emission of propagating SWs from confined modes at specific frequencies, uniform field excitation of an anisotropy-modulated ferromagnetic film continuously generates SWs at the anisotropy boundaries if the effective magnetic fields of neighboring domains differ. The simulation data of Fig. 6.5 illustrate this effect. Dissimilar forced magnetization precessions in the neighboring domains trigger the emission of SWs. The excitation process can be explained by the Schlömann mechanism [47], which postulates that translational symmetry breaking at a magnetic nonuniformity causes the emission of propagating SWs in a uniform excitation field. Other examples of this excitation mechanism such as SW emission from the edge of a patterned film [29] or an interface between yttrium iron garnet (YIG) and a ferromagnetic metal [48, 49] can be found in the literature. In all these geometries, an analytical theory describes the properties of the emitted SWs [50]. In our anisotropy-modulated multiferroic heterostructure, we find that the character of emitted SWs changes gradually from magnetostatic to exchange dominated with increasing excitation frequency (Fig. 6.5d).

6.4 Spin-Wave Manipultion by a Programmable Domain Wall

Transmission of SWs through a magnetic domain wall can be used for amplitude or phase manipulation [51, 52]. The transmission efficiency depends on several micromagnetic factors. For instance, SW transport through an infinitely extended 1D Bloch wall induces a finite phase shift without reflection [53]. In contrast, interactions between SWs and 180° Néel walls or magnetic domain walls in confined geometries are more complex [54]. Dynamic stray fields in such domain walls reduce the transmission of SWs if their wavelength exceeds the wall width [55, 56]. Additionally, domain wall resonances limit the transmission of SWs at specific frequencies [57–60]. This effect, known as resonant reflection, depends strongly on the spatial inhomogeneity of the effective magnetic field inside the wall. Because of dynamic stray fields and resonance modes, narrow domain walls tend to reflect SWs more than broad walls.

Active control of the SW amplitude or phase requires reprogramming of the domain wall spin structure. The strain-coupled multiferroic heterostructures that we discuss in this chapter are ideally suited for this. For example, if a ferromagnetic film is strain-coupled to an a_1–a_2 domain pattern of a $BaTiO_3$ substrate, its magnetic microstructure can be switched reversibly between

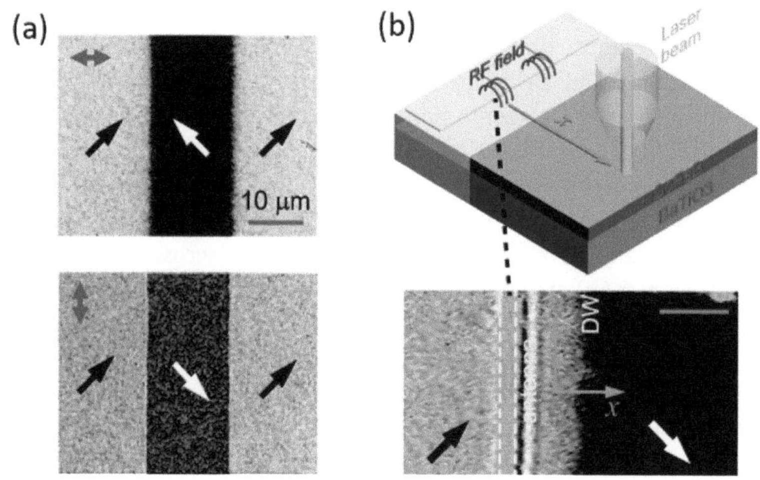

Figure 6.6 (a) Kerr microscopy images of the magnetization configuration in a $CoFeB/BaTiO_3$ sample after the application of a magnetic field along the domain boundaries (top) and perpendicular to the domain boundaries (bottom). (b) Schematic and Kerr microscopy image of the μ-BLS measurement geometry. Adapted from Ref. [20] (https://creativecommons.org/licenses/by/4.0/).

one comprising alternating 90° head-to-head/tail-to-tail domain walls (Fig. 6.6a, top image) and another containing 90° head-to-tail domain walls (Fig. 6.6a, bottom image). The widths of these two domain walls are very different. The head-to-head and tail-to-tail domain walls exhibit a large magnetostatic energy, and consequently, their width increases with ferromagnetic film thickness [7–9]. In contrast, the width of head-to-tail domain walls is almost entirely determined by a competition between ferromagnetic exchange and the strength of magnetic anisotropy, and consequently, it varies much less with film thickness [7]. A related attractive feature of strain-coupled multiferroic heterostructures is strong domain wall pinning, which prevents domain walls from moving under the action of propagating SWs [61, 62]. In this chapter, we summarize results on active SW manipulation by programming domain walls in a 50-nm-thick $Co_{40}Fe_{40}B_{20}$ film on top of a $BaTiO_3$ substrate with a_1–a_2 stripe domains [20]. The widths of the 90°

head-to-head/tail-to-tail and 90° head-to-tail domain walls in this system are approximately 1.6 μm and 50 nm, respectively.

Microfocused Brillouin light scattering (μ-BLS) was used to characterize SW transmission through pinned magnetic domain walls. The SWs were excited by 500-nm-wide microwave antennas that were patterned on top of the sample at a distance of about 2 μm from a pinned domain wall and the BLS signal was recorded either locally or by scanning the laser spot across the domain wall (Fig. 6.6b). Figure 6.7a shows μ-BLS spectra measured at a fixed location 2.5 μm beyond the center of the pinned domain wall and by scanning the excitation frequency from 5 to 15 GHz. The solid and dashed curves depict BLS intensity data for a broad 90° head-to-head and narrow 90° head-to-tail domain wall, respectively. The resonances ranging from ±7.7 to ±12.2 GHz in the μ-BLS spectra illustrate the intensity of propagating SWs after transmission through the magnetic domain wall. The local μ-BLS measurements of Fig. 6.7a thus indicate that the broad head-to-head domain wall transmits the SWs well. Yet, switching of the magnetization to the narrow head-to-tail domain wall significantly suppresses the SW signal. The same conclusions can be drawn from μ-BLS line scans recorded at 9.85 GHz (Fig. 6.7b) and 10.85 GHz (Fig. 6.7c). At an excitation frequency of 10.85 GHz, the BLS intensity behind the head-to-tail domain wall almost reduces completely to the noise level of the measurement.

Micromagnetic simulations confirm full transparency of the broad 90° head-to-head domain wall and strong reflection at the narrow 90° head-to-tail domain wall around a frequency of 10.85 GHz (Fig. 6.8a,c). The profiles of the effective magnetic field inside the domain wall explain the different transmission characteristics. In the head-to-head configuration, the effective magnetic field reduces gradually (Fig. 6.8b). This produces a finite SW phase shift upon transmission (see the change of slope in the contour plot of Fig. 6.8a), but it hardly affects the SW amplitude. In contrast, the effective magnetic field inside the head-to-tail domain wall peaks sharply at its center (Fig. 6.8d). Two field minima at $x = \pm 20$ nm surround this peak. This non-uniform field profile produces a resonance mode that is characterized by two oscillatory out-of-phase antinodes on opposite sides of the domain wall center

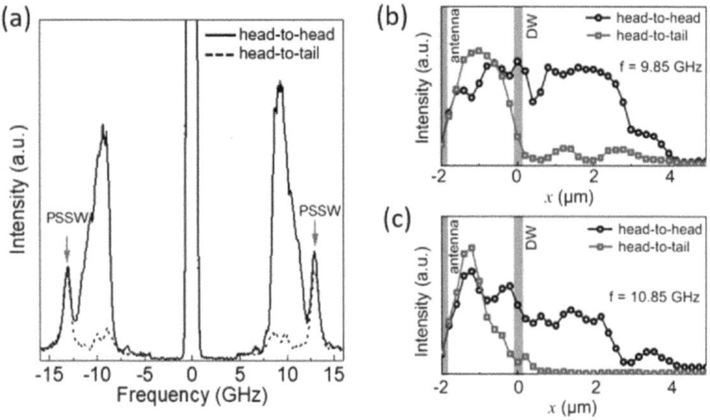

Figure 6.7 (a) μ-BLS spectra measured on a 50 nm CoFeB/$BaTiO_3$ sample at a fixed position, 2.5 μm beyond the center of the pinned domain wall. (b, c) BLS line scans recorded from the antenna edge across the pinned domain wall at excitation frequencies of 9.85 and 10.85 GHz. Adapted from Ref. [20] (https://creativecommons.org/licenses/by/4.0/).

(see inset in Fig. 6.8c). The domain wall resonance mode absorbs and reflects propagating SWs, as illustrated by the undulations of SW maxima and minima in the contour graph of Fig. 6.8c. The SW intensity is suppressed by the domain wall resonance mode over a >2 GHz frequency range around 10.85 GHz [20]. Reversible switching between different 90° configurations of a pinned magnetic domain wall thus allows for active control over the amplitude and phase of propagating SWs. The operation frequency of the programmable SW filter depends on the strength of the strain-induced magnetic anisotropy. This parameter is tunable by variation of the deposition parameters, the insertion of a seed layer at the multiferroic interface, or the use of ferromagnetic materials with different magnetostriction constants.

6.5 Electric Field Control of Spin Waves

Electric field control of magnetism in $BaTiO_3$-based multiferroic heterostructures enables low-power SW manipulation. $BaTiO_3$

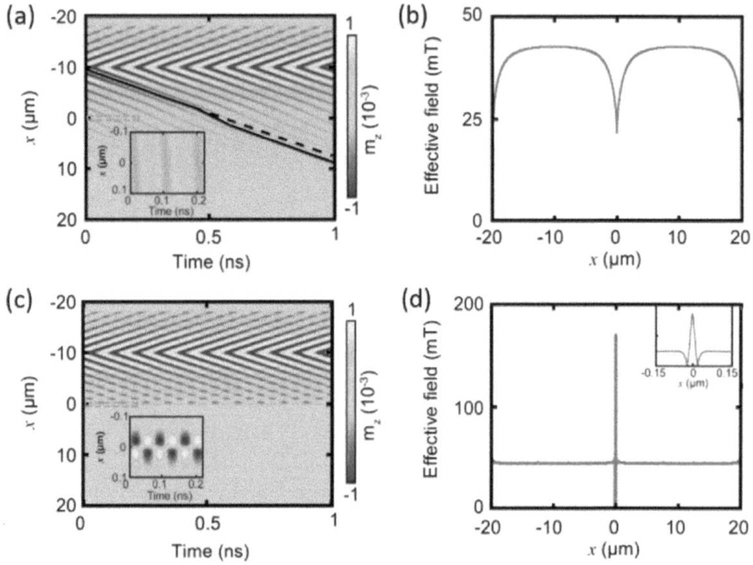

Figure 6.8 Micromagnetic simulations of spin wave transmission through a broad 90° head-to-head domain wall (a) and a narrow 90° head-to-tail domain wall (c) at a frequency of 10.85 GHz. The effective magnetic fields of the two domain walls are plotted in (b, d). The inset in (c) illustrates strong out-of-phase oscillations of two anti-nodes inside the head-to-tail domain wall. Adapted from Ref. [20] (https://creativecommons.org/licenses/by/4.0/).

substrates with a–c stripe domains are particularly interesting for this purpose. In this geometry, out-of-plane electric fields reversibly and gradually change the ferroelectric and ferromagnetic domain patterns by lateral domain wall motion. As an example, we show experimental results on Fe/$BaTiO_3$ in Fig. 6.9. The Kerr microscopy images in Fig. 6.9a show the remanent magnetic microstructure of a 20-nm-thick Fe film on top of a $BaTiO_3$ substrate with a ferroelectric a–c domain configuration. Reversible motion of the ferroelectric boundary and the pinned magnetic domain wall is achieved by the application of out-of-plane electric field pulses with a magnitude of 2 kV/cm. If the electric field aligns along the direction of ferroelectric polarization in the c domain, the c domain grows at the expense of the a domain by lateral wall motion toward the bottom of the images. An electric field in the opposite direction shrinks the c domain by

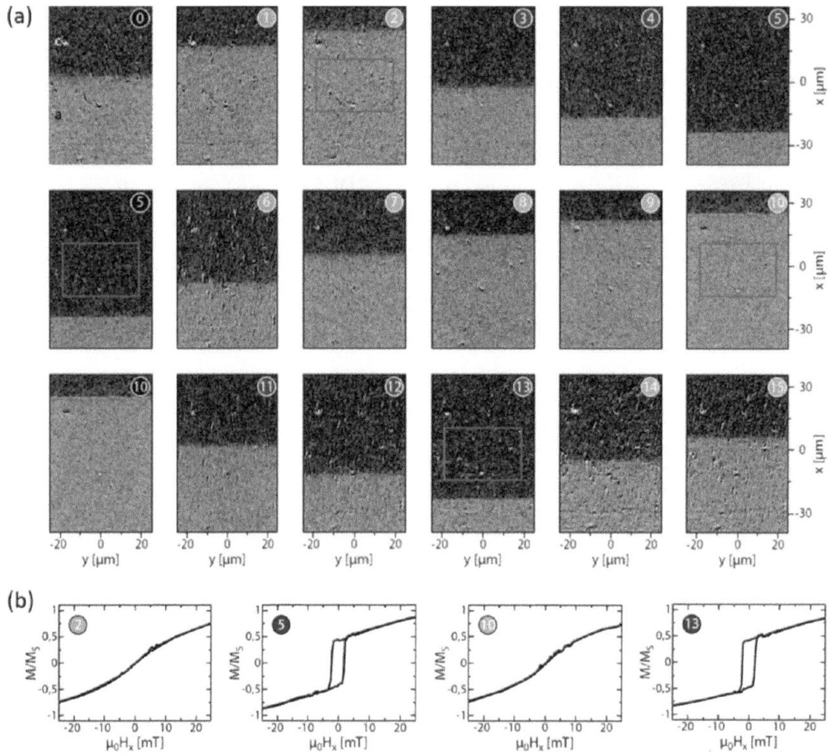

Figure 6.9 (a) Kerr microscopy images illustrating the motion of a magnetic domain wall in an Fe film during the application of positive and negative voltage pulses across the $BaTiO_3$ substrate. (b) Magnetic hysteresis curves for a selected sample area—boxes in (a). The data demonstrate reversible electric-field-controlled switching between uniaxial (*a* domain) and biaxial (*c* domain) magnetic anisotropy via lateral domain wall motion. Adapted from Ref. [6] (https://creativecommons.org/licenses/by/3.0/).

moving the ferroelectric boundary and pinned magnetic domain wall back up. The hysteresis curves of Fig. 6.9b are recorded on the same area (indicated by the boxes in Fig. 6.9a) after a series of positive and negative voltage pulses. Following the passing of the domain wall, the magnetic anisotropy in the area corresponds to that of an *a* domain (numbers 2 and 10) or *c* domain (numbers 5 and 13). In other words, the application of electric field pulses changes the local magnetic anisotropy reversibly and deterministically.

While electric field control of SW propagation has not been demonstrated yet, we anticipate strong nonvolatile effects in multiferroic heterostructures. The transmission of SWs through a strain-coupled ferromagnetic film can either be changed by electrically moving a magnetic domain wall away or into the area of interest (see previous section on SW filtering) or a shift of the SW dispersion relation by a reversible change of the local magnetic anisotropy. Future experiments should confirm whether electric field manipulation of SWs is indeed feasible.

Acknowledgments

We are thankful to the following collaborators for their contributions to the presented work: K. J. A. Franke, B. Van de Wiele, P. Baláž, F. Montoncello, F. Brandl, D. Grundler, M. Madami, G. Gubbiotti, Y. Shirahata, and T. Taniyama. The Academy of Finland and the European Research Council provided financial support for this work.

References

1. Lahtinen, T. H. E., Tuomi, J. O. and van Dijken, S. (2011). Pattern transfer and electric-field induced magnetic domain formation in multiferroic heterostructures, *Adv. Mater.*, **23**, 3187–3191.
2. Lahtinen, T. H. E., Tuomi, J. O. and van Dijken, S. (2011). Electrical writing of magnetic domain patterns in ferromagnetic/ferroelectric heterostructures, *IEEE Trans. Magn.*, **47**, 3768–3771.
3. Lahtinen, T. H. E., Franke, K. J. A. and van Dijken, S. (2012). Electric-field control of magnetic domain wall motion and local magnetization reversal, *Sci. Rep.*, **2**, 258.
4. Hall, R. C. (1960). Magnetic anisotropy and magnetostriction of ordered and disordered cobalt-iron alloys, *J. Appl. Phys.*, **31**, 157S–158S.
5. Lahtinen, T. H. E., Shirahata, Y., Yao, L., Franke, K. J. A., Venkataiah, G., Taniyama, T. and van Dijken, S. (2012). Alternating domains with uniaxial and biaxial magnetic anisotropy in epitaxial Fe films on $BaTiO_3$, *Appl. Phys. Lett.*, **101**, 262405.

6. Franke, K. J. A., Van de Wiele, B., Shirahata, Y., Hämäläinen, S. J., Taniyama, T. and van Dijken, S. (2015). Reversible electric-field-driven magnetic domain-wall motion, *Phys. Rev. X*, **5**, 011010.
7. Franke, K. J. A., López González, D., Hämäläinen, S. J. and van Dijken, S. (2014). Size dependence of domain pattern transfer in multiferroic heterostructures, *Phys. Rev. Lett.*, **112**, 017201.
8. Casiraghi, A., Rincón Domínguez, T., Rößler, S., Franke, K. J. A., López González, D., Hämäläinen, S. J., Frömter, R., Oepen, H. P. and van Dijken, S. (2015). Influence of elastically pinned magnetic domain walls on magnetization reversal in multiferroic heterostructures, *Phys. Rev. B*, **92**, 054406.
9. López González, D., Casiraghi, A., Kronast, F., Franke, K. J. A. and van Dijken, S. (2017). Influence of magnetic field and ferromagnetic film thickness on domain pattern transfer in multiferroic heterostructures, *J. Magn. Magn. Mater.*, **441**, 404–408.
10. Chopdekar, R. V., Malik, V. K., Fraile Rodríguez, A., Le Guyader, L., Takamura, Y., Scholl, A., Stender, D., Schneider, C. W., Bernhard, C., Nolting, F. and Heyderman, L. J. (2012). Spatially resolved strain-imprinted magnetic states in an artificial multiferroic, *Phys. Rev. B*, **86**, 014408.
11. Chopdekar, R. V., Heidler, J., Piamonteze, C., Takamura, Y., Scholl, A., Rusponi, S., Brune, H., Heyderman, L. J. amd Nolting, F. (2013). Strain-dependent magnetic configuration in manganite–titanate heterostructures probed with soft X-ray techniques, *Eur. Phys. J. B*, **86**, 241.
12. Streubel, R., Köhler, D., Schäfer, R. and Eng, L. M. (2013). Strain-mediated elastic coupling in magnetoelectric nickel/barium-titanate heterostructures, *Phys. Rev. B*, **87**, 054410.
13. Fackler, S. W., Donahue, M. J., Gao, T., Nero, P. N. A., Cheong, S.-W., Cumings, J. and Takeuchi, I. (2014). Local control of magnetic anisotropy in transcritical permalloy thin films using ferroelectric $BaTiO_3$ domains, *Appl. Phys. Lett.*, **105**, 212905.
14. Xie, Y., Zhan, Q., Liu, Y., Dai, G., Yang, H., Zuo, Z., Chen, B., Wang, B., Zhang, Y., Rong, X. and Li, R.-W. (2014). Electric-field control of magnetic anisotropy in $Fe_{81}Ga_{19}/BaTiO_3$ heterostructure films, *AIP Adv.*, **4**, 117113.
15. Phillips, L. C., Cherifi, R. O., Ivanovskaya, V., Zobelli, A., Infante, I. C., Jacquet, E., Guiblin, N., Ünal, A. A., Kronast, F., Dkhil, B., Barthélémy, A., Bibes, M. and Valencia, S. Local electrical control of magnetic order and orientation by ferroelastic domain arrangements just above room temperature, *Sci. Rep.*, **5**, 10026.

16. Zhang, X., Hashimoto, T. and Joy, D. C. (1992). Electron holographic study of ferroelectric domain walls, *Appl. Phys. Lett.*, **60**, 784–786.
17. Hlinka J. and Márton, P. (2006). Phenomenological model of a 90° domain wall in $BaTiO_3$-type ferroelectrics, *Phys. Rev. B*, **74**, 104104.
18. Zhang Q. and Goddard, W. A. (2006). Charge and polarization distributions at the 90° domain wall in barium titanate ferroelectric, *Appl. Phys. Lett.*, **89**, 182903.
19. Franke, K. J. A., Lahtinen, T. H. E. and van Dijken, S. (2012). Field tuning of ferromagnetic domain walls on elastically coupled ferroelectric domain boundaries, *Phys. Rev. B*, **85**, 094423.
20. Hämäläinen, S. J., Madami, M., Qin, H., Gubbiotti, G. and van Dijken, S. (2018). Control of spin-wave transmission by a programmable domain wall, *Nat. Commun.*, **9**, 4853.
21. Shirahata, Y., Shiina, R., López González, D., Franke, K. J. A., Wada, E., Itoh, M., Pertsev, N. A., van Dijken, S. and Taniyama, T. (2015). Electric-field switching of perpendicular magnetized multilayers, *NPG Asia Mater.*, **7**, e198.
22. López González, D., Shirahata, Y., Van de Wiele, B., Franke, K. J. A., Casiraghi, A., Taniyama, T. and van Dijken, S. (2017). Electric-field-driven domain wall dynamics in perpendicularly magnetized multilayers, *AIP Adv.*, **7**, 035119.
23. Demidov, V. E., Kostylev, M. P., Rott, K., Krzysteczko, P., Reiss, G. and Demokritov, S. O. (2009). Excitation of microwaveguide modes by a stripe antenna, *Appl. Phys. Lett.*, **95**, 112509.
24. Demidov, V. E., Urazhdin, S. and Demokritov, S. O. (2010). Direct observation and mapping of spin waves emitted by spin-torque nano-oscillators, *Nat. Mater.*, **9**, 984–988.
25. Madami, M., Bonetti, S., Consolo, G., Tacchi, S., Carlotti, G., Gubbiotti, G., Mancoff, F. B., Yar, M. A. and Åkerman, J. (2011). Direct observation of a propagating spin wave induced by spin-transfer torque, *Nat. Nanotechnol.*, **6**, 635–638.
26. Locatelli, N., Cros, V. and Grollier, J. (2014). Spin-torque building blocks, *Nat. Mater.*, **13**, 11–20.
27. Urazhdin, S., Demidov, V. E., Ulrichs, H., Kendziorczyk, T., Kuhn, T., Leuthold, J., Wilde, G. and Demokritov, S. O. (2014). Nanomagnonic devices based on the spin-transfer torque, *Nat. Nanotech.*, **9**, 509–513.
28. Demidov, V. E., Kostylev, M. P., Rott, K., Münchenberger, J., Reiss, G. and Demokritov, S. O. (2011). Excitation of short-wavelength spin waves in magnonic waveguides, *Appl. Phys. Lett.*, **99**, 082507.

29. Au, Y., Davison, T., Ahmad, E., Keatley, P. S., Hicken, R. J. and Kruglyak, V. V. (2011). Excitation of propagating spin waves with global uniform microwave fields, *Appl. Phys. Lett.*, **98**, 122506.

30. Sklenar, J., Bhat, V. S., Tsai, C. C., DeLong, L. E. and Ketterson, J. B. (2012). Generating wave vector specific Damon-Eshbach spin waves in Py using a diffraction grating, *Appl. Phys. Lett.*, **101**, 052404.

31. Yu, H., Duerr, G., Huber, R., Bahr, M., Schwarze, T., Brandl, F. and Grundler, D. (2013). Omnidirectional spin-wave nanograting coupler, *Nat. Commun.*, **4**, 2702.

32. Davies, C. S., Sadovnikov, A. V., Grishin, S. V., Sharaevskii, Y. P., Nikitov, S. A. and Kruglyak, V. V. (2015). Generation of propagating spin waves from regions of increased dynamic demagnetising field near magnetic antidots, *Appl. Phys. Lett.*, **107**, 162401.

33. Davies, C. S. and Kruglyak, V. V. (2016). Generation of propagating spin waves from edges of magnetic nanostructures pumped by uniform microwave magnetic field, *IEEE Trans. Magn.*, **52**, 2300504.

34. Au, Y., Ahmad, E., Dmytriiev, O., Dvornik, M., Davison, T. and Kruglyak, V. V. (2012). Resonant microwave-to-spin-wave transducer, *Appl. Phys. Lett.*, **100**, 182404.

35. Yu, H., D'Allivy Kelly, O., Cros, V., Bernard, R., Bortolotti, P., Anane, A., Brandl, F., Heimbach, F. and Grundler, D. (2016). Approaching soft x-ray wavelengths in nanomagnet-based microwave technology, *Nat. Commun.*, **7**, 11255.

36. Mozooni, B. and McCord, J. (2015). Direct observation of closure domain wall mediated spin waves, *Appl. Phys. Lett.*, **107**, 042402.

37. Van de Wiele, B. Hämäläinen, S. J., Baláž, P., Montoncello, F. and van Dijken, S. (2016). Tunable short-wavelength spin wave excitation from pinned magnetic domain walls, *Sci. Rep.*, **6**, 21330.

38. Wintz, S., Tiberkevich, V., Weigand, M., Raabe, J., Lindner, J., Erbe, A., Slavin, A. and Fassbender, J. (2016). Magnetic vortex cores as tunable spin-wave emitters, *Nat. Nanotechnol.*, **11**, 948–953.

39. Voto, M., Lopez-Diaz, L. and Martinez, E. (2017). Pinned domain wall oscillator as a tuneable direct current spin wave emitter, *Sci. Rep.*, **7**, 13559.

40. Holländer, R. B., Muller, C., Schmalz, J., Gerken, M. and McCord, J. (2018). Magnetic domain walls as broadband spin wave and elastic magnetisation wave emitters, *Sci. Rep.*, **8**, 13871.

41. Vansteenkiste, A., Leliaert, J., Dvornik, M., Helsen, M., Garcia-Sanchez, F. and Van Waeyenberge, B. (2014). The design and verification of MuMax3, *AIP Adv.*, **4**, 107133.

42. Saitoh, E., Miyajima, H., Yamaoka, T. and Tatara, G. (2004). Current-induced resonance and mass determination of a single magnetic domain wall, *Nature*, **432**, 203–206.

43. Bedau, D., Kläui, M., Krzyk, S., Rüdiger, U., Faini, G. and Vila, L. (2007). Detection of current-induced resonance of geometrically confined domain walls, *Phys. Rev. Lett.*, **99**, 146601.

44. Baláž, P., Hämäläinen, S. J. and van Dijken, S. (2018). Static properties and current-induced dynamics of pinned 90° magnetic domain walls under applied fields: an analytic approach, *Phys. Rev. B*, **98**, 064417.

45. Brandl, F., Franke, K. J. A., Lahtinen, T. H. E., van Dijken, S. and Grundler, D. (2014). Spin waves in CoFeB on ferroelectric domains combining spin mechanics and magnonics, *Solid State Commun.*, **198**, 13–17.

46. Hämäläinen, S. J., Brandl, F., Franke, K. J. A., Grundler, D. and van Dijken, S. (2017). Tunable short-wavelength spin-wave emission and confinement in anisotropy-modulated multiferroic heterostructures, *Phys. Rev. Appl.*, **8**, 014020.

47. Schlömann, E. (1964). Generation of spin waves in nonuniform magnetic field. I. Conversion of electromagnetic power into spin-wave power and vice versa, *J. Appl. Phys.*, **35**, 159–166.

48. Qin, H., Hämäläinen, S. J. and van Dijken, S. (2018). Exchange-torque-induced excitation of perpendicular standing spin waves in nanometer-thick YIG films, *Sci. Rep.*, **8**, 5755.

49. Klingler, S., Amin, V., Geprägs, S., Ganzhorn, K., Maier-Flaig, H., Althammer, M., Huebl, H., Gross, R., McMichael, R. D., Stiles, M. D., Goennenwein, S. T. B. and Weiler, M. (2018). Spin-torque excitation of perpendicular standing spin waves in coupled YIG/Co heterostructures, *Phys. Rev. Lett.*, **120**, 127201.

50. Poimanov, V. D., Kuchko, A. N. and Kruglyak, V. V. (2018). Magnetic interfaces as sources of coherent spin waves, *Phys. Rev. B*, **98**, 104418.

51. Hertel, R., Wulfhekel, W. and Kirschner, J. (2004). Domain-wall induced phase shifts in spin waves, *Phys. Rev. Lett.*, **93**, 257202.

52. Buijnsters, F. J., Ferreiros, Y., Fasolino, A. and Katsnelson, M. I. (2016). Chirality dependent transmission of spin waves through domain walls, *Phys. Rev. Lett.*, **116**, 147204.

53. Bayer, C., Schultheiss, H., Hillebrands, B. and Stamps, R. L. (2005). Phase shift of spin waves traveling through a 180° Bloch domain wall, *IEEE Trans. Magn.*, **41**, 3094–3096.

54. Chang, L.-J., Liu, Y.-F., Kao, M.-Y., Tsai, L.-Z., Liang, J.-Z. and Lee, S.-F. (2018). Ferromagnetic domain walls as spin wave filters and the interplay between domain walls and spin waves, *Sci. Rep.*, **8**, 3910.

55. Macke, S. and Goll, D. (2010). Transmission and reflection of spin waves in the presence of Néel walls, *J. Phys. Conf. Ser.*, **200**, 042015.

56. Wang, X.-G., Guo, G.-H., Zhang, G.-F., Nie, Y.-Z. and Xia, Q.-L. (2013). An analytical approach to the interaction of a propagating spin wave and a Bloch wall, *Appl. Phys. Lett.*, **102**, 132401.

57. Han, D.-S., Kim, S.-K., Lee, J.-Y., Hermsdoerfer, S. J., Schultheiss, H., Leven, B. and Hillebrands, B. (2009). Magnetic domain-wall motion by propagating spin waves, *Appl. Phys. Lett.*, **94**, 112502.

58. Kim, J.-S., Stärk, M., Kläui, M., Yoon, J., You, C.-Y., Lopez-Diaz, L. and Martinez, E. (2012). Interaction between propagating spin waves and domain walls on a ferromagnetic nanowire, *Phys. Rev. B*, **85**, 174428.

59. Wang, X.-G., Guo, G.-H., Nie, Y.-Z., Zhang, G.-F. and Li, Z.-X. (2012). Domain wall motion induced by the magnonic spin current, *Phys. Rev. B*, **86**, 054445.

60. Wang, X.-G., Guo, G.-H., Zhang, G.-F., Nie, Y.-Z. and Xia, Q.-L. (2013). Spin-wave resonance reflection and spin-wave induced domain wall displacement, *J. Appl. Phys.*, **113**, 213904.

61. Hinzke, D. and Nowak, U. (2011). Domain wall motion by the magnonic spin Seebeck effect, *Phys. Rev. Lett.*, **107**, 027205.

62. Yan, P., Wang, X. S. and Wang, X. R. (2011). All-magnonic spin-transfer torque and domain wall propagation, *Phys. Rev. Lett.*, **107**, 177207.

Chapter 7

Patterned Spin Textures for Magnonics

E. Albisetti,[a,b] D. Petti,[a] and R. Bertacco[a]
[a] *Dipartimento di Fisica, Politecnico di Milano, P.zza Leonardo da Vinci 32, Milano, 20133 Italy*
[b] *Advanced Science Research Center, CUNY Graduate Center, 85 St. Nicholas Terrace, New York, NY 10031, USA*
edoardo.albisetti@polimi.it

In this chapter, we present our recent work on the control of the excitation and propagation of spin waves in ferromagnetic thin films by using nanopatterned spin-textures. The chapter is divided into three sections, which cover both the principles and technical aspects related to magnetic patterning as well as its applications to magnonics.

In the first part, we discuss thermally assisted magnetic scanning probe lithography (tam-SPL), focusing on its main features and patterning capabilities. In particular, we discuss the working principle of tam-SPL and its use for stabilizing 2D magnetic domains, 1D domain walls, and 0D topological spin textures with deterministically controlled properties.

In the second part, we discuss the spatial control of spin waves via patterned spin textures. We experimentally show that by controlling point by point the orientation of the magnetization and the effective fields via tam-SPL, it is possible to spatially control the

Three-Dimensional Magnonics: Layered, Micro- and Nanostructures
Edited by Gianluca Gubbiotti

ISBN 978-981-4800-73-0 (Hardcover), 978-0-429-29915-5 (eBook)
www.jennystanford.com

spin-wave excitation and propagation in continuous magnetic films. Furthermore, we present a strategy for realizing reconfigurable magnonic crystals based on periodic magnetization patterns.

In the third section, we shift our focus from micron-size magnetic domains, to nanoscopic spin textures. In particular, we discuss our results on the use of tailored domain walls for confining at the nanoscale and steering propagating spin waves. We visualize and study, via space- and time-resolved X-ray microscopy and micromagnetic simulations, the properties of confined spin-wave modes. Finally, we demonstrate a tunable spin-wave circuit based on domain walls for finely controlling the interference of confined spin-wave modes.

7.1 Introduction

Spin textures in magnetic materials, such as magnetic domains, domain walls (DWs), or vortices, are gaining attention for their potential as active components in data storage and processing devices. The interest arises from their scalability toward nanoscale dimensions, resilience against external perturbations, and high degree of tunability via external stimuli. For example, racetrack memories are based on storing and manipulating information in DWs [1] or skyrmions [2], which are controllably stabilized and manipulated in ferromagnetic nanowires. The dynamic properties of vortices, that is, their excitation, the core reversal, and their coupling [3–5], have been extensively studied and are at the basis, for example, of magnetic vortex oscillators [6, 7].

In magnonics, the possibility of using spin textures for controlling spin waves is extremely appealing due to the strong influence of the spin configuration on the generation, propagation, and spatial confinement of spin waves down to the nanoscale. In this framework, nanoscale spin textures such as DWs and vortices represent an active field of study [8]. Spin-wave transmission and reflection at DWs has been investigated [9–11], together with the interaction of spin waves with vortices [12] and the generation of spin waves via vortex core-reversal [13]. The channelling [14] and steering [15] of spin waves along DWs has been shown, together with the possibility of using vortices [16] and DWs [17] as localized spin-wave emitters.

With the rising of spin textures as device components, the importance of controlling their properties and configuration has become crucial. The most widely used methods for doing so are via physical patterning, for example, via optical or electron beam lithography, by exploiting the shape anisotropy of the system, or via ion beam irradiation. However, these methods pose limitations on the patterning capabilities and reconfigurability of the system.

In this chapter, we present our results on the development of thermally assisted magnetic scanning probe lithography (tam-SPL) [18], which we use for reversibly patterning spin textures in magnetic multilayers characterized by exchange bias. We discuss the features of tam-SPL in terms of patterning capabilities, spatial resolution, and reversibility, with reference to the patterning of 2D magnetic domains, 1D DWs [19], and 0D topological solitons [20], with deterministically controlled properties. Then, we discuss several applications in magnonics, where the enhanced tunability and control of spin textures by tam-SPL give rise to novel functionalities. In this framework, we show the spatial control of spin-wave properties via magnetic domains written through tam-SPL. In particular, we show, via microfocused Brillouin light scattering (μ-BLS), the spatial modulation of the spin-wave excitation and propagation, by stripe-shaped 90° and 0° domains [18], which can be tuned by external fields [21]. Then, we present a strategy for developing reconfigurable and tunable magnonic crystals based on periodic nanoscale magnetization patterns. In the subsequent section, we focus on the use of nanoscale spin textures for controlling and confining spin waves at the nanoscale [15]. In particular, we directly visualize via scanning transmission X-ray microscopy (STXM) the channelling and steering of localized spin-wave modes in patterned DWs, and we study the geometry and properties of the modes via micromagnetic simulations. Finally, we present the realization of a nanoscale spin-wave circuit based on DWs allowing for the tunable superposition and interference of confined spin-wave modes.

7.2 Thermally Assisted Magnetic Scanning Probe Lithography

Since the invention of writing around 3200 BC, there has always been a trade-off between two key features in the development of every patterning method: the endurance of the patterning and the rewritability. Magnetism intrinsically brings in a very efficient solution to face this problem. This is the basis for the large success of magnetic information technology, where information is written at the micro- and nanoscale via magnetic and electric fields. In this framework, there is an increasing need of patterning techniques for fabricating novel reconfigurable magnetic metamaterials, implementing complex memory and logic functionalities at the nanoscale. Metamaterials are composite materials displaying intriguing functionalities that arise from engineered combinations of nanoscale building blocks. In particular, magnetic metamaterials are highly appealing for the realization of novel device architectures, depending on the controlled propagation of electromagnetic radiation [22], surface plasmons [23], and spin waves [24, 25].

Conventionally, the fabrication of magnetic micro- and nanostructures and the stabilization of spin textures are mainly carried out using conventional lithography [26] or ion irradiation [27, 28]. In the former case, the micromagnetic configuration arises from the physical shape of the structure (e.g., rings, nanowires, dots). In the latter, the selective irradiation of a continuous film modifies magnetism via chemical, microstructural modifications or via thermal effects, while keeping unaltered the film topography. In parallel, advanced scanning probe lithography (SPL) is gaining attention in the scientific community for its flexibility and broad applicability, which is revealing extremely useful in the development of research lines based on novel materials [29].

Recently, we introduced a novel technique, thermally assisted magnetic scanning probe lithography (tam-SPL) [18], to write nanoscale reconfigurable spin textures in a continuous and homogeneous film. tam-SPL is the nanoscale analog of the uniform field cooling, which is used to set a unidirectional magnetic anisotropy in an exchange-biased ferromagnetic layer due to the interface exchange interaction with an antiferromagnet [30, 31]. In tam-SPL, we use the heatable tip of a scanning probe microscope (SPM), in the presence of a magnetic field, to heat the portion of the film in contact to the tip and then we exploit the cooling due to heat dispersion when the tip is moved to another position. This confined "field cooling" causes a local pinning of the magnetization of the ferromagnet along the direction of the applied magnetic field, which reflects in a shift of its hysteresis loop. This method allows writing of arbitrary-shaped spin textures such as magnetic domains, DWs, or solitons, with controlled magnetic anisotropy and spin orientation. The patterned spin textures can be modulated via external fields but, importantly, are stable at remanence and cannot be permanently erased with magnetic fields. On the other hand, the pattern can be reconfigured by simultaneous application of a magnetic field and heating, thus providing an excellent compromise between endurance and rewritability.

7.2.1 Concept of tam-SPL

The concept of tam-SPL is sketched in Fig. 7.1. Before writing, the exchange-biased bilayer is initialized by uniform heating above the critical temperature at which the exchange bias disappears, that is, the blocking temperature (T_B), followed by cooling in a uniform magnetic field (see Fig. 7.1A). This sets a uniform unidirectional anisotropy axis in the ferromagnetic film, whose hysteresis loop is shifted by the exchange bias field (H_e) (see Fig. 7.1D).

Then, by sweeping the hot tip of an SPM [32, 33], we locally heat the exchange-biased multilayer above T_B, while applying a uniform static magnetic field H_W. When the tip is displaced over the film, the heated region of the film undergoes a "local field cooling," which resets the exchange bias direction according to the orientation of the magnetization of the ferromagnet, which is set by the external field H_W (Fig. 7.1B).

Upon removal of the writing field H_W, the direction of the spins within the ferromagnet is determined by the new, spatially modulated, unidirectional magnetic anisotropy set during writing. In the specific case shown in Fig. 7.1, the writing field is antiparallel to the initialization field. Therefore, the magnetization in the

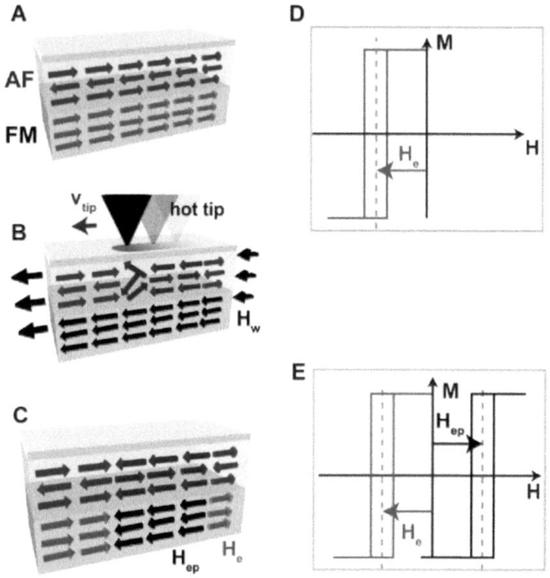

Figure 7.1 (A) Initial state: Magnetization of the ferromagnet (FM) is uniformly pinned in one direction by the exchange interaction with the antiferromagnet (AF). (B) Writing by sweeping the heated SPM tip on the sample in the presence of an external magnetic field H_W. An opposite exchange bias direction is set according to the underlying ferromagnetic spins, aligned with H_W. (C) Upon removal of the external magnetic field H_W, the spin texture in the ferromagnet is stabilized by the local exchange bias. The bottom panels show the hysteresis loops before (D) and after patterning (E). Adapted from Ref. [18].

area scanned by the tip (Fig. 7.1C) is antiparallel to that of the surrounding film portion, still in the initial state. Noteworthy, the local hysteresis loop in the written area is shifted to the right by a new exchange bias field H_{ep} (Fig. 7.1E), opposite to that set during initialization.

7.2.2 Patterning 2D Domains, 1D Domain Walls, and 0D Solitons

To demonstrate the patterning capabilities of our technique we chose a prototypical exchange bias system made of a thin film of $Co_{40}Fe_{40}B_{20}$ (CoFeB) in contact with an $Ir_{20}Mn_{80}$ (IrMn) layer.

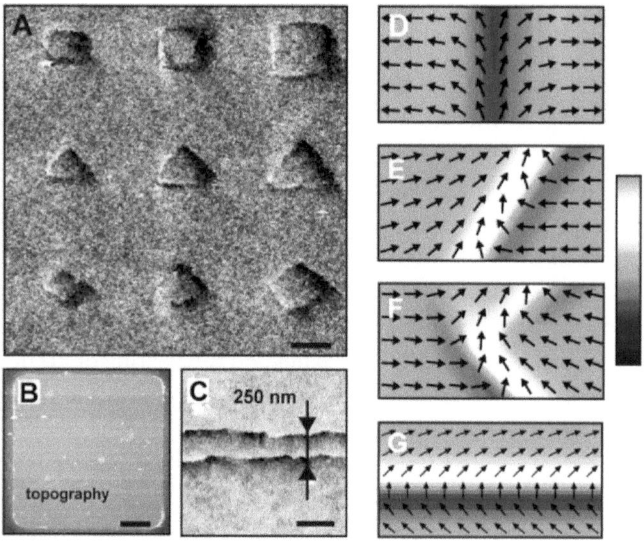

Figure 7.2 MFM image of square-, triangle-, and diamond-shaped magnetic domains with different sizes (A) and topographies (B) of the same pad after tam-SPL. (C) High-resolution MFM image of a stripe domain. (D–G) Micromagnetic simulation of the patterned domain walls on square, triangles, diamonds, and single lines, respectively. Adapted from Ref. [18].

Ru(2)/IrMn(7)/CoFeB(5)/SiO2(1000)/Si (thickness in nanometers) multilayers were grown via magnetron sputtering in an in-plane magnetic field $H_G = 300$ Oe, used to set the uniaxial anisotropy axis of the ferromagnetic CoFeB layer. In this system, the blocking temperature is well above the room temperature ($\sim$160°C), allowing to perform local field cooling by using the heatable SPM tip. In this example, the exchange bias is about 75 Oe, thus ensuring robust stabilization of one of the two possible remanent states.

Magnetic domains with different shapes (squares, diamonds, and triangles) were patterned by raster-scanning the hot tip of the SPM, as shown in Fig. 7.2. The writing speed in this specific example was 3 μm s^{-1}. A writing field of 700 Oe was applied in the direction opposite to the initialization field to pattern domains with antiparallel remanent magnetization. The heater temperature (T_H) on the tip was set to 560°C, 675°C, and 765°C for squares, triangles, and diamonds, respectively. These values ensured that the writing temperature at the sample surface T_W, which is lower than the heater temperature due to thermal dissipation, was larger than the blocking temperature. The magnetic contrast observed in the magnetic force microscopy (MFM) analysis reported in Fig. 7.2A shows the patterned magnetic domain structures. Noteworthy, no topographical counterpart of the patterned areas was present, as clearly seen in the topography image of the same pad reported in Fig. 7.2B. This proves that the patterning is "purely magnetic" and is not accompanied by a physical modification of the film.

The capabilities of tam-SPL go beyond the writing of 2D domains with uniform magnetization. This is evident from the micromagnetic simulations reported in Fig. 7.2D–G, where the spin texture of the DWs stabilized at the boundaries of the patterned areas is shown. Black arrows indicate the local spin direction, while the dark/white contrast shows the calculated force on the SPM tip, for directly comparing simulations with experimental data. Due to the peculiar orientation of the 180° DWs defining the edges of the shapes, different spin textures in the walls are promoted. Head-to-head and tail-to-tail DWs are found on the left and right sides of square domains (Fig. 7.2D), respectively, where the spins at the center of the DW are parallel to the wall. On the top and bottom sides, instead, the spins at the center of the DW are perpendicular to the wall itself. For different orientations of the DW, such as in triangles (Fig. 7.2E) and diamond shapes (Fig. 7.2F), the arrangement of the spins within the wall is in between the two cases above. These examples show that by controlling the shape of the pattern, in combination with the magnetization direction, DWs with tailored spin configuration and position can be obtained.

To investigate the minimum domain size achievable by tam-SPL, we patterned a set of parallel lines with a spacing of 2 μm, each of them obtained by a single sweep of the heated tip, in an applied field antiparallel to the initialization field. From the high-resolution MFM image of Fig. 7.2C, the structure of a single line emerges: it presents two 180° DWs at a distance of 250 nm, which define an inner domain with the minimum size achievable in this system. The corresponding spin texture is reported in the simulation of Fig. 7.2G, displaying one of the two Néel DWs corresponding to the experimental MFM contrast. Noteworthy, the minimum domain size is limited by the DW width, rather than by the spatial thermal profile on the film generated by the tip, which has been proven to give rise to a sub-10-nm resolution [34]. In fact, the actual patterning resolution is largely due to the micromagnetic properties of the system (e.g., the exchange bias strength, magnetocrystalline anisotropy, exchange stiffness), as well as the grain size distribution.

The possibility of changing vectorially and point by point the exchange bias direction opens up intriguing possibilities, which go beyond the patterning of domains and DWs. Next, we show that tailored topological solitons, such as vortices, antivortices, and Bloch lines, can be stabilized via tam-SPL in continuous films.

In Fig. 7.3 we show the controlled nucleation of vortex-antivortex pairs (panels A–C) and Bloch lines (panels D–F). The patterned exchange bias field configuration (panels A and D), the MFM image (panels B and E), and the results of micromagnetic simulations (panels C and F) are presented. Gray arrows indicate the direction of the patterned exchange bias field and black arrows the direction of the magnetization. In the micromagnetic simulation, the gray contrast indicates the divergence of the magnetization, which is related to the experimental MFM contrast. In panels A–C, we show that a vortex-antivortex pair can be stabilized by patterning sequentially four triangular-shaped areas with exchange bias oriented in a counterclockwise direction. Such a non-uniform configuration of the exchange bias field, which, in turn, stabilizes a non-uniform magnetization configuration, generates a local potential minimum for the vortex core in the center of the patterned region (dark circle in panel A) and an antivortex core at the corner of the patterned region (indicated with a bright circle). This can be seen in the MFM image (panel B) and micromagnetic simulation (panel C) of the pattern at remanence.

The properties of the vortex-antivortex pairs can be tailored by setting the direction of the patterned exchange bias field with respect to the geometrical shape of the patterned area. In particular, by changing the direction of the patterned exchange bias field, the chirality of the vortex and the location of the antivortex can be controlled.

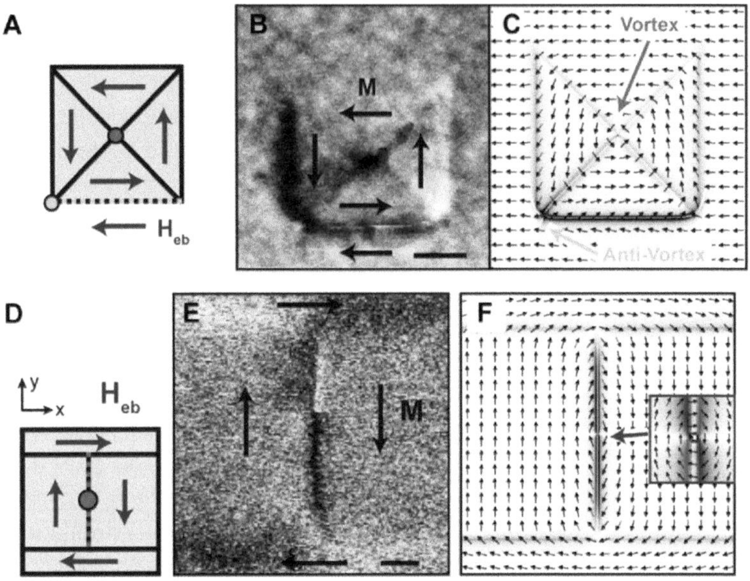

Figure 7.3 (A, D) Sketch of the exchange bias directions (black arrows) for stabilizing vortex-antivortex pairs (A) and vortex Bloch lines in a straight Néel wall (D). (B, E) MFM images of the vortex-antivortex pair (B) and Bloch line (E). Black arrows indicate the magnetization direction. Scale bar = 2 μm. (C, F) Corresponding simulated micromagnetic configuration showing the location of the vortex and antivortex (C) and the structure of the vortex Bloch line (F). The inset shows a zoom-in of the central region of the simulation. Adapted from Ref. [20], with the permission of AIP Publishing.

In panels D–F, we show the stabilization of a vortex Bloch line within a patterned 180° Néel wall. In panel D, the sketch of the configuration of the patterned exchange bias is shown. The 180° wall is defined by rectangular regions with magnetization oriented in opposite directions. In such a system, the orientation of the magnetization within the DW, that is, the chirality of the DW, is determined by the boundary conditions, hence by the direction of the magnetization at the edges of the DW. The Bloch line is stabilized by imposing two opposite chiralities at the two ends of the wall by defining two regions with opposite magnetization. It is worth noting that both the chirality of the Bloch lines and their vorticity, that is, vortex or antivortex, can be tailored by controlling the orientation of the magnetization of the central domains. In particular, in panel E, we show the MFM image of a vortex Bloch line with clockwise chirality within a straight 180° Néel wall. The magnetization within the wall points toward opposite directions in the bottom (top) part of the wall, as indicated by the opposite MFM contrast, in good agreement with the micromagnetic simulations of panel F. The bright/dark colors at the top and bottom sides of the MFM image confirm the opposite direction of the magnetization in the two domains.

7.2.3 Tunability and Reconfigurability

tam-SPL offers the possibility to tune point by point the strength and direction of the unidirectional anisotropy related to the exchange bias by varying the writing temperature during patterning. Figure 7.4A reports the MFM image taken at remanence from square areas patterned with different heater temperatures, T_H = 240°C (central square) and 435°C (left square), while applying a writing field opposite to H_i. For the nonpatterned sample (right square) no magnetic contrast is seen, in agreement with the fact that the local hysteresis loop measured via micro-magneto-optical Kerr effect (μ-MOKE) microscopy (Fig. 7.4C, left loop) is completely shifted from the 0, showing an exchange bias field of −50 Oe, as in the surrounding area. By increasing T_H up to 240°C (central square) the patterned area starts to show a magnetic contrast and the corresponding local MOKE signal (Fig. 7.4B) displays a loop centered around 0. By further increasing T_H to 435°C (left square) the contrast further increases as the full inversion of the exchange bias field up to +50 Oe (right loop in Fig. 7.4C) is achieved. This behavior is related to the distribution of blocking temperatures within the polycrystalline film, arising from the different sizes of the grains [35]. Therefore, by finely tuning the writing temperature, it is possible to modulate the unidirectional anisotropy strength, eventually leading to a "continuous tuning" of the anisotropy strength.

The reversibility of tam-SPL is illustrated in Fig. 7.4D. In the top panel, the MFM image from two squares written upon application of a writing field H_W pointing to the left is reported. The central

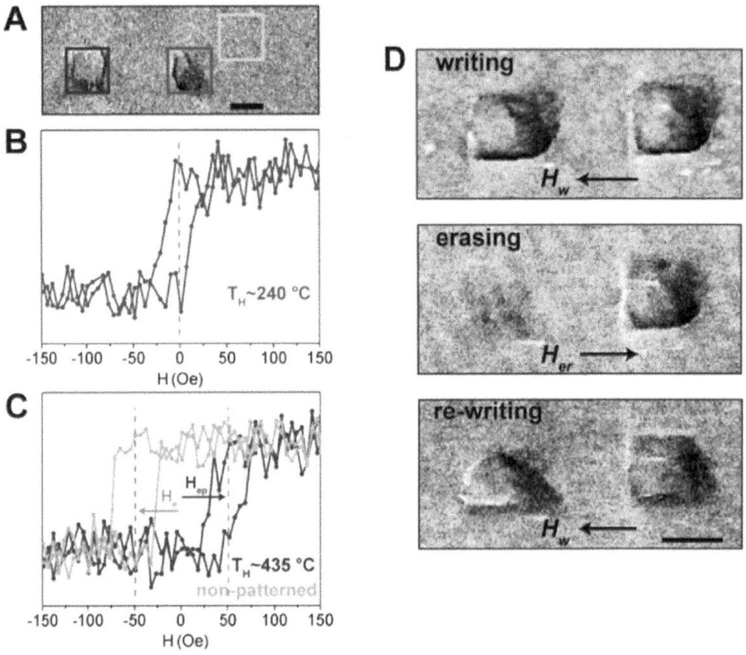

Figure 7.4 (A) MFM image from different areas: nonpatterned (right square) and patterned with $H_W = 700$ Oe and $T_H = 240°C$ (central square) and 435°C (left square). (B) μ-MOKE hysteresis loop corresponding to the central square of panel A ($T_H = 240°C$). (C) μ-MOKE hysteresis loops corresponding to the left square of panel A ($T_H = 435°C$, dark line) and to the nonpatterned area (bright line). (D) Top panel: MFM image of two squares patterned with $T_H = 600°C$ and $H_w = 700$ Oe. Central panel: the very same area after erasure of the left square by scanning it with $T_H = 600°C$ and an erasing field $H_{er} = -700$ Oe, opposite to H_W and in the same direction of H_i. Lower panel: a triangle is rewritten with $T_H = 600°C$ and $H_W = 700$ Oe. Adapted from Ref. [18].

panel shows the MFM image of the same area, after erasing the left square by sweeping the tip at the same temperature, with an erasing field (H_{er}) pointing to the right. The lower panel shows the patterning of a triangular-shaped domain on the same area. This example underlines the possibility of reconfiguring the patterned spin textures, at variance with conventional irreversible patterning techniques.

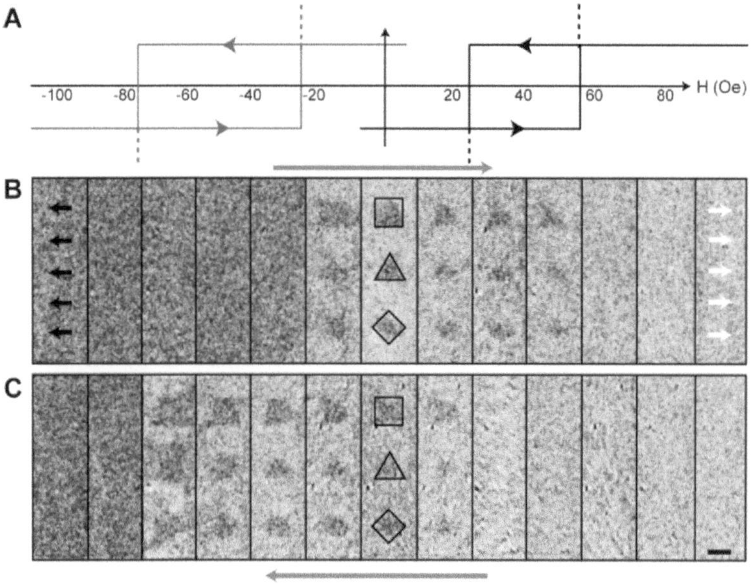

Figure 7.5 (A) Hysteresis loops from the nonpatterned (left loop) and patterned (right loop) areas. (B, C) μ-MOKE images of patterned domains at different magnetic fields. Adapted from Ref. [18].

7.2.4 Pattern Stability and Modulation with External Fields

The evolution of the magnetization in the patterned areas as a function of the external magnetic field, as investigated by μ-MOKE, is shown in Fig. 7.5. We report μ-MOKE images of the patterned area of Fig. 7.2B,C. The top sequence of images is taken while sweeping a field parallel to the unidirectional anisotropy axis, from −100 Oe to +100 Oe, while the bottom sequence refers to the second part of the loop (from +100 Oe to −100 Oe).

Due to the opposite exchange bias of patterned and unpatterned areas, the magnetic contrast appears only for the applied fields stabilizing an antiparallel configuration in the two regions. In particular, the contrast is visible at remanence but disappears at positive and negative fields large enough to saturate the magnetization in the whole film. Noteworthy, however, the same magnetic state at remanence is always recovered after 2000 hysteresis loops. This points to the high stability of the pattern, which arises from the resilience of exchange bias to external magnetic fields. Furthermore, Fig. 7.5 shows that the patterned spin textures can be tuned via external fields, due to the sensitivity of the DWs to external magnetic perturbations. In particular, the size of the patterned domains extends when the external field is applied in the same direction as the magnetization of the domains and shrinks when the direction is the opposite. Importantly, the process is reversible as, upon removing the external field, the spin texture is restored to the configuration at remanence. This provides an efficient way to controllably reconfigure the spin texture with external stimuli, without permanently altering the patterned magnetic anisotropy landscape.

7.3 Controlling Spin Waves with Patterned Domains

7.3.1 Spin Waves in Patterned 90° Stripe Domains

Figure 7.6B shows the MFM image of 2.5-μm-wide stripe-shaped magnetic domains patterned via tam-SPL in an exchange bias IrMn 7 nm/CoFeB 5 nm system. For patterning the domains, after initializing the continuous film via uniform field cooling in an external field H_i, the tip, heated above the blocking temperature of the system, was raster-scanned in the presence of a magnetic field applied in the direction perpendicular to H_i. The presence of sharp 90° DWs is confirmed by the alternated black-white contrast in the MFM image, where the black arrows indicate the direction of the magnetization. For spin-wave excitation, coplanar waveguides (CPWs) were patterned on top of the film by electron beam lithography and lift-off of Ti 7 nm/Au 200 nm after depositing a SiO_2 insulating layer. The optical image of the sample is shown in Fig. 7.6A, where the tam-SPL patterned area is indicated by the outer square and the DWs separating the alternated 90°-oriented magnetic domains are indicated by dashed lines.

For studying the spin-wave intensity with high spatial resolution, μ-BLS was employed. In BLS, the incident photons are scattered

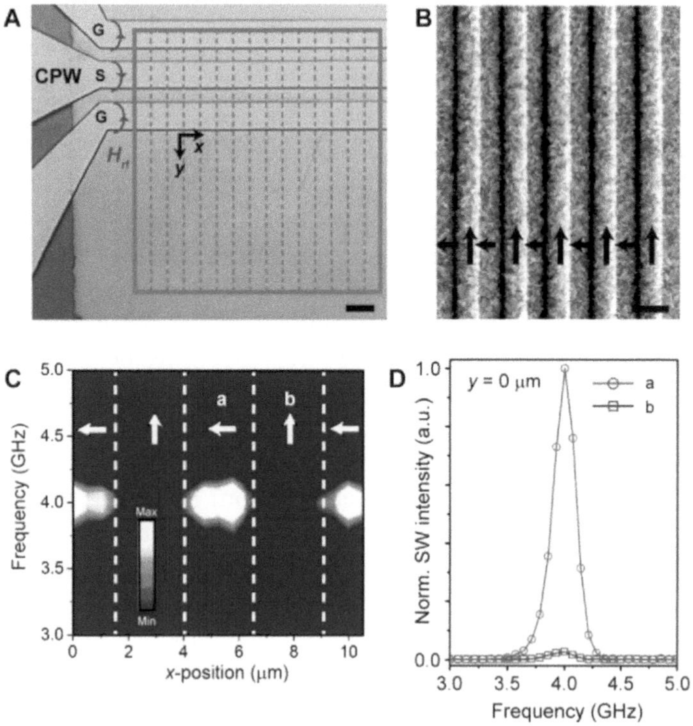

Figure 7.6 (A) Optical image of the sample, showing the coplanar waveguide (CPW) and the rectangle indicating the patterned area. Scale bar = 4 μm. (B) MFM image of the patterned domains consisting of 2.5-μm-wide stripes with alternated 0°- and 90°-oriented remanent static magnetization (black arrows). The track boundaries are marked by dashed lines in panel A. Scale bar = 4 μm. (C) Spin-wave intensity measured at remanence via μ-BLS along the *x* direction, right outside the CPW. (D) Spin-wave spectra measured in track *a* (circles) and track *b* (squares), right outside the CPW. Adapted from Ref. [18].

inelastically by the interaction with magnons in the film. Therefore, by analyzing the frequency shift, the wave vector and the intensity of the scattered photons, it is possible to obtain information on the dispersion and intensity of the spin waves propagating in the film, with a spatial resolution that is determined by the laser spot size (~250 nm in the case presented here). Figure 7.6C shows a line map of the spin-wave intensity in the patterned film, acquired by scanning the laser spot along the x direction (parallel to the CPW) right outside the CPW. Spin waves were excited by running a microwave current at 3–6 GHz in the CPW. The measured spin-wave intensity shows pronounced maxima in correspondence of the regions where the magnetization points in the x direction (regions marked with the letter a) and remarkably lower intensity in the region where the magnetization points in the y direction (letter b).

Figure 7.6D shows the intensity profile as a function of frequency, displaying a 36 times intensity difference between the two regions. This can be explained considering the different spin-wave excitation efficiency in the two regions. Spin waves are excited by the coupling of the Oersted field generated by the CPW with the dynamical magnetization of the spin-wave mode. In stripe of type a, the magnetization and the spin-wave wave vector form a 90° angle with each other (Damon–Eshbach configuration). Here, the spin-wave dynamical magnetization is characterized by a strongly elliptical precession, with the major axis in the plane of the film along the y direction and the minor axis along the out-of-plane direction. In this configuration, the efficient spin-wave excitation is determined by the efficient coupling of the Oersted field component in the y direction with the corresponding dynamical magnetization along y. In stripes of type b, the magnetization and the wave vector are parallel (backward-volume configuration), and the corresponding elliptical precession of the magnetization has its major axis along x. In this case, no efficient coupling with the Oersted field generated by the CPW is possible, which has no x component. This demonstrates that by controlling point by point the magnetization vector, it is possible to fabricate magnonic metamaterials for spatially modulating the spin-wave excitation efficiency without physically patterning the film.

Figure 7.7A shows the linear μ-BLS spin-wave intensity map acquired along the y direction, that is, in the direction perpendicular to the CPW. The top panel, related to stripe a, shows the detection of propagating spin waves up to about 4 μm from the CPW, while in stripe b, spin waves are detected up to roughly 2 μm from the CPW. Figure 7.7B shows the corresponding frequency profiles at a 2.25 μm distance from the CPW, displaying that while spin waves are still detectable in the domain in the Damon–Eshbach configuration,

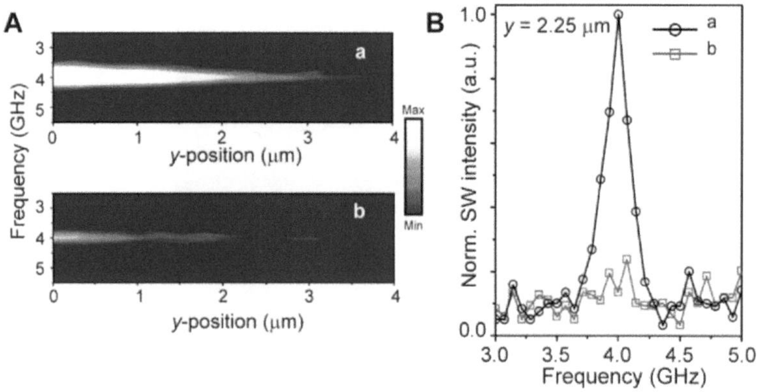

Figure 7.7 (A) One-dimensional map of the spin-wave intensity measured at remanence by μ-BLS as a function of the distance *y* from the CPW. (B) Spin-wave spectra measured in track *a* (circles) and track *b* (squares) at distance $y = 2.25$ μm from the CPW. Adapted from Ref. [18].

spin waves in the backward-volume configuration are below the measurement noise. These results show that the patterning of micrometric magnetic domains leads to a reconfigurable and tunable spatial modulation of the intensity of propagating spin waves within the film and to the spatial confinement of spin-wave modes.

7.3.2 Reversible Tuning with External Fields

Due to the fact that tam-SPL patterns are imprinted in the exchange bias, the application of external fields can be used for changing reversibly the magnetic configuration and therefore tuning the magnonic response of the system.

Next, we show the tuning of the properties of spin waves excited in patterned micrometric 2.5-μm-wide stripe domains with alternated 90° direction of the magnetization. Figure 7.8 shows the μ-BLS linear spatial map of the spin-wave intensity in the 2.7 to 8.3 GHz frequency range, acquired scanning along the *x* direction, across the magnetic domains. In panel A, the measurement was performed at remanence, while in panel B, the measurement was performed in a 300 Oe magnetic field applied along the *y* direction.

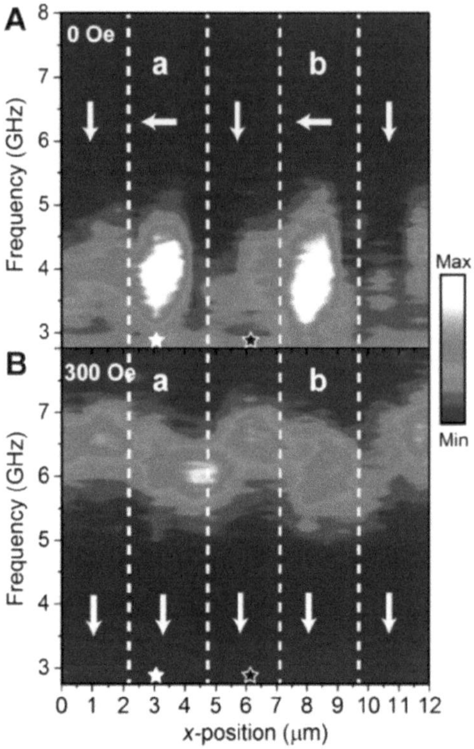

Figure 7.8 One-dimensional μ-BLS mapping of the spin-wave intensity across patterned stripe-shaped domains with alternated 0° −90° magnetization direction. (A) Measurement at remanence and (B) measurement performed while applying a 300 Oe external magnetic field in the vertical direction for uniformly saturating the magnetization. Upon removal of the field, the configuration of panel A is retrieved. Adapted from Ref. [21] (https://creativecommons.org/licenses/by/4.0/).

In the measurement at remanence (top panel), in analogy to what is discussed above, a strong spin-wave excitation is observed in the Damon–Eshbach configuration, while a markedly lower spin-wave excitation is observed in the backward-volume configuration.

Upon application of the external field, the magnetization of the film is uniformly oriented along y, and as a consequence, the same low spin-wave intensity is observed in the whole system. Interestingly, the spin-wave spectrum within the type a domains is shifted toward lower frequencies by roughly 0.5 GHz with respect to the spectrum of type *b* domains. This is consistent with the fact that despite the uniform magnetization orientation, the two domain types are characterized by a different exchange bias field, which, combined with the external field, determines the spin-wave spectrum of the material. It is worth noting that the initial configuration is restored by the exchange bias field upon removal of the external field, therefore ensuring the complete reversibility of the process. These interesting features suggest that the vectorial tuning of the internal field represents an interesting avenue for controlling spin waves, in addition to the patterning of the magnetization. In this sense, realizing magnonic crystals based on finely tuned internal fields controlled at the nanoscale via tam-SPL is a promising application.

7.3.3 Magnonic Metamaterials Based on Patterned Spin Textures

A particularly appealing application of tam-SPL is the creation of reconfigurable magnonic metamaterials, that is, manipulating the spin-wave propagation in magnetic thin films via the local tuning of their magnetic properties with high spatial resolution. In particular, as shown above, the local exchange bias field can change the dispersion of the propagating spin wave. This is the basis for designing magnonic elements, such as spin-wave phase shifters or magnonic crystals described in Fig. 7.9, which reports the results of static and dynamic simulations carried out with the Object Oriented MicroMagnetic Framework (OOMMF; see below for details). The working principle of these devices is based on the strong anisotropy of spin-wave propagation. Spin waves excited by a point source in a multilayer with uniform exchange bias propagate anisotropically, depending on the relative angle of the wave vector ($\mathbf{k}$) and the local magnetization (M), whose orientation is determined by exchange bias. In particular, if we consider a point excitation of spin waves at 10 GHz frequency, the wavelength is about 1.6 times larger (the refractive index is 1.6 times smaller) if the wave vector is perpendicular to the magnetization direction compared to the parallel case.

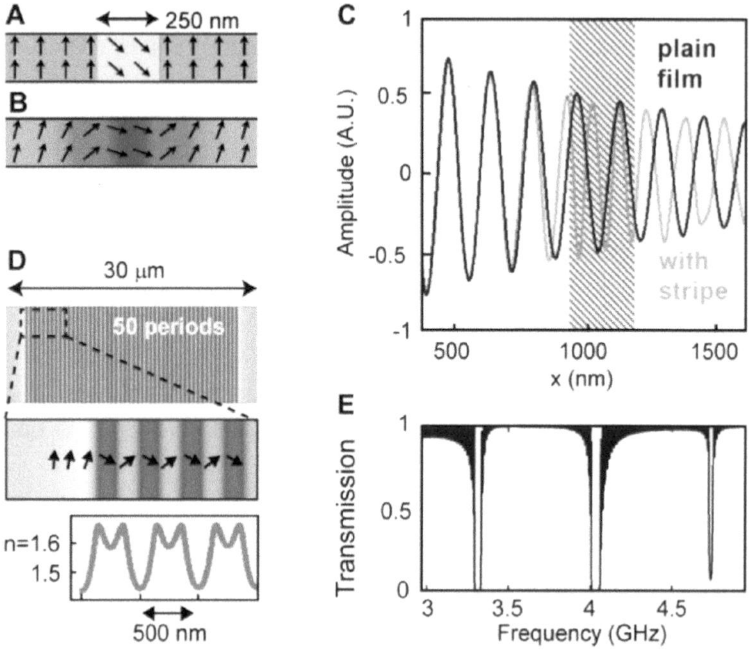

Figure 7.9 (A) Exchange bias field plot for a single 250 nm stripe patterned via tam-SPL. (B) Micromagnetic configuration of the stripe after relaxation. (C) Simulated amplitude of a spin-wave propagating through the stripe. (D) Structure of a simulated magnonic crystal resulting from 50 repetitions of the stripe above. Bottom: the local micromagnetic configuration and effective refractive index profile is shown for a short segment of the crystal. (E) Simulated transmission of the magnonic crystal. Adapted from Ref. [18].

This change in the refractive index can be exploited for designing a phase shifter or a magnonic crystal, as shown in Fig. 7.9. In panel A, we plot the spatially modulated orientation of the exchange bias field, upon patterning via tam-SPL a vertical stripe, 250 nm wide, with alternated direction of the exchange bias field of 0° and 120°. The corresponding ground-state micromagnetic configuration calculated by OOMMF is plotted in panel B. Panel C shows that the presence of the patterned stripe causes a phase shift of about 3π in the travelling spin wave, excited at 10 GHz, with respect to the case of free propagation in a film with uniform magnetization direction.

In this sense, a single stripe with misoriented magnetization orientation acts as an ideal phase shifter that can constitute the basis for realizing spin-wave based Mach–Zehnder-type interferometers. Moreover, it is possible to envision magnonic devices consisting of arrays of stripes. For example, in panel D, we simulated a magnonic crystal made of 50 periods of stripes with identical configuration with respect to the ones presented in panel A, spaced by 500 nm. The micromagnetic configuration shown in the zoom on the right displays a spin rotation between adjacent stripes, which gives rise to a modulation of the refractive index for magnons (from 1.45 to 1.65), as seen in the profile shown in the inset. As a result, the transmission profile, presented in panel E, calculated with 1D transfer matrix methods, shows the emergence of the characteristic bandgaps of a magnonic crystal.

The bandwidth and position of the bandgaps can be tuned by controlling the stripe width, the number of periods, the separation between the stripes, and the width of the regions of spin rotation between the stripes. In particular, this last feature can be deterministically set by controlling the exchange bias strength profile set during patterning. Interestingly for applications, after patterning, the magnonic crystal bandwidth and the entity of the phase shift in the single stripe could be continuously and reversibly tuned by applying external magnetic fields, which determine the size of the magnetic domains, and the reorientation of the magnetization, as shown in Fig. 7.8.

7.4 Spin-Wave Circuits Based on Nanoscale Spin Textures

In the previous section, we focused on the control of spin waves via micron-sized patterning of magnetic domains. Here, we show that spin-wave confinement and guidance at the nanoscale can be obtained with tailored nanoscale spin textures such as shaped magnetic DWs.

In particular, we report on the realization of a nanoscale spin-wave circuitry for confining and steering spin waves using patterned spin textures. First, we show the stabilization of arbitrarily shaped magnetic DWs, which act as magnonic waveguides. Then we realize a prototypic spin-wave circuit allowing for the tunable superposition of signals propagating in two converging waveguides. We directly visualize the channeling and steering of localized spin-wave modes propagating along straight and curved DWs via space and time-resolved scanning transmission X-ray microscopy (STXM), with no need for an applied bias magnetic field. Furthermore, we show that by applying a small external field, we can controllably displace the position of the DWs and, therefore, change the spatial superposition and interference of confined spin-wave modes propagating within the two converging nanoscale waveguides. The properties and geometry of the confined spin-wave modes are also studied via micromagnetic simulations.

7.4.1 Patterning, Characterization, and Static Simulations of Nanoscale Spin Textures

The different spin textures were patterned via tam-SPL in an exchange bias ferromagnet/antiferromagnet bilayer. The $Co_{40}Fe_{40}B_{20}$ 20 nm/$Ir_{22}Mn_{78}$ 10 nm/Ru 2 nm stack, shown in Fig. 7.10A was grown on Si/Si_3N_4 200-nm-thick membranes via magnetron sputtering, as described previously. Then, tam-SPL patterning was performed by sweeping in a raster-scan fashion the scanning probe, heated above the blocking temperature of the exchange bias system $T_B \approx 300°C$, in the presence of an external magnetic field. Two rotatable permanent magnets were employed for generating a uniform external magnetic field applied in the sample plane during patterning. After patterning, 2 μm × 30 μm microstrip antennas were fabricated via maskless optical lithography and lift-off, after depositing a 50-nm-thick SiO_2 insulating layer via magnetron sputtering. The stripline consisted of a Cr 5 nm/Cu 200 nm bilayer deposited by means of thermal evaporation. Figure 7.10B shows the optical image of the final sample, where a straight DW (dashed line) perpendicular to the stripline is located in the middle of the stripline.

STXM was used for acquiring static images of the magnetization configuration of the spin textures patterned via tam-SPL. In this

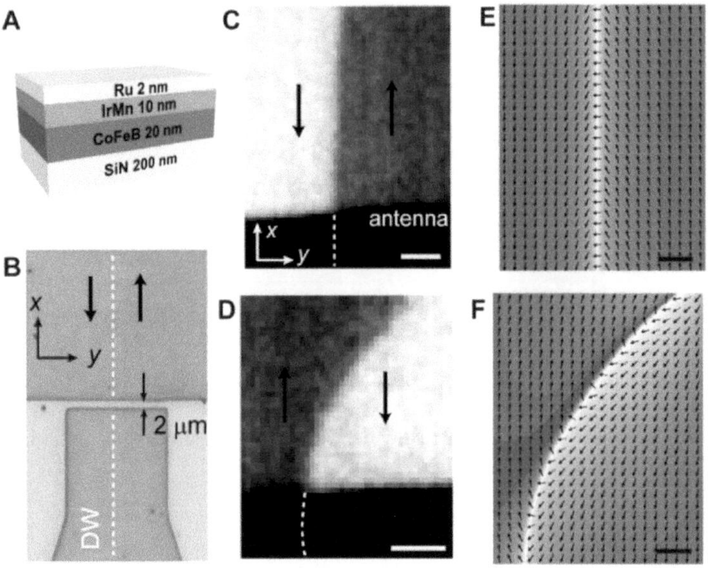

Figure 7.10 (A) Sample structure. (B) Optical image of the sample, showing the 2-μm-wide microstrip antenna for spin-wave excitation. The white line indicates the position of a patterned straight domain wall. (C, D) Static STXM images of a straight 180° Néel domain wall (C) and a parabola-shaped 180° Néel wall (D). The dark arrows indicate the direction of the magnetization within the domains. The dashed lines indicate the continuation of the domain walls under the antenna. (E, F) Micromagnetic simulations corresponding to the central panels. The black arrows indicate the local spin configuration. Scale bar = 500 nm. Adapted from Ref. [15] (https://creativecommons.org/licenses/by/4.0/).

technique, monochromatic X-rays, tuned to the Co L_3 absorption edge (photon energy of about 781 eV), are focused onto a spot on the sample, and the transmitted photons are recorded using an avalanche photodiode. To form the image, the sample is scanned using a piezoelectric stage, and the transmitted X-ray intensity is recorded for each pixel in the image. Magnetic contrast in the images arises from the X-ray magnetic circular dichroism (XMCD) by illuminating the sample with circularly polarized light. In the particular measurement configuration we employed, the black-white contrast is related to the in-plane component of the magnetization.

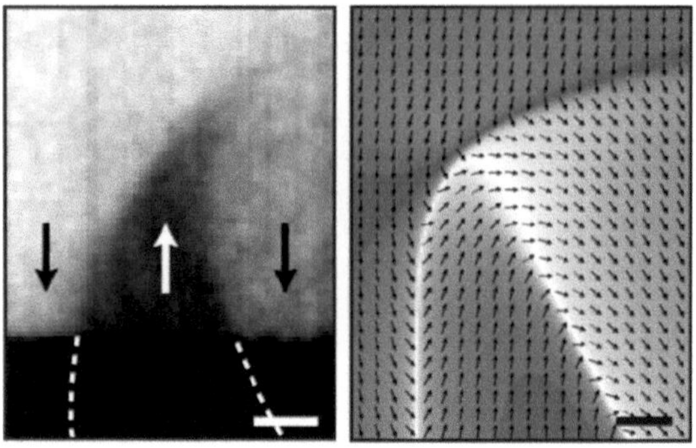

Figure 7.11 Left: STXM image of a complex spin texture comprising two converging 180° Néel walls sharing a common apex. Right: corresponding micromagnetic simulation. Scale bar = 500 nm. Adapted from Ref. [15] (https://creativecommons.org/licenses/by/4.0/).

Figure 7.10C,D shows STXM images of straight and curved 180° Néel DWs, respectively, stabilized by patterning two magnetic domains with antiparallel remanent magnetization. The dark (bright) contrast corresponds to M_x>0 (<0); the dark arrows indicate the direction of the magnetization within the domains. The black region at the bottom of the figure shows the boundary of the microstrip used for spin-wave excitation. For both straight and parabolic DWs, the images, acquired at zero external field, display sharp 180° Néel walls. Panels E and F show the corresponding micromagnetic simulations, which display a 180° spin rotation within the sample plane, with the central spins, indicated by the white region, defining the DW profile.

By placing in close proximity two shaped DWs, it is possible to stabilize more complex spin textures. Figure 7.11 (left panel) shows the static image of a structure comprising two 180° Néel walls tilted by a 30° angle from each other, sharing a common apex. The corresponding micromagnetic simulation, in the right panel, shows that the two DWs merge at the intersection, where the magnetization orientation is determined by the spin configuration of the two walls. After the intersection, a "transition" region is formed, where the magnetization rotates continuously toward the uniform magnetization direction of the domain.

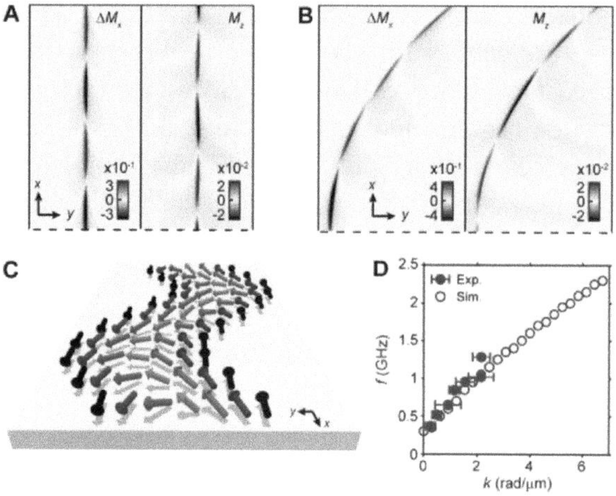

Figure 7.12 (A, B) Micromagnetic simulations of the ΔM_x and M_z spin-wave modes propagating in a Néel 180° domain wall (A) and in a curved Néel 180° wall (B), respectively. The excitation is generated by a horizontal line antenna placed below the dashed line. (C) Sketch of the spin-wave mode in a straight wall. The mode is characterized by the flexural motion of the domain wall in the plane of the film and by the precession of the spins along the domain wall. (D) Experimental (filled circles) and simulated spin-wave dispersion (empty circles) in the straight domain wall. Adapted from Ref. [15] (https://creativecommons.org/licenses/by/4.0/).

7.4.2 Simulations of Spin-Wave Propagation along Straight and Curved Domain Walls

Micromagnetic simulations of the spin-wave excitation in the presence of 180 Néel DWs are presented in Fig. 7.12 for straight (panel A) and curved (panel B) walls. For exciting spin waves, a horizontal line source of sinusoidal magnetic field applied in the x direction at 1.28 GHz (straight) and 1.11 GHz (curved wall) was used, placed in correspondence of the dashed line on the bottom part of the image.

The M_z component is reported in the right panel, while the magnetic deviation $\Delta M_x(t)$ from the time averaged value $\langle M_x(t)\rangle$ is shown in the left panel. The flexural motion of the wall, associated with the spin precession within the wall, can be observed for both straight and curved geometries. The 3D configuration of the spin-wave modes is better evidenced in the sketch shown in panel C. Spin waves are confined in the transverse direction of the wall by the reduced local effective field arising from the inhomogeneous magnetization profile and propagate freely along the wall [14, 36]. Such modes, called Winter magnons [37], are characterized by the elliptical precession of the spins along the wall, with the major axis lying in the film plane, associated with a propagating flexural motion of the wall profile. Interestingly, such propagating modes are observed both in the straight and curved geometry, indicating that shaped DWs can be indeed used for confining and steering spin waves.

Figure 7.12D reports the simulated (empty dots) and experimental dispersion relation (filled dots) for spin waves confined at a straight 180° Néel DW. In these simulations, a sinc-shaped field pulse was used as excitation. The experimental values were extracted from experiments performed at different excitation frequencies.

Both simulations and experiments confirm the propagating character of the excitations, showing a positive dispersion and the presence of a small bandgap below 0.3 GHz, which can be ascribed to the residual effective field within the DW due to exchange bias and uniaxial anisotropies. The simulated (experimental) spin-wave group velocity $\mathbf{v}_g = \partial\omega/\partial\mathbf{k}$, extracted at $\mathbf{k} \to 0$, is $\mathbf{v}_{gSim} = 2.30 \pm 0.34$ km s^{-1} ($\mathbf{v}_{gExp} = 2.77 \pm 1.4$ km s^{-1}). Noteworthy, the demonstration of the propagating character and of the positive dispersion confirms that such guided modes (see Section 7.4.3) can be used for transporting information within integrated nanomagnonic circuits. Furthermore, interestingly, wall-bound modes are the only ones supported in this frequency range, because it corresponds to the bandgap of the propagating spin-wave modes (backward-volume in this case) of the uniform film. This guarantees the confinement of spin-wave excitations at the DW.

7.4.3 Experiments of Spin-Wave Guidance and Steering in Patterned Domain Walls

Spin waves are imaged stroboscopically via time-resolved STXM, in a pump-probe scheme, using a radiofrequency (RF) magnetic field

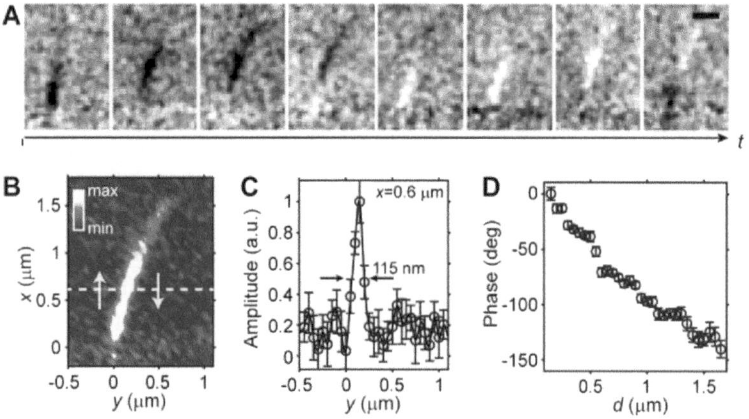

Figure 7.13 (A) Normalized time-resolved STXM images of the propagation of localized spin waves along a curved 180° Néel wall. (B) Spatial maps of the spin-wave amplitude, showing localization at the domain wall. The arrows indicate the direction of the magnetization within the domains. (C) Horizontal profile extracted from (B) in correspondence of the white dashed lines. The FWHM is 115 nm. (D) Phase shift as a function of distance from the antenna along the curved domain wall. Scale bar = 500 nm. Adapted from Ref. [15] (https://creativecommons.org/licenses/by/4.0/).

generated by the microstrip antenna as a pumping signal and the X-ray flashes generated by the synchrotron light source as a probing signal. In our experimental configuration, a phase resolution of about 15° in the time-resolved images was obtained. A pixel size between 40 and 75 nm was employed during acquisition.

In Fig. 7.13A, the propagation of spin waves along a curved DW is visualized by a sequence of STXM images. In particular, the normalized M_x contrast for the curved wall, calculated as the magnetic deviation $\Delta M_x(t)$ from the time-averaged state $\langle M_x(t)\rangle$, was acquired at different times within a single period of sinusoidal excitation. The excitation frequency was 1.11 GHz, and no static external magnetic field was applied. The normalized M_x contrast shows spin waves confined at the DW and propagating away from the antenna located at the bottom of the panels. The corresponding spatial map of the amplitude of the spin-wave excitation is reported in Fig. 7.13B. The map is obtained by fitting the time trace of each acquired pixel (raw data) with a sinusoidal function and plotting for each pixel the amplitude of the corresponding fit. The horizontal profile extracted from it at $x = 0.6$ μm from the stripline, shown in panel C, confirms that the mode is confined at the DW, with a lateral extension (full width at half maximum [FWHM]) of 115 nm. To demonstrate the propagating character of the spin waves, the time traces (sinusoidal fitting) related to pixels located within the DW at different distances from the antenna are plotted as a function of time (panel D). The linear phase shift with distance observed when moving away from the stripline along the wall, shown in panel D, clearly confirms the propagating character of the mode. Noteworthy, propagating distances exceeding 4 μm, combined with a lateral confinement <200 nm, were observed along curved walls, confirming the potential of shaped spin textures for spin-wave guidance.

7.4.4 Tunable Spin-Wave Circuits Based on Domain Walls

Figure 7.14A,B shows static STXM images of a spin texture comprising two DWs sharing a common apex (see also Fig. 7.11). By applying a small static magnetic field in the $-x$ direction, ranging from 2 to 1.68 mT, the distance between the two DWs (dashed white lines) can be controlled. In particular, in panel A, the two DWs are spatially separate. By decreasing the field, the two walls are brought closer and finally converge at the common apex (panel B). Since, as shown in the previous section, the spin waves are confined to the DWs, by displacing the walls we can spatially control the location of the spin-wave excitation with high accuracy.

To better visualize the progressive overlapping of the two modes, in panels C and D we show the amplitude of the spin-wave excitation along the horizontal profiles of panels A and B (dark dashed lines), in the two configurations. The two peaks, with an FWHM of around 200 nm, correspond to the two guided modes. For an applied magnetic field of 2 mT, the two modes are separated by 810 nm and do not overlap (panel C). By decreasing the field down to 1.68 mT, the two modes are brought closer and partially overlap, with a peak-to-peak distance of 340 nm (panel D). For better studying the properties of the confined spin-wave modes, in panels E and F we plot the time

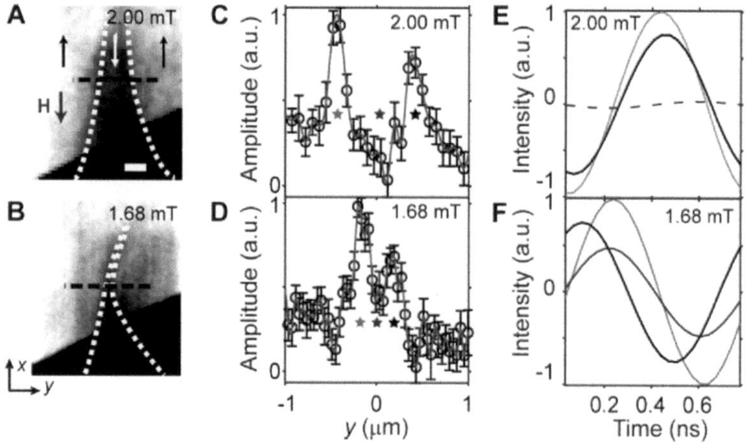

Figure 7.14 (A, B) STXM images of a spin texture consisting of two 180° Néel domain walls sharing a common apex for different external magnetic fields **H**. The arrows on the top indicate the direction of the magnetization within the domains. (C, D) Spin-wave excitation amplitude as a function of the distance along the horizontal profiles indicated by the dashed lines in panels A and B, respectively. (E, F) Time traces extracted in correspondence of the stars in panels C and D. For 2 mT the modes are spatially separate. By reducing the external field, a tunable spatial superposition of the two modes is achieved. Scale bar = 500 nm. Adapted from Ref. [15] (https://creativecommons.org/licenses/by/4.0/).

traces (sinusoidal fits) extracted in correspondence of the stars in panels C and D for different external fields.

The higher amplitude curves show the magnetization dynamics in correspondence of the maximum amplitude of the two guided modes. As expected, their phase difference depends on the applied field because of the modulation of the waveguide geometry and spin configuration. The lower amplitude curves show the dynamics in the region where the two modes overlap. For 2.00 mT (panel E), the two modes are spatially separate; therefore no excitation is measured (dashed line). On the contrary, for lower fields we observe a sizeable modulation of the excitation amplitude and phase in the overlap region, which arises from the tuning of the spatial superposition of the two guided modes. We anticipate that the control of the superposition, phase difference, and amplitude of the guided modes via external stimuli, such as fields or current, allows to envision the implementation of logic functions in spin texture–based devices, such as Mach–Zehnder-type spin-wave interferometers.

7.5 Conclusions and Perspectives

In this chapter, we presented our recent results on the development of tam-SPL for reversibly controlling the ferromagnetic spin texture with nanoscale precision. We discussed the different aspects regarding the technique and presented the first applications in the field of magnonics, which comprise the spatial modulation of spin-wave excitation and propagation in tailored magnetic domains and the development of magnonic crystals based on periodic magnetization structures and vectorial nanopatterning of the effective field. Then, we switched our focus toward anisotropic spin textures, demonstrating the channeling and steering of spin waves in DWs and the realization of a tunable nanoscale circuit allowing for the control of the superposition and interference of confined spin waves.

Since the rise of spin textures as active device components, plenty of theoretical and experimental work has been done for identifying the most promising systems and realizing proof-of-principle demonstrations. Nevertheless, the study of spin-wave dynamics in complex spin textures is still at its infancy and offers virtually limitless possibilities for engineering novel functionalities in next-generation spintronic and magnonic devices. One of the crucial limitations in this sense is the experimental stabilization of such spin structures. With tam-SPL we made a substantial step in this direction. However, a lot of work is still to be done for unveiling the full potential of engineered spin textures. In magnonics, one of the most interesting future directions is the study and control of spin waves in fully 3D spin textures. Furthermore, the study and control of nonreciprocal and topologically protected spin-wave modes is highly appealing for their potential applications. Finally, the study of spin textures in different magnetic systems such as ferrimagnets, antiferromagnets, and synthetic antiferromagnet films and structures represents an extremely promising avenue for obtaining a tailored magnonic response exploiting the full spin-wave spectrum up to the THz frequency range.

Acknowledgments

The authors would like to acknowledge the following contributions to the results reported in this chapter: Matteo Pancaldi and Paolo Vavassori for μ-MOKE measurements; Marco Madami and Silvia Tacchi for μ-BLS measurements; Raffaele Silvani for micromagnetic simulations of propagating spin waves; Simone Finizio, Sebastian Wintz, and Jörg Raabe for STXM measurements; Adam Papp, Gyorgy Csaba, and Wolfgang Porod for micromagnetic simulations of magnonic crystals; Giacomo Sala for sample preparation; William P. King, Martin Spieser, and Armin W. Knoll for providing thermal nanolithography probes; and Elisa Riedo for coordination, discussions, and the thermal nanolithography tool.

References

1. Parkin, S. S. P., Hayashi, M. and Thomas, L. (2008). Magnetic domain-wall racetrack memory, *Science*, **320**(5873), 190–194.
2. Fert, A., Cros, V. and Sampaio, J. (2013). Skyrmions on the track, *Nat. Nanotechnol.*, **8**(3), 152–156.
3. Kaka, S., Pufall, M. R., Rippard, W. H., Silva, T. J., Russek, S. E. and Katine, J. A. (2005). Mutual phase-locking of microwave spin torque nano-oscillators, *Nature*, **437**(7057), 389–392.
4. Manfrini, M., Kim, J.-V., Petit-Watelot, S., Van Roy, W., Lagae, L., Chappert, C. and Devolder, T. (2013). Propagation of magnetic vortices using nanocontacts as tunable attractors, *Nat. Nanotechnol.*, **9**(20), 121–125.
5. Choe, S.-B. (2004). Vortex core-driven magnetization dynamics, *Science*, **304**(5669), 420–422.
6. Pribiag, V. S., Krivorotov, I. N., Fuchs, G. D., Braganca, P. M., Ozatay, O., Sankey, J. C., Ralph, D. C. and Buhrman, R. A. (2007). Magnetic vortex oscillator driven by d.c. spin-polarized current, *Nat. Phys.*, **3**(7), 498–503.

7. Mistral, Q., van Kampen, M., Hrkac, G., Kim, J.-V., Devolder, T., Crozat, P., Chappert, C., Lagae, L. and Schrefl, T. (2008). Current-driven vortex oscillations in metallic nanocontacts, *Phys. Rev. Lett.*, **100**(25), 257201.
8. Duerr, G., Huber, R. and Grundler, D. (2012). Enhanced functionality in magnonics by domain walls and inhomogeneous spin configurations, *J. Phys.: Condens. Matter*, **24**(2), 24218.
9. Macke, S. and Goll, D. (2010). Transmission and reflection of spin waves in the presence of Néel walls, *J. Phys.: Conf. Ser.*, **200**(4), 42015.
10. Buijnsters, F. J., Ferreiros, Y., Fasolino, A. and Katsnelson, M. I. (2016). Chirality-dependent transmission of spin waves through domain walls, *Phys. Rev. Lett.*, **116**(14), 147204.
11. Hämäläinen, S. J., Madami, M., Qin, H., Gubbiotti, G. and van Dijken, S. (2018). Control of spin-wave transmission by a programmable domain wall, *Nat. Commun.*, **9**(1), 4853.
12. Park, J. P. and Crowell, P. A. (2005). Interactions of spin waves with a magnetic vortex, *Phys. Rev. Lett.*, **95**(16), 167201.
13. Kammerer, M., Weigand, M., Curcic, M., Noske, M., Sproll, M., Vansteenkiste, A., Van Waeyenberge, B., Stoll, H., Woltersdorf, G., Back, C. H. and Schuetz, G. (2011). Magnetic vortex core reversal by excitation of spin waves, *Nat. Commun.*, **2**(1), 279.
14. Wagner, K., Kákay, A., Schultheiss, K., Henschke, A., Sebastian, T. and Schultheiss, H. (2016). Magnetic domain walls as reconfigurable spin-wave nanochannels, *Nat. Nanotechnol.*, **11**(5), 432–436.
15. Albisetti, E., Petti, D., Sala, G., Silvani, R., Tacchi, S., Finizio, S., Wintz, S., Calò, A., Zheng, X., Raabe, J., Riedo, E. and Bertacco, R. (2018). Nanoscale spin-wave circuits based on engineered reconfigurable spin-textures, *Commun. Phys.*, **1**(1), 56.
16. Wintz, S., Tiberkevich, V., Weigand, M., Raabe, J., Lindner, J., Erbe, A., Slavin, A. and Fassbender, J. (2016). Magnetic vortex cores as tunable spin-wave emitters, *Nat. Nanotechnol.*, **11**(11), 948–953.
17. Holländer, R. B., Müller, C., Schmalz, J., Gerken, M. and McCord, J. (2018). Magnetic domain walls as broadband spin wave and elastic magnetisation wave emitters, *Sci. Rep.*, **8**(1), 13871.
18. Albisetti, E., Petti, D., Pancaldi, M., Madami, M., Tacchi, S., Curtis, J., King, W. P., Papp, A., Csaba, G., Porod, W., Vavassori, P., Riedo, E. and Bertacco, R. (2016). Nanopatterning reconfigurable magnetic landscapes via thermally assisted scanning probe lithography, *Nat. Nanotechnol.*, **11**(6), 545–551.

19. Albisetti, E. and Petti, D. (2016). Domain wall engineering through exchange bias, *J. Magn. Magn. Mater.*, **400**, 230–235.

20. Albisetti, E., Calò, A., Spieser, M., Knoll, A.W., Riedo, E. and Petti, D. (2018). Stabilization and control of topological magnetic solitons via magnetic nanopatterning of exchange bias systems, *Appl. Phys. Lett.*, **113**(16), 162401.

21. Albisetti, E., Petti, D., Madami, M., Tacchi, S., Vavassori, P., Riedo, E. and Bertacco, R. (2017). Nanopatterning spin-textures: a route to reconfigurable magnonics, *AIP Adv.*, **7**(5), 55601.

22. Schurig, D., Mock, J. J., Justice, B. J., Cummer, S. A., Pendry, J. B., Starr, A. F. and Smith, D. R. (2006). Metamaterial electromagnetic cloak at microwave frequencies, *Science*, **314**(5801), 977–980.

23. Temnov, V. V. (2012). Ultrafast acousto-magneto-plasmonics, *Nat. Photonics*, **6**(11), 728–736.

24. Lenk, B., Ulrichs, H., Garbs, F. and Münzenberg, M. (2011). The building blocks of magnonics, *Phys. Rep.*, **507**(4–5), 107–136.

25. Chumak, A. V., Vasyuchka, V. I., Serga, A. A. and Hillebrands, B. (2015). Magnon spintronics, *Nat. Phys.*, **11**(6), 453–461.

26. Ross, C. (2001). Patterned magnetic recording media, *Annu. Rev. Mater. Res.*, **31**(1), 203–235.

27. Chappert, C. (1998). Planar patterned magnetic media obtained by ion irradiation, *Science*, **280**(5371), 1919–1922.

28. Holzinger, D., Zingsem, N., Koch, I., Gaul, A., Fohler, M., Schmidt, C. and Ehresmann, A. (2013). Tailored domain wall charges by individually set in-plane magnetic domains for magnetic field landscape design, *J. Appl. Phys.*, **114**(1), 13908.

29. Garcia, R., Knoll, A. W. and Riedo, E. (2014). Advanced scanning probe lithography, *Nat. Nanotechnol.*, **9**(8), 577–587.

30. Sharma, P. P., Albisetti, E., Monticelli, M., Bertacco, R. and Petti, D. (2016). Exchange bias tuning for magnetoresistive sensors by inclusion of non-magnetic impurities, *Sensors*, **16**(7), 1030.

31. Nogués, J. and Schuller, I. K. (1999). Exchange bias, *J. Magn. Magn. Mater.*, **192**(2), 203–232.

32. Wei, Z., Wang, D., Kim, S.-Y. S. S.-Y., Kim, S.-Y. S. S.-Y., Hu, Y., Yakes, M. K., Laracuente, A. R., Dai, Z., Marder, S. R., Berger, C., King, W. P., de Heer, W. A., Sheehan, P. E. and Riedo, E. (2010). Nanoscale tunable reduction of graphene oxide for graphene electronics, *Science*, **328**(5984), 1373–1376.

33. Pires, D., Hedrick, J. L., De Silva, A., Frommer, J., Gotsmann, B., Wolf, H., Despont, M., Duerig, U. and Knoll, A. W. (2010). Nanoscale three-dimensional patterning of molecular resists by scanning probes, *Science*, **328**(5979), 732–735.

34. Ryu Cho, Y. K., Rawlings, C. D., Wolf, H., Spieser, M., Bisig, S., Reidt, S., Sousa, M., Khanal, S. R., Jacobs, T. D. B. and Knoll, A. W. (2017). Sub-10 nanometer feature size in silicon using thermal scanning probe lithography, *ACS Nano*, **11**(12), 11890–11897.

35. O'Grady, K., Fernandez-Outon, L. E. and Vallejo-Fernandez, G. (2010). A new paradigm for exchange bias in polycrystalline thin films, *J. Magn. Magn. Mater.*, **322**(8), 883–899.

36. Garcia-Sanchez, F., Borys, P., Soucaille, R., Adam, J.-P., Stamps, R. L. and Kim, J.-V. (2015). Narrow magnonic waveguides based on domain walls, *Phys. Rev. Lett.*, **114**(24), 247206.

37. Winter, J. M. (1961). Bloch wall excitation. Application to nuclear resonance in a Bloch Wall, *Phys. Rev.*, **124**(2), 452–459.

Chapter 8

Spin Textures as Sources for Magnons with Short Wavelengths and 3D Mode Profiles

Volker Sluka[a] and Sebastian Wintz[b]
[a] *Helmholtz-Zentrum Dresden-Rossendorf, 01328 Dresden, Germany*
[b] *Paul Scherrer Institut, 5232 Villigen PSI, Switzerland*
v.sluka@hzdr.de, sebastian.wintz@psi.ch

8.1 Introduction

In this chapter, we will give an overview of contemporary methods for spin-wave excitation and propagation. We will focus on the exploitation of spin textures for the generation and propagation of spin waves with very short wavelengths as well as on the 3D nature of such excited waves.

8.2 Spin-Wave Excitation and Propagation

8.2.1 Spin Waves

Spin waves are collective excitations of spin systems, with the magnon being their fundamental quantum of excitation [7, 29, 47]. Therefore spin waves are also referred to as magnons.

Three-Dimensional Magnonics: Layered, Micro- and Nanostructures
Edited by Gianluca Gubbiotti

ISBN 978-981-4800-73-0 (Hardcover), 978-0-429-29915-5 (eBook)
www.jennystanford.com

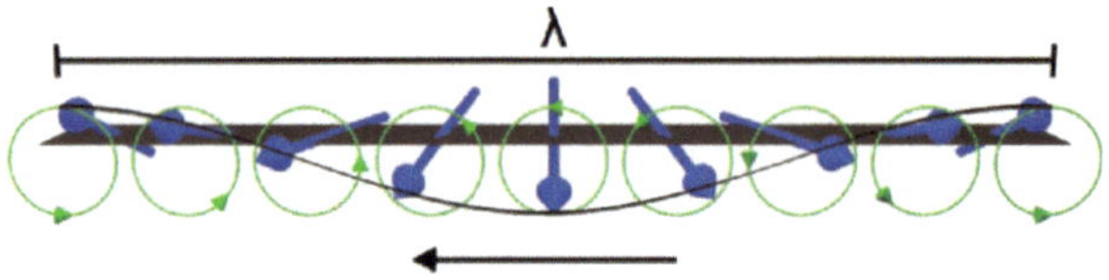

Figure 8.1 Schematic representation of a spin wave. Magnetic moments (blue arrows) precess with a lateral phase shift, as indicated by the green arrows. Thereby a lateral wave of the wavelength λ is formed, which propagates in the direction of the black arrow to the left.

In Fig. 8.1 a spin wave is illustrated schematically. Magnetic moments precess around their equilibrium position with a harmonic spatial phase shift. This spatial phase shift defines the spin-wave wavelength λ and the reciprocal wavenumber k is given by $k = 2\pi/\lambda$. The precession of the magnetic moments with frequency f causes the spin wave to propagate in space with a corresponding time period $T = 1/f$. In terms of the angular frequency $\omega = 2\pi f$, the phase velocity of the spin wave is $v_{ph} = \omega/k$, while the group velocity reads $v_{gr} = \mathrm{d}w/\mathrm{d}k$. The functional relation between the frequency and the wavenumber of a spin wave is the spin-wave dispersion relation $f(k)$.

In this section we will consider a mean field situation of a ferromagnet with a spatially homogeneous magnetization $\mathbf{M}$ the modulus of which (saturation magnetization M_s) is furthermore constant in time. The most relevant magnetic energy contributions here are exchange energy, magnetostatic energy, anisotropy energy, and Zeeman (external field) energy. While the latter two are local values, the first two have to be treated as self-interactions of the ferromagnet. The exchange interaction is typically very strong on short-range scales (nearest atomic neighbor), whereas the magnetostatic (dipole–dipole) interaction is relevant rather for long-range scales. Therefore, depending on the wavelength of a spin wave, different interactions are dominant for determining their dispersion relations, that is, one distinguishes between magnetostatic and exchange spin waves with the intermediate range being referred to as dipole exchange waves.

Spin waves occur with wavelengths from macroscopic down to atomic scales and their frequencies range up to the THz regime. Their dispersion relation not only depends on the magnetic properties of the hosting medium, such as exchange constant (A), saturation magnetization, and magnetic anisotropy but also particularly on the relative orientation of **M** and the wave vector $\mathbf{k} = k\mathbf{e}_\mathbf{k}$, with $\mathbf{e}_\mathbf{k}$ being the propagation direction vector of unity. In a 3D bulk system one can therefore distinguish two principal propagation geometries, that is, collinear or perpendicular to **M**, with all the intermediate directions being a superposition of the two principal ones. For the situation of a 2D ferromagnetic film, the film plane provides an additional symmetry breaking, separating the cases of in-plane and perpendicular magnetization, thereby leading to three basic propagation geometries with distinct dispersion relations. For **M** being perpendicular to the film plane, **k** is always perpendicular to **M** and the corresponding waves are called forward-volume waves. For the case of in-plane magnetization there is further a differentiation between backward-volume waves ($\mathbf{k} \updownarrow \mathbf{M}$) and so-called Damon–Eshbach waves ($\mathbf{k} \perp \mathbf{M}$) [18]. Again, intermediate cases can be seen as a superposition of these three different basic geometries. In addition to the continuous spectra of laterally propagating waves, there can be discrete exchange-dominated perpendicular standing waves along the thickness profile of the film with propagating spin-wave branches at higher frequencies [27, 35, 48]. However, the vast majority of existing works on spin waves in thin films address the case a quasi-uniform spin-wave amplitude distribution over the film thickness, which means such waves can be essentially considered 2D.

From a fundamental point of view, spin waves have been studied already for more than 100 years. More recently they have been also proposed for being used in applications [15, 52], giving rise to the field of magnonics [16, 50]. In particular spin waves are seen as high-potential signal carriers for future spintronic information and communication devices. This is mainly for two reasons. On the one hand, spin waves are not necessarily associated with the motion of charges, which means that ohmic losses can be omitted in the signal transfer, potentially allowing for reduced power consumption of spin-wave devices compared to standard complementary metal-oxide semiconductor (CMOS) technology. On the other hand, device miniaturization could strongly benefit from spin-wave technology

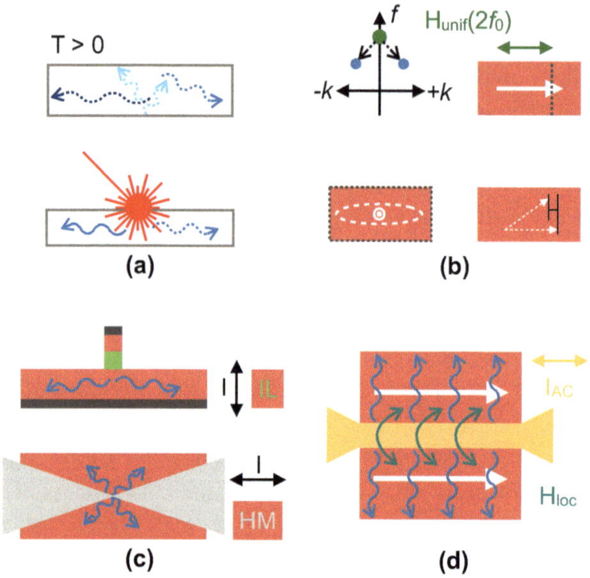

Figure 8.2 Schematic representation of spin-wave excitation mechanisms. (a) Top: Incoherent thermal excitation of magnons (blue arrows). Bottom: Optically driven coherent and incoherent magnon excitation. (b) Parametric excitation of magnons ($+k$ and $-k$) of frequency f by a spatially uniform magnetic field H of double frequency $2f$. The field (green arrow) and equilibrium magnetization (M_{eq}, white arow) are collinear. In the continuum model, the field couples into the difference of the projection of M onto M_{eq} during the elliptical precession. (c) Top: Magnon excitation by spin-transfer torques stemming from an electric (spin-poarized) current flowing through a nanocontact trilayer stack, consisting of two ferromagnetic layers (red) that are separated by a nonferromagnetic interlayer (green). Bottom: Spin–orbit torque excitation of magnons in a ferromagnetic layer interfaced with a heavy metal constriction on top. A current flowing in the heavy metal acquires transversal spin polarization via the spin–orbit interaction, which then provides a torque to the ferromagnet. (d) Excitation of magnons by inhomogeneous/local magnetic fields originating from an alternating current flowing in patterned metallic conductors, here a metal strip (yellow).

for cases where the wavelength sets a limit to the device's footprint (e.g., interference based applications), as spin waves exhibit orders of magnitudes shorter wavelengths than electromagnetic waves at GHz frequencies [38, 74].

Yet for enabling the implementation of competitive magnonic devices, novel concepts for the creation, propagation, manipulation, and detection of spin waves need to be developed, while also the underlying physics has to be addressed. One important aspect of this effort relates to spin waves with nanoscale or sub-μm wavelengths, which are highly desirable in terms of magnonic applications.

In this chapter we will review the state of the art of spin-wave generation and propagation methods in thin films, with the focus being on short-wavelength spin waves and their significant 3D properties. For that purpose we will at first address a variety of existing spin-wave generation and propagation methods in general in the remaining part of this section (8.2). In the following two sections we will provide a more detailed review of recent works reporting short-wavelength excitation (8.3) and directional propagation (8.4) in vortex spin textures, involving significant 3D spin-wave characteristics.

8.2.2 Spin-Wave Excitation

In this section, different mechanisms and methods for the excitation of spin waves will be discussed, as schematically represented in Figs. 8.2 and 8.3.

Figure 8.2a (top) shows a mechanism that is present at finite temperatures of the spin system, namely thermal excitation of magnons. Depending on the temperature, a certain distribution $(f, \mathbf{k})$ of magnons is excited. However, the excited magnons are fully incoherent and therefore typically not useful for most kinds of applications. Illuminating a ferromagnetic thin film by a laser, on the other hand, can excite fractions of both coherent and incoherent magnons (Fig. 8.2a, bottom).

Figure 8.2b shows a spin-wave excitation method that is referred to as parametric pumping [26]. Here a microwave photon corresponding to a spatially uniform, alternating magnetic field $\mathbf{H}$ of double frequency $2f$ upon annihilation creates two magnons with opposing k-values of frequency f. The process relies on the ellipticity of precession in a thin film where different projections of **M** along the equilibrium orientation $\mathbf{M}_{\text{eq}}$ occur, depending on whether the precession state is on either of its two in-plane or perpendicular turning points. Note that for the process to set-on, thermal activation is necessary as otherwise no torques were applied to the magnetization because $\mathbf{M}_{\text{eq}}$ and **H** are collinear. Second this process requires typically a certain threshold value of the magnitude of the magnetic field. While in principle there is no restriction for the wavelengths that can be excited by parametric pumping, the spatial control over the emission is rather limited for this method.

A method for a more controlled excitation of coherent spin waves is by exploiting the spatially inhomogeneous Oersted magnetic fields from an alternating current flowing in a metallic conductor patterned adjacent to the magnetic film (see Fig. 8.2d). Here the excitation mechanism is linear in frequency and amplitude, and it has proven very efficient for exciting spin waves of moderate wavelengths. Yet for generating nanoscale wavelengths, this method suffers from two severe difficulties: As a consequence of the spatial magnetic field distribution it is very inefficient for exciting wavelengths that are smaller than the width of the metallic conductor. And, while it is nowadays possible to pattern structures on the sub-100 nm scale, it is on the other hand very problematic to impedance-match these structures to achieve a sufficient transmission of high-frequency microwaves.

Another method for the local excitation of coherent spin waves is to exploit the effects from spin-polarized currents [6, 66, 67, 80] (see Fig. 8.2c). These can be spin-transfer torques in a perpendicular sandwich stack of two ferromagnetic layers and a nonferromagnetic interlayer (Fig. 8.2c, top). An alternating electric current passing through a nanocontact of such a stack (only one ferromagnetic layer is extended) may acquire spin polarization, which in turn exerts a local torque on the magnetization of the extended layer. This local torque then causes the excitation of coherent spin waves [8, 9, 24, 25, 45, 55, 56, 65, 81]. A similar effect can be achieved in a bilayer stack of an extended ferromagnetic film and a patterned heavy metal film on top laid out as a lateral nanoconstriction (Fig. 8.2c, bottom). Here a lateral alternating current flowing mostly in the heavy metal layer acquires spin polarization locally as a consequence of the spin–orbit interactions in the heavy metal material. This spin-polarized current then acts on the ferromagnetic film in the interface region, having the highest torques at the nanoconstriction (highest current densities) and therefore leading to spin-wave generation. Interestingly these effects do not only occur for alternating currents but may also arise for direct currents as spin-wave autoemission processes. While spin-transfer/spin–orbit torque excitation of spin waves has been proven very useful in many cases, similar restrictions apply as for the patterned field antennas discussed above, namely a limit for the smallest wavelength excitable following from the patterning sizes and impedance-matching issues.

Further methods to excite spin waves involve, for example, magnetoelastic coupling and short wavelengths, can, for example, also be achieved postexcitation by adapting the dispersion relation locally in the propagation path of the waves [23, 75].

A different route for the excitation of spin waves is to convert external, uniformly alternating magnetic fields into inhomogeneous local fields and hence spin waves by exploiting the internal magnetic fields of the sample itself. One way to do so is by making use of the Schlömann mechanism [62]. This mechanism relies on a variation of the static or dynamic effective fields over the sample's volume or area in the case of a thin film. A consequence from this variation is that the ferromagnetic resonance frequency and the spin-wave dispersion relation becomes a local property. Typically the uniform external field can then couple directly to a sample region, where the ferromagnetic resonance condition is fulfilled and the resulting spin excitation may enter and propagate through adjacent regions where the corresponding wavelength is already finite according to the local spin-wave dispersion. An example of such a spin-wave excitation based on the Schlömann mechanism is shown in Fig. 8.3a. Here the dynamic dipolar fields from an originally nonresonant-driven precession at the edges of a rectangular patterned thin film change the magnonic landscape locally. Thereby spin waves can be efficiently excited at the sample edge [58] or other internal field variations in general [20–22].

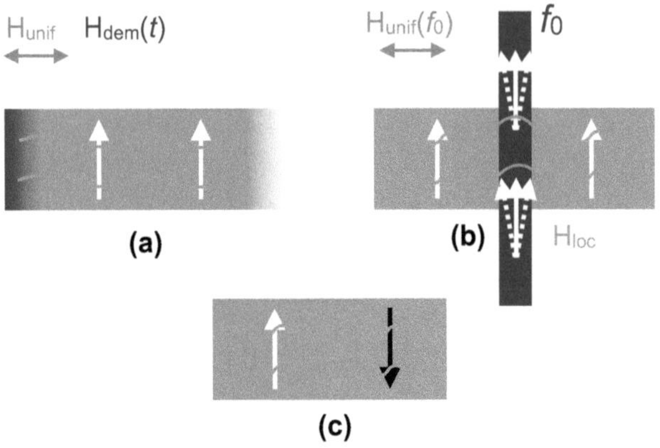

Figure 8.3 Schematic representation of spin-wave excitation mechanisms. (a) Example of the Schlömann mechanism [62] converting a spatially homogeneous external magnetic field via static or dynamic local demagnetization fields into spin waves. Here the dynamic demagnetization fields arise from the driven precession at the sample edge [58]. (b) Example of a system with a Fano resonance [30] where a discrete resonance interacts with a frequency-continuous mode (e.g., a spin-wave dispersion relation $f(k)$). Here the discrete resonance corresponds to the precessional eigenfrequency (f_0) of a magnetic wire [4], which by stray field interaction causes the excitation of spin waves in an adjacent film. (c) Example of spin textures as naturally formed local antennas for spin waves; in this case a 180° magnetic domain wall acts as a source for spin waves. Such excitation can be originating from the Schlömann mechanism [92] or from Fano resonances [71] or from both at the same time.

The second prominent mechanism for converting uniform external magnetic fields into finite-wavelength spin waves relies on Fano resonances [30]. In general a Fano resonance describes the situation where a discrete resonance is interacting with a continuous dynamic mode. In terms of magnonics this could be a situation where the external field excites a discrete local magnetic resonance of the sample, which then hybridizes with the continuous spin-wave dispersion relation, causing spin-wave excitation. An example for Fano resonance spin-wave excitation is schematically shown in Fig. 8.3b. The external field excites the precessional eigenfrequency of a ferromagnetic wire on top of a ferromagnetic film. The dynamic dipolar fields from this precession in turn interact with the film, locally exciting spin waves [4]. The so-called magnonic grating coupler is another example for a Fano resonance [101, 102].

A third mechanism relying on the action of internal magnetic fields is based on the use of spin textures such as domain walls, vortices, or skyrmions. In the example shown in Fig. 8.3c a magnetic domain wall acts as an emitter for spin waves [41, 43, 46, 57, 72, 88, 93]. Such an effect can either be dominated by the Schlömann mechanism [92] or a Fano resonance [71] or even a mixture of the two.

8.2.3 Spin-Wave Propagation

As mentioned earlier, the directional propagation and controlled waveguiding of spin waves plays an important role in realizing magnonic devices. In this section, examples of the state of the art of spin-wave waveguiding will be reviewed, as also schematically shown in Fig. 8.4.

Figure 8.4a shows a situation where directional control of spin-wave propagation is achieved by patterning elements out of a magnetic film [40]. Here, rectangles are arranged in a chain with the major axis of the element (and thus M_{eq}) being perpendicular to the path of the chain. The elements are dipolarly coupled to each other, thereby allowing for spin-wave propagation (apart from certain bandgaps) in this effective Damon–Eshbach geometry. If the path of the chain is now fabricated to turn a corner, the spin waves can follow this path to a certain extent, while remaining in the Damon–Eshbach geometry [78].

The second example is that of the realization of a magnonic multiplexer (Fig. 8.4b) [85]. Here, in a patterned y-shaped structure the initial magnonic signal can be routed into either of the y-branches. The routing can be controlled by the path of a direct current flowing in a conductor beneath the ferromagnetic structure along either y-branch. The reason for the selective spin-wave propagation lies in the action of the Oersted field of the current: Wherever the current is flowing, the magnetization aligns perpendicular to the magnetic wire, thereby realizing the Damon–Eshbach geometry. In contrast to that, **M** aligns collinear to the wire in the branch where no current is flowing and the spin waves

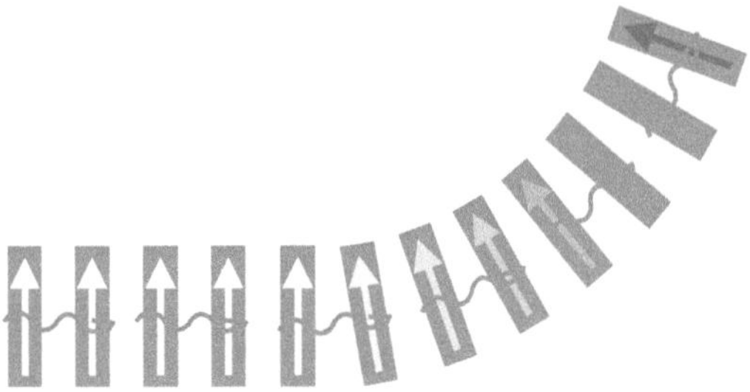

(a)

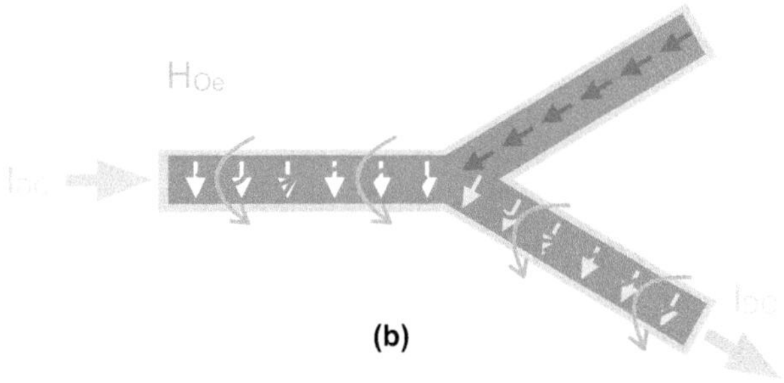

(b)

(c)

Figure 8.4 Schematic representation of exemplary spin-wave guiding mechanisms. A chain of slightly tilted rectangular thin-film elements with equilibrium magnetization ($\mathbf{M_{eq}}$, solid arrows) along their major axis, allowing to guide spin waves (wiggly arrows) around the corner, while almost keeping the Damon–Eshbach geometry ($\mathbf{M_{eq}} \perp \mathbf{k}$). (b) Spin-wave multiplexer [85]: Spin waves can be selectively guided into either branch of a Y-structure, depending on along which branch a direct current is flowing in a patterned metallic conductor below the ferromagnetic layer. The Oersted field from the current locally enforces the Damon–Eshbach geometry, while M_{eq} aligns parallel to the wire where there is no such current. (c) Domain walls as natural waveguides for spin waves. The local effective fields in domain walls can confine spin waves to the region of the walls [32] and the spatial distribution of such waveguide can be engineered [2].

are blocked due to dispersion mismatch and/or the noncollinear magnetization transition. Using a direct current, spin waves can also be guided around a simple corner in a similar way [86]. While this method was shown to work very well for spin-wave routing, concerns may arise about the electric power consumption in an efficiency-sensitive framework. Other examples of spin-wave routing are by making use of antenna emission characteristics [37] or by varying the magnonic index of a film [19].

Figure 8.4c shows an example where naturally formed and reconfigurable spin textures are used as magnonic waveguides, here a 1D 180° magnetic domain wall [94]. While ferromagnetic films exhibit a magnonic frequency gap, depending, for example, on their anisotropy, this frequency gap may vanish in corresponding domain walls, thus leading to a situation where spin waves are naturally confined to a domain wall that can be straight or arbitrary shaped in its path [3, 32, 71, 72, 76, 89, 100], even in a controlled way [1, 2].

8.3 Spin Vortex Cores for Short-Wavelength Magnon Excitation

8.3.1 Introduction

In this section, a detailed review will be given on a recently reported spin-wave excitation mechanism using the driven oscillation of spin vortex cores [5, 27, 71, 72, 99], with the focus being on Ref. [99] where also the full details of the experiments can be found. This method falls into the category of Fano resonances described before, applied to vortex spin textures with the potential to generate coherent spin waves with very short wavelengths.

A spin vortex is a nontrivial topological spin distribution [79], where the magnetization arranges in a planary flux-closing way in order to minimize the magnetodipolar energy [49]. At the center of this vortex, driven by the exchange interaction the magnetization rotates out of the plane, forming a perpendicular vortex core with a size of the order of 10 nm diameter only [31, 64]. For both, the in-plane circulation sense (clockwise or counterclockwise) and the core polarity (up or down), two configuration states are possible and they are typically degenerated in energy, that is, equally distributed. Spin vortices are found as the ground state of thin-film microelements, but they also exist in more complex domain patterns of extended film geometries.

From a magnetodynamic perspective, spin vortices exhibit a variety of eigenmodes. The lowest-frequency gyration mode corresponds to an orbiting of the vortex core around its equilibrium position [13, 77, 82, 87]. At higher frequencies, confined radial and azimuthal spin-wave eigenmodes exist [11, 54, 59, 84], as well as higher-order gyrotropic dynamics [28]. Furthermore, the dynamic switching of the vortex core polarity may occur as a nonlinear process, accompanied by the emission of incoherent spin waves [14, 44, 53].

8.3.2 System Description

For the situation of a ferromagnet/nonferromagnetic/ferromagnetic trilayer, different relative vortex pair configurations are possible with respect to vorticity and polarity [10, 68–70, 95–98]. Here a Co(47.8)/Ru(0.8)/$Ni_{81}Fe_{19}$(42.8) (nm) trilayer is considered, as shown in Fig. 8.5a, where the Ru interlayer mediates an antiferromagnetic interlayer exchange interaction [36]. The therefore expected vortex pair ground state is that of opposite vorticities (interlayer exchange) and parallel polarities (dipolar interaction overcoming interlayer exchange), as schematically shown in Fig. 8.5b, for a square-shaped element. Note that instead of a continuous vorticity exhibited in circular elements, square-shaped elements exhibit a discrete four-domain vortex pattern, yet from a topological point of view both types of vortices are identical.

To confine vortex pairs, samples were patterned into microelements, here, for example, squares with 4 μm base size. Scanning transmission X-ray microscopy (STXM) [60] was performed to directly image the magnetic state of the samples in a layer-specific way with a lateral resolution of about 25 nm. Results from these measurements are shown in Fig. 8.5c: top line (NiFe), bottom line (Co), left column (perpendicular contrast), and right column (in-plane contrast). By means of these images it gets clear that there is

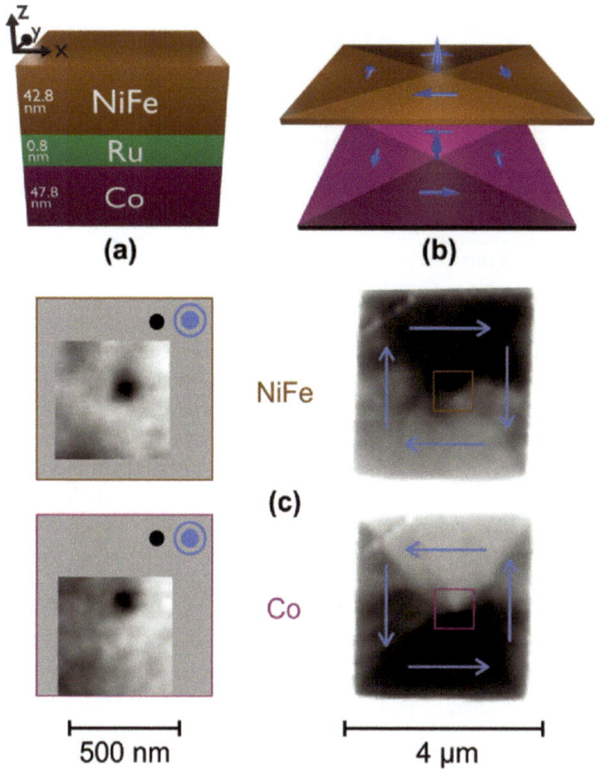

Figure 8.5 Spin vortex pairs in magnetic multilayers. (a) Schematic of a trilayer stack with two ferromagnetic layers (Co and NiFe) and a nonferromagnetic interlayer (Ru) mediating antiferromagnetic interlayer exchange coupling. (b) Schematic of a corresponding square-shaped vertical vortex pair with opposite in-plane vorticities and parallel vortex cores. (c) STXM images: NiFe layer (top) and Co layer (bottom). Left: Parallel vortex cores as black dots. Right: Opposite in-plane vorticities as indicated by blue arrows.

each a vortex in both ferromagnetic layers and that their vorticities are opposite, while their core polarities remain parallel.

8.3.3 Spin-Wave Excitation

When such kinds of samples are exposed to a uniformly alternating magnetic field, short-wavelength spin waves are excited by the

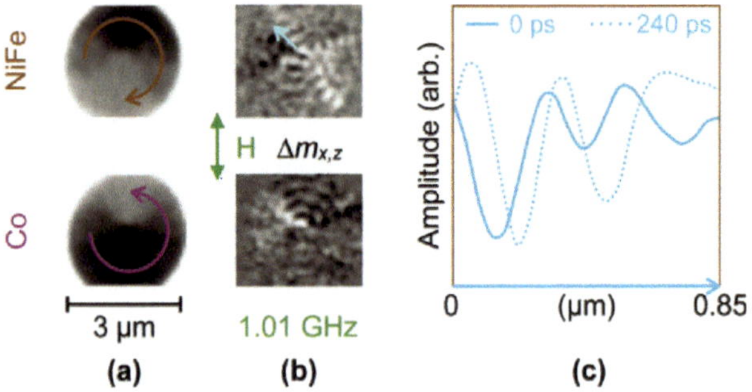

Figure 8.6 Spin-wave emission from vortex cores. (a) STXM images of a vortex pair in a disk structure showing opposite vorticities: top NiFe and bottom Co. (b) Corresponding TR-STXM snapshot of the emitted wave pattern upon excitation by a 1.01 GHz alternating magnetic field, showing the magnetic deviation of both m_x and m_z with respect to the time-average state. (c) Line profile along the blue arrow in (b), for two different time steps revealing the outward-propagating nature of the spin waves recorded.

driven gyration of the vortex cores (see Fig. 8.6), as measured by time-resolved STXM (TR-STXM). Here a 3μm circular disk of the same trilayer, as discussed above, is considered, containing a vortex pair, as shown in Fig. 8.6a. Note that here the vorticities are continuous, as expected. The response of the vortex pair to a field excitation of 1.01 GHz frequency and $\mu_0 H \approx 1$ mT in amplitude is shown in Fig. 8.6b, μ_0 being the vacuum permeability. These images show a snapshot in time with normalized magnetic contrast, that is, the combined magnetic deviation in x, z at a certain time compared to the time average state: top (NiFe) and bottom (Co). A circular spin-wave emission pattern is clearly visible, and the waves excited at the cores are propagating outward to the rim of the disk. This outward-propagating nature is further illustrated by means of line profiles, as indicated for two different time steps of 240 ps delay, as shown in Fig. 8.6c.

The underlying excitation process is that of a Fano resonance [30]. The driven gyration of the coupled vortex core mode [12, 39, 42] excites spin waves via a local interaction with the continuous spin-wave dispersion [27].

It was further shown that the spin-wave excitation mechanism observed can be tuned by varying the driving frequency in an

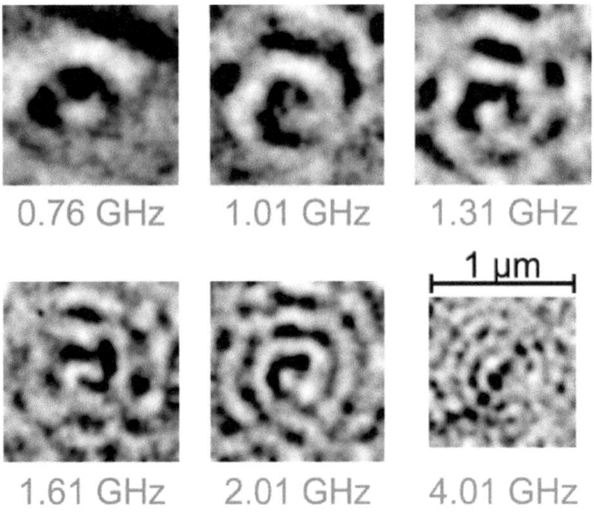

Figure 8.7 Tunability of vortex core spin-wave emission. TR-STXM snapshot images of the sample shown in Fig. 8.5 upon different excitation frequencies, as indicated (perpendicular magnetic deviation Δm_z of the NiFe layer). A clear decrease of the resulting wavelength can be observed with increasing excitation frequency.

experimentally verified range from 750 MHz to 4 GHz. Figure 8.7 shows the response of the center region of the sample from Fig. 8.5 for six different excitation frequencies as snapshots of the NiFe layer in normalized view (Δm_z). Thereby it was found that the resulting wavelength scales with the driving frequency, that is, longer waves are excited at lower frequencies and shorter waves are excited at higher frequencies.

8.3.4 Spin-Wave Dispersion Relation

It is noticeable that the spin waves observed exhibit relatively short wavelengths compared to waves in single-layer films. This result was verified by comparing the experiment to both micromagnetic simulations [61] and an analytic theory. Figure 8.8 shows the experimentally measured dispersion relation $f(k)$ together with the results from micromagnetic simulations revealing very good agreement with each other. The micromagnetic simulations were

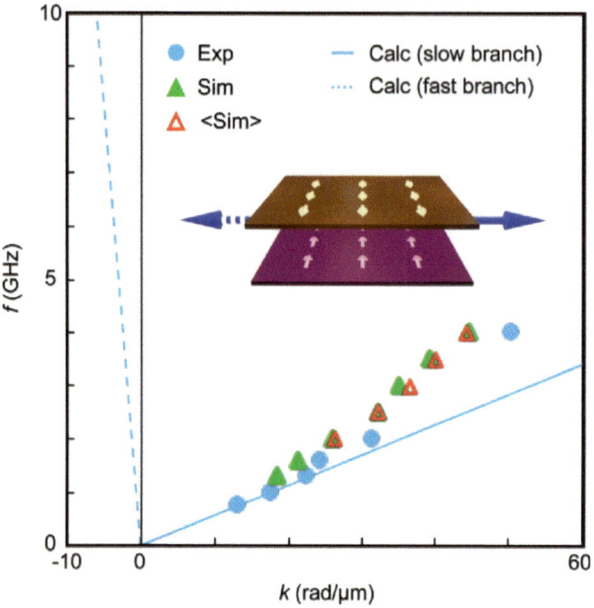

Figure 8.8 Spin-wave dispersion relation $f(k)$ for the antiparallel magnetic bilayer in the collective Damon–Eshbach geometry (as sketched). Blue solid dots: Experimental data points. Green solid triangles: Micromagnetic simulations. Red hollow triangles: Micromagnetic simulations of identical layers with mean magnetic properties. Blue solid line: Analytic calculation for the slow spin-wave branch $(+k)$. Blue dashed line: Analytic calculation for the fast spin-wave branch $(-k)$, the two of which being highly nonreciprocal.

separately performed for the vortex pair in the trilayer stack as measured as well as for a symmetric stack with averaged magnetic properties, essentially proving that both situations match equally well with the experimental data. In addition, the spin-wave dispersion for a corresponding symmetric antiparallel bilayer continuous film was analytically calculated for the layer-collective Damon–Eshbach geometry. Also this calculated dispersion, in the approximation of relatively long wavelengths and thin interlayers, is in agreement with the experiment to a wide extend. Furthermore the calculations revealed that the dispersion in this system is strongly nonreciprocal, that is, the wavelengths of spin waves of identical

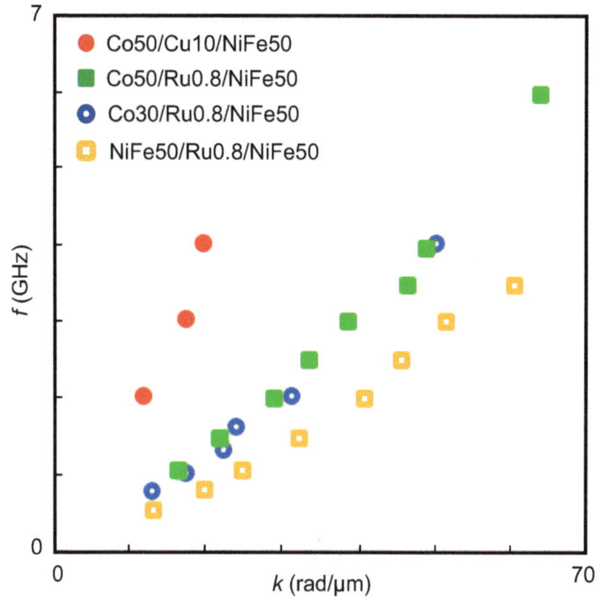

Figure 8.9 Antiparallel bilayer spin-wave dispersion in the collective Damon–Eshbach geometry for different multilayers, as measured in the experiment. Green solid squares: Co50/Ru0.8/NiFe50 (nominal nm) stack. Blue hollow circles: Co30/Ru0.8/NiFe50 (nominal nm) stack. Red solid circles: Co50/Cu10/NiFe50 (nominal nm) stack. Yellow hollow squares: NiFe50/Ru0.8/NiFe50 (nominal nm) stack. A higher thickness of the interlayer leads to a steeper dispersion relation, while a lower average saturation magnetization leads to a more gradual dispersion.

frequencies propagating into opposite directions are different by a factor of about 10. This nonreciprocity is a consequence of the different interlayer dynamic field interaction or, more abstractly, of the chiral nature of the two magnetization vectors and the direction of propagation, as schematically shown. In the experiment, yet only the slow (and therefore short-wavelength) spin-wave branch was observed, which is because the expected amplitudes for the fast branch are relatively low and the wavelength is easily exceeding the size of the whole disk structure.

The influence of the magnetic material parameters and thicknesses on the antiparallel bilayer collective Damon–Eshbach spin-wave dispersion can be seen in Fig. 8.9. Here the experimentally measured dispersion data points in vortex pairs are displayed for four different trilayer stacks, as defined in the figure by their nominal thicknesses. Note that the nominal Co50/Ru0.8/NiFe50 stack corresponds to the one discussed above. All four spin-wave dispersion relations are approximately linear in the given frequency range. In agreement with analytic predictions the slope of the dispersion, and thus the group and phase velocities depend on the interlayer thickness. The trilayer comprising a 10-nm-thick Cu interlayer exhibits a significantly higher slope in its dispersion than the same stack with a Ru 0.8 nm interlayer. A thickness asymmetry of the ferromagnetic layers, on the other hand, does not strongly influence the spin-wave dispersion relation as long as the thicknesses are of the same order of magnitude. This can be seen by means of comparing the dispersions of the nominal Co30/Ru0.8/NiFe50 (nm) and Co50/Ru0.8/NiFe50 (nm) stacks being almost identical. Besides the interlayer thickness also the saturation magnetization influences the spin-wave dispersion. Figure 8.9 also displays the data points for a NiFe50/Ru0.8/NiFe50 stack (lower average M_S), resembling a smaller dispersion slope than that of the Co50/Ru0.8/NiFe50 (nm) stack (higher average M_S).

8.3.5 Spin-Wave Nonreciprocity

With respect to the analytically calculated spin-wave nonreciprocity discussed along Fig. 8.8, it was not possible to directly observe spin waves belonging to the fast dispersion branch experimentally. Nevertheless it was possible to indirectly confirm the spin-wave nonreciprocity by measuring a vortex pair with reversed vorticities compared to the cases discussed above. Figure 8.10a shows a static STXM image with magnetic in-plane contrast of such a reversed vortex pair in a Co/Ru/NiFe disk. Figure 8.10b shows the dynamic response at 2.01 GHz excitation frequency as absolute and normalized TR-STXM snapshots. Here the spin waves of the detectable slow spin-wave branch are now propagating inward from the disk edge to the center. This inward propagating nature can be visualized by means of the line profile shown in Fig. 8.10c, taken at two different delay times along the inward-pointing arrow in Fig. 8.10b. The slow branch is now propagating inward because

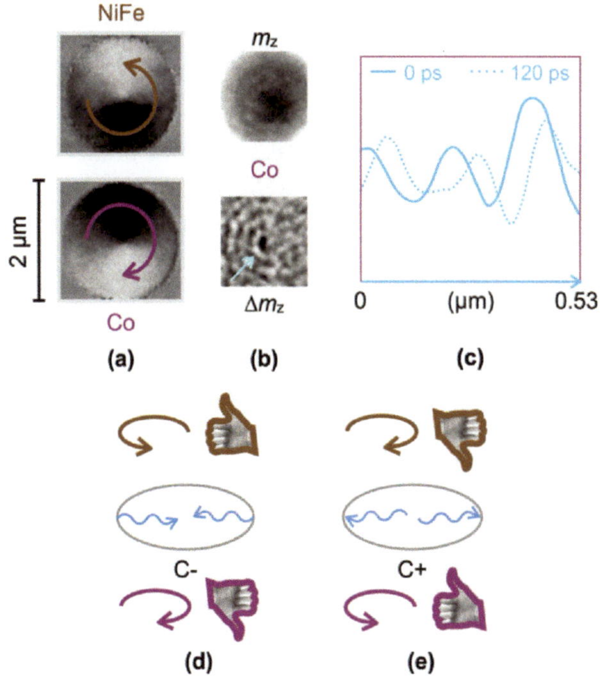

Figure 8.10 Nonreciprocal spin-wave dispersion in antiparallel magnetic bilayers. (a) STXM image of a spin vortex with antiparallel in-plane vorticities, opposite to the previous cases. Top: NiFe; bottom: Co. (b) TR-STXM snapshot of the spin-wave emission pattern in the Co layer. Top: absolute perpendicular contrast; bottom: normalized perpendicular contrast. (c) Line profiles along the blue arrow at different time steps revealing the inward-propagating nature of the spin waves. (d and e) Schematics of the two different relative handednesses (C- and C+) of vortex pairs with opposite vorticities and the resulting direction of propagation for the slow branch of collective Damon–Eshbach spin waves.

this direction corresponds to the configuration for which the solid dispersion line was calculated in Fig. 8.8, whereas for the earlier-discussed vortex pairs this solid line corresponded to outward-propagating waves. This effect can be understood by attributing a handedness to the vortex pair system with opposite vorticities. As shown in the bottom panels of Fig. 8.10 both a $C-$ (d) and a $C+$ (e) handedness can be defined in such way. For $C-$ the observable slow nonreciprocal spin-wave branch is propagating inward, while it is outward propagating for the $C+$ case. Note that even for the situation that the slow branch is inward propagating, the original wave excitation occurs in the core region at the center. As verified by micromagnetic simulations, in this case spin waves are propagating outward via the fast nonreciprocal spin-wave branch to the edge of the structure, where they get reflected and then subsequently propagate inward via the observable slow spin-wave branch.

8.3.6 3D Nature of Spin Waves

As it was mentioned already in the Introduction section, for most experiments so far spin waves of moderate wavelengths can essentially be considered a 2D phenomenon in thin films, that is, their amplitude and phase distribution is quasi-uniform over the film thickness profile. This is different to the 3D bulk sample case, for which Damon–Eshbach spin waves were initially considered and where they exhibit a surface wave character [18]. On the other hand it could also be expected that a relevant 3D profile reoccurs for Damon–Eshbach waves in thin films when the wavelength starts to approach the length scale of the film thickness, as schematically shown in Fig. 8.11a, or for the situation of precessional nodes along the film's normal direction [27, 35, 48]. By means of Fig. 8.11a the nonreciprocal character of Damon–Eshbach wave propagation becomes obvious. The spin-wave amplitude decays from each surface exponentially, and the propagation at that surface occurs in the direction to follow a right-handed sense with respect to the equilibrium magnetization.

For the spin waves observed in the experiment, it was found that 3D effects clearly need to be taken into account. At first it is important that the propagating waves in the antiparallel magnetic bilayer resemble a collective Damon–Eshbach geometry, that is, the lateral wave characteristics are present in both layers at all times. There are in principle two different coupling modes, namely the acoustic and the optical one, as it will be further discussed in Section 8.4. Since the magnetization precession is right handed in both layers, however, this mode character can only be descriptive for one of the two dynamic components (lateral or perpendicular). Here the definition is used with respect to the

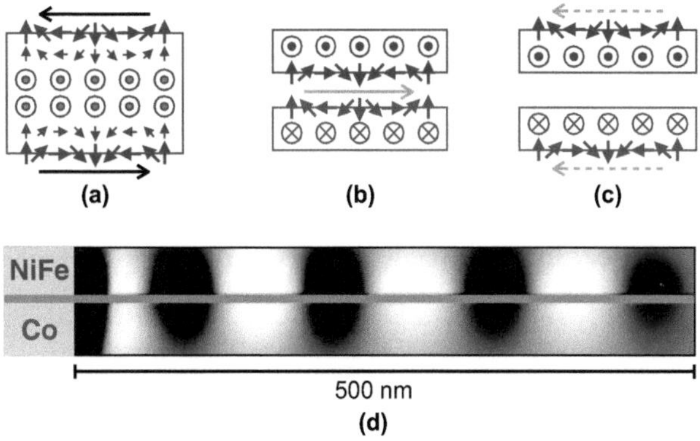

Figure 8.11 3D nature of short-wavelength spin waves in thin films. (a–c) Schematics of spin waves in thin films. (a) Damon–Eshbach waves in a thin film, propagating on the surface in the directions indicated by the black arrows. (b) Slow collective Damon–Eshbach spin-wave branch in an antiparallel bilayer propagating along the interlayer interfaces. (c) Fast collective spin-wave branch propagating along the top and bottom surfaces. (d) Micromagnetic simulation of a snapshot of the dynamic m_z component during propagation of the slow collective Damon–Eshbach branch in an antiparallel magnetic bilayer in a cross-sectional view for the two layers Co and NiFe.

perpendicular dynamic component, meaning for the acoustic mode, the perpendicular dynamic components are in phase between the two layers, while the in-plane components are antiphased. For the optical mode the situation is vice versa. Yet for the experiment only the acoustic mode is relevant (with its two nonreciprocal branches $\pm k$) since the optical mode resides at significantly higher frequencies (see Section 8.4). With respect to the mixed symmetry of the perpendicular and lateral dynamic components between the two layers, the spin-wave dynamics in the antiparallel bilayer is of 3D nature in itself.

Moreover, as the wavelength of the spin waves is approaching the order of magnitude of the film thickness, also within each layer the spin-wave amplitude profile becomes strongly position dependent along the perpendicular axis. This is schematically shown in Fig. 8.11b for the slow branch and Fig. 8.11c for the fast branch. The strong 3D nature of the slow branch was further verified by means of micromagnetic simulations (Fig. 8.11d). Here, a snapshot of the perpendicular dynamic magnetic component is shown in a cross-sectional view with the grayscale being normalized to the maximum. A clear 3D localization of the spin-wave amplitude around the interlayer can be observed.

8.4 Short-Wavelength Magnon Generation and Propagation in 1D and 2D Spin Textures

8.4.1 Introduction

In the previous section, we discussed the generation of submicrometer wavelength spin waves from field-driven vortex motion in a system of two exchange-coupled ferromagnetic layers. In the described case, the spin waves emitted from the vortex pair travel in the radial direction and, apart from the Gilbert damping [33], are therefore subject to a geometrical damping of their amplitude. Due to the circular symmetry, the amplitude is proportional to $r^{-1/2}$, r being the distance from the source (Fig. 8.12a). To circumvent such unnecessary losses, concepts are needed to guide and control the way spin waves propagate. One such concept is the idea to harness magnetic domain walls as spin-wave guides [2, 3, 32, 52, 72, 76, 89, 94, 100]. In a recent experiment, it has actually been shown that

Figure 8.12 Schematics of spin waves propagating under various geometrical conditions. (a) Spin waves are generated by a point source and travel in a 2D medium. Accordingly, apart from the Gilbert damping, the spin-wave amplitude must decay with distance r from the source as $r^{-1/2}$. This is avoided in (b) and (c), where source and medium dimensions differ by one. (b) The source has dimension $d = 1$, while for the propagation medium $d = 2$. (c) The spin-wave is generated by a point defect in the magnetization texture, and a domain wall acts as a waveguide, so source and propagation medium dimensions are 0 and 1, respectively.

domain walls can support localized modes, which can be excited by external alternating current magnetic fields [89]. Furthermore, as the lateral position of the excited region corresponds to the position of the domain wall, the former could be controlled by tuning static magnetic fields. However, the magnetic excitations observed in Ref. [89] decayed along the domain wall coordinate within subwavelength distances from the exciting antenna.

In this section, we review a recent work (first published as [71]) addressing the generation and propagation of short-wavelength spin waves. Using STXM in combination with X-ray magnetic circular dichroism (XMCD) [63], it has been shown that using an anisotropic bilayer system of interlayer exchange-coupled ferromagnetic layers, submicrometer spin waves can be excited and propagate across distances corresponding to multiples of the wavelength. These waves can be plane waves in a 2D medium (Fig. 8.12b), or they can be quasi-1D waves traveling along domain walls (Fig. 8.12c). The latter can be curved, so in this case, spin waves can be guided to change their direction of propagation [71].

8.4.2 System Description

The samples of Ref. [71] are based on the $Ni_{81}Fe_{19}(44.9)/Ru(0.8)/Co_{40}Fe_{40}B_{20}(46.6)$ multilayer system (Fig. 8.13a), where the numbers in brackets refer to the layer thicknesses in nanometers. The thicknesses of NiFe and CoFeB were determined from transmission electron microscopy measurements. The multilayer stack was deposited by magnetron sputtering on X-ray transparent silicon nitride membrane substrates and includes an Al capping layer on the CoFeB for oxidation protection. The Ru layer gives rise to an antiferromagnetic interlayer exchange coupling (IEC) between the NiFe and the CoFeB layer, as confirmed by magneto-optical Kerr effect (MOKE) measurements carried out on extended films. In addition to the IEC, the ferromagnetic films exhibit collinear uniaxial in-plane anisotropies, which was confirmed by MOKE measurements on corresponding single-layer NiFe and CoFeB films. The anisotropy strengths K_u for the single-layer films are $0.2\,\mathrm{kJ/m^{-3}}$ and $1.1\,\mathrm{kJ/m^{-3}}$, for NiFe and CoFeB, respectively. The extended multilayer was patterned into disk- and square-shaped

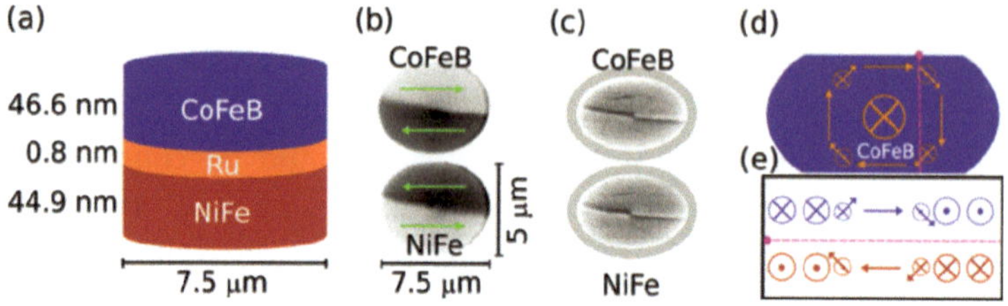

Figure 8.13 Sample details. (a) Schematic of an elliptically shaped microelement also displaying the individual layers' thicknesses. The NiFe and CoFeB layers couple antiferromagnetically across the Ru spacer. STXM was used to image the spatial distribution of the XMCD signal and thus obtain information of the magnetization distribution in the sample. (b) The signal from the in-plane magnetization components of Co and Ni, showing that the magnetization distribution is a pair of antiferromagnetically coupled vortices, distorted into an elongated pattern due to anisotropy. The green arrows indicate the magnetization direction. (c) The out-of-plane contrast revealing that the out-of-plane magnetization components in the domain wall actually couple ferromagnetically, which can by understood as a result from the dipolar interaction. The information about the magnetization distribution is schematically summarized in panels (d) and (e), where (d) is a top view of the CoFeB layer and (e) is a cross section that can be imagined as taken along the pink line in (d).

microstructures (Fig. 8.13) using electron beam lithography (EBL) and ion beam etching. Subsequently, a 200 nm-thick Cu strip was fabricated on top of the microelements. This was done using EBL and the lift-off technique.

Figure 8.13b–e illustrates the magnetic ground-state configuration of an elliptically shaped microelement. The configuration is a pair of stacked vortices whose vorticities, due to the antiferromagnetic IEC, are opposite. The vortex pattern is significantly deformed as a result of the uniaxial anisotropies, so effectively in each layer, there are two homogeneously in-plane magnetized domains separated by a 180° domain wall (Fig. 8.13b). In each disk, this domain wall spans the microelement and contains the vortex core. As micromagnetic simulations confirm, the walls are actually mixtures of Néel and Bloch domain walls [51]. Their in-plane components couple antiferromagnetically across the Ru layer, while, due to the diploar interaction, the out-of-plane components are ferromagnetically aligned; in particular, this holds for the vortex cores. The magnetization distribution is illustrated in Fig. 8.13d,e and is also visible in the STXM images in Figs. 8.13b and Fig. 8.13c, which display XMCD contrast arising from the in-plane and out-of-plane magnetization components, respectively. In what follows, we will refer to the system of coupled domain walls simply as "the domain wall," and when we actually mean the individual walls, this will be clear from the context. The micromagnetic simulations performed in Ref. [71] also indicate that the actual CoFeB uniaxial anisotropy in the sample is significantly higher than in the single-layer film. To reproduce the experimentally found magnetization distribution, a value of about $3\,\mathrm{kJ/m^3}$ for K_u was required. This elevation of the CoFeB K_u can be attributed to strain due to the presence of the waveguide on top of the structure. Indeed, CoFeB is known for its sensitivity to strain [90]. The orientation of the observed magnetization distribution is also consistent with this picture.

8.4.3 Generation of Spin Waves

Using the Oersted field generated by an alternating current sent through the Cu strip, spin waves could be excited on top of the above-described magnetic background. The spin waves were recorded using TR-STXM. The method allows for a time resolution of about 100 ps. Figure 8.14a displays a snapshot of spin waves measured at the Ni L_3 edge excited at 0.74 GHz. Similarly, Fig. 8.14b shows the excitations at 1.1 GHz in the CoFeB layer (measured at the Co L_3 edge). In both cases, plane spin waves with wave fronts parallel to the domain wall are excited and travel from the domain wall to the rim of the microelement. The Oersted field is oriented parallel to the short axis of the elliptical sample and thus aligned perpendicularly to the background magnetization in the domains. The Oersted field mainly excites dynamics of the domain wall. Considering the domain walls in the NiFe and CoFeB layers separately, the Oersted field excites an antiphase width oscillation of the two walls. This oscillation in turn leads to highly localized torques that result in the generation of the observed plane spin waves [43, 72, 88].

Similar to the plane spin waves excited by the domain wall oscillations, the Oersted field–induced dynamics of the vortex cores

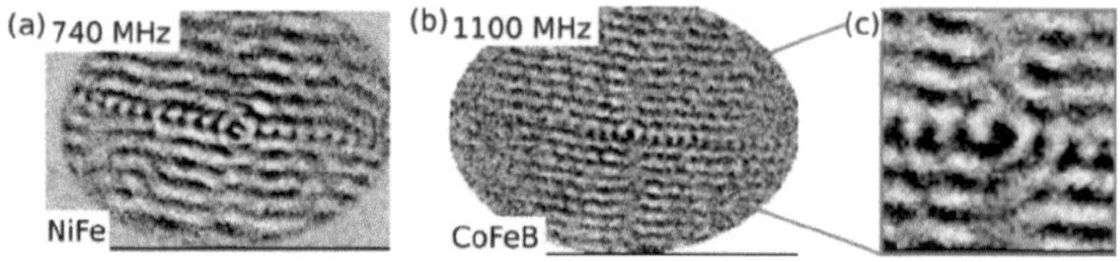

Figure 8.14 Spin-wave excitation at different frequencies. (a) Spin waves excited at 0.74 GHz. (b) Excitations at 1.1 GHz. (a) and (b) are taken at the Ni and the Co absorption edge, respectively. The plane spin waves visible are excited by the Oersted field–induced dynamics of the domain wall. Similarly, the combined motion of the vortex cores gives rise to spin waves with radial wave fronts [99], as shown in (c) as an enlargement of the boxed region in (b).

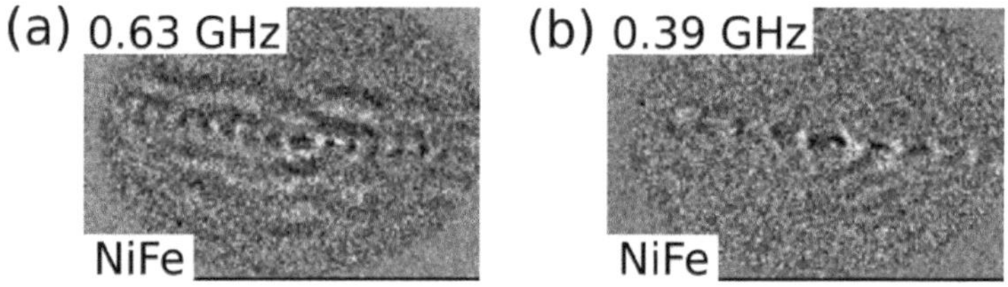

Figure 8.15 At lower frequencies, the magnetization dynamics changes qualitatively. While at 0.63 GHz (a), plane spin waves are still visible, such dynamics is absent at 0.39 GHz (b). However, magnons inside the domain wall are visible. These are excited by the vortex core dynamics and, similarly to the plane waves at higher frequencies, travel outward toward the rim of the microelement.

acts as a point source for radial spin waves, which also travel outward and give rise to interference with the plane waves excited by the domain walls. These are visible in Fig. 8.14c, which is an enlargement of the highlighted region in Fig. 8.14b. From such images, the wavelength and *k*-vector for the given frequency can be extracted.

When going to excitation frequencies below values of about 500 MHz, a qualitative change in the magnetization dynamics is observed, as can be seen in Fig. 8.15. The figure shows snapshots of excitations at 0.63 GHz (Fig. 8.15a) and 0.39 GHz (Fig. 8.15b). While in Fig. 8.15a, plane spin waves are still visible, at the low frequency used in Fig. 8.15b, no such plane waves are excited anymore.

However, one can clearly see waves in and traveling along the domain wall. Such domain wall waves are excited by the combined motion of the vortex cores and are also present in Fig. 8.15a and in the measurements displayed in Fig. 8.14, where they coexist with the plane spin waves excited by the domain wall dynamics.

8.4.4 Micromagnetic Simulations

For further investigation of these results, micromagnetic simulations were performed using the MuMax^3 code [83]. The performed simulations can be divided into two sets: The first kind of simulation was aimed at obtaining the plane spin-wave dispersion relation in the coupled layers. For this purpose, the bilayer system was modeled using a discretization of the x, y, and z directions into 4096, 16, and 115 cells, respectively. The y direction here corresponds to the easy-axis direction and thus to the direction along which the static part of the magnetization in the domains is aligned. Therefore, periodic boundary conditions are applied along that direction in the simulations. The material parameters for NiFe were chosen as follows: Saturation magnetization $M_s^{\text{NiFe}} = 800\,\text{kAm}^{-1}$, exchange stiffness $A_{\text{xc}}^{\text{NiFe}} = 7.5$, pJm^{-1} [91], and $K_u^{\text{NiFe}} = 200\,\text{Jm}^{-3}$. Corresponding parameters [17] for the CoFeB layer were chosen to $M_s^{\text{CoFeB}} = 1250\,\text{kAm}^{-1}$, $A_{\text{xc}}^{\text{CoFeB}} = 12\,\text{pJm}^{-1}$, and $K_u^{\text{CoFeB}} = 3000\,\text{Jm}^{-3}$. The IEC between the two ferromagnetic layers was set to $J = -0.1\,\text{mJm}^{-2}$, and the Gilbert damping to 0.01 for NiFe and 0.008 for CoFeB. These values were increased linearly to 0.065 for both of the layers in order to reduce the reflection of the waves from the edges. The exciting alternating magnetic field was applied out of plane in a 100-nm-wide region in the center. Once the system reached dynamic equilibrium, the k-vector was obtained via a spatial fast Fourier transform, performed along the x direction. As will be discussed below, the chosen set of parameters yields plane spin-wave dispersion relations that fit well to the experimentally obtained ones. Additional simulations comparing excitation efficiencies by in-plane and by out-of-plane alternating magnetic fields (always oriented perpendicularly to the magnetization) were performed. As a result, it was observed that out-of-plane fields are suited better for spin-wave excitation in this system, which can be understood as follows: The spin-wave modes actually exhibit an in-phase oscillation of the out-of-plane magnetization component; therefore the out-of-plane external field tends to drive the individual layer magnetizations in a way that is consistent with the mode. The in-plane magnetization component, however, exhibits an antiphase oscillation. As we will discuss next, there exists an additional mode in the system that indeed has an in-phase oscillation of the in-plane magnetization components. However, that mode appears at frequencies higher than the ones considered here.

The second type of simulation was performed to investigate the spin-wave excitation inside the domain wall. For this purpose, an elliptically shaped microelement (short axis 2.16 μm, long axis 3.24 μm) was discretized into cells as (648, 432, 115). Some of the simulation parameters chosen in this case were slightly different compared to the plane-wave simulations described above: $A_{\text{xc}}^{\text{NiFe}} = 10.5\,\text{pJm}^{-1}$, $A_{\text{xc}}^{\text{CoFeB}} = 13\,\text{pJm}^{-1}$, $K_{\text{u}}^{\text{CoFeB}} = 5000\,\text{Jm}^{-3}$, and $J = -0.3\,\text{mJm}^{-2}$. Before excitation, the magnetization distribution found in the experiment was qualitatively reproduced in the simulated system. For the excitation, a sinc(t) time profile for the magnetic field was used with a cut-off frequency of 10 GHz. By filtering, the response at the desired frequency could be obtained. As a result, these simulations qualitatively confirm the excitation and propagation of spin waves inside the domain wall.

8.4.5 Dispersion Relations

An analytical model was developed that allows the calculation of the plane spin-wave dispersion relations. These model calculations are much faster to perform compared to the micromagnetic simulations and therefore can be used to explore the parameter space for the micromagnetic computations. The core of the model considers plane-wave spin-wave excitations in homogeneously magnetized coupled thin layers. In each thin layer, the spatial dependence of the dynamic component of the magnetization along the coordinate perpendicular to the layer system is neglected. Since in the experiment, the layer thicknesses were beyond the respective materials' exchange lengths, the spatial dependence of the magnetization modes along the perpendicular coordinate was modeled by dividing each magnetic layer into a number of thin layers, for which the thin-film approximation holds. These individual layers are coupled to each other by an effective intralayer exchange coupling whose strength is determined by the respective value of A_{xc} and the number of layers. As a result, the model can be used to calculate the spin-wave dispersion relations in coupled bilayer systems with thicknesses higher than the exchange lengths and in particular in the system considered in the experiment of Ref. [71]. Figure 8.16a displays the main result, summarizing the measured dispersion relations for spin waves in the domains (squares) and the domain walls (diamond symbols) and comparing them to the dispersion relations obtained via the model calculations and the micromagnetic simulations. Note that dispersion relations from the theory and micromagnetic simulations are only for the plane waves in the domains. As Fig. 8.16a shows, theory and micromagnetics agree well with each other and with the experimental results for the spin waves in the domains. The chosen values for the IEC and the CoFeB anisotropy in the simulation and model calculation are $J = -0.1\,\mathrm{mJm^{-2}}$ and $K_u^{\mathrm{CoFeB}} = 3\,\mathrm{kJm^{-3}}$, respectively. As mentioned before, the anisotropy value is elevated compared to the measurement results for single CoFeB films; however, as CoFeB reacts sensitively to strain, the higher value can be explained taking into account the presence of the Cu strip line on top of the microelement.

The plane-wave magnon modes in the domains observed in the experiment correspond to the slow acoustic branch of spin waves. The model calculation result and the data from the micromagnetics displayed in Fig. 8.16a show spin-wave dispersion relations of two different branches. The branch low in frequency is the acoustic branch and includes the experimentally observed modes on the side of positive k. Above the acoustic branch, there is the optical branch, whose frequency has a much stronger k dependence. Both of these branches exhibit a minimum in frequency at finite energy—a so-called frequency gap. The gap is around 500 MHz for the acoustic branch and about 3 GHz for the optical mode. Furthermore, both the acoustic and optical branches are nonreciprocal, as they exhibit an asymmetric k dependence.

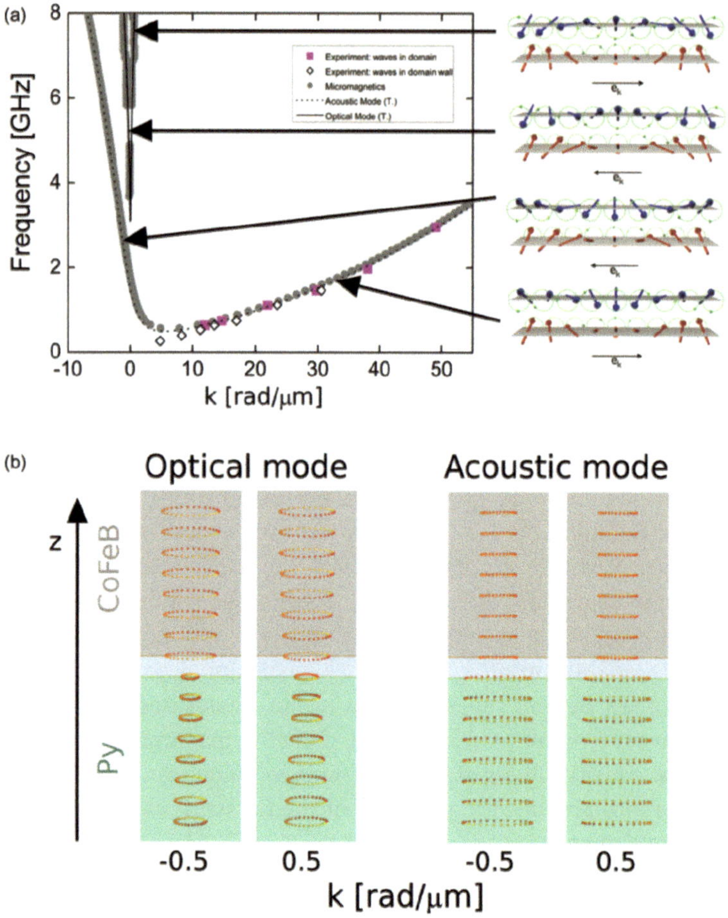

Figure 8.16 Spin-wave modes and dispersion relations. (a) The graph shows the calculated dispersion relations of spin waves in the domains. The gray dots refer to the results of micromagnetic simulations, and the dotted/solid lines represent the model calculations. The square and diamond-shaped elements correspond to the experimentally measured dispersion relations of the plane wave and the wave in the domain wall, respectively. The results from theory and micromagnetics comprise parts of the optical and acoustic branches of the spin waves in the domains, which are both clearly asymmetric with respect to $k = 0$. The associated phase relations are schematically depicted on the right-hand side of the graph. (b) This shows how amplitude and phase for the four cases of different phase relationships (optical/acoustic) and positive or negative sign of k depend on the coordinate perpendicular to the layer plane. The modes are the plane-wave modes inside the domains.

This is easily visible for the acoustic branch, which exhibits a higher group velocity $d\omega/dk$ for negative than for positive k. Therefore, the negative and positive k polarity parts of the dispersion relation are called the fast and the slow branch, respectively. Next to the graph in Fig. 8.16a, the phase relationships corresponding to the particular combination of propagation direction (determined by the k polarity) and mode (acoustic or optical) are illustrated. Basically, when looking at each layer's magnetization from the perspective such that the static part of the magnetization points directly at the viewer, one sees that the phase relationship corresponds to a shift of π for the optical mode, while in the acoustic mode, the layers are in phase. (Strictly speaking, the clear separation of the optical and acoustic modes is only possible for two identical layers; however, for the layers and frequencies considered here, the phase deviations from such pure optical and acoustic modes are small and therefore neglected.) Figure 8.16b shows the corresponding mode profiles along the coordinate perpendicular to the layer system for $k = \pm 0.5\,\text{rad}/\mu\text{m}$.

Similar to the spin waves in the domains, the dispersion relations of the domain wall modes also exhibit modes of acoustic and optical character [73]. The modes measured in the experiment and displayed in Fig. 8.16a as white diamonds correspond to the slow acoustic branch. Interestingly, the mode does not exhibit any notable frequency gap. As a result, it is possible to excite the domain wall spin waves selectively by tuning the excitation frequency to values below $\approx$500 MHz, where the domain modes are not populated (c.f. Fig. 8.15b).

8.4.6 Discussion

As described before, the observed selective excitation of spin-wave modes in the domain wall is possible due to the existence of a frequency gap for the modes in the domains, and the fact that the acoustic domain wall mode with its near linear frequency dependence on k, in the limit of long wavelengths, runs below the plane-wave dispersion and intercepts the frequency axis close to zero. This particular property of the acoustic domain wall wave dispersion relation can be explained by the Goldstone theorem

[34], which states that a system with a continuous symmetry spontaneously broken by the ground state exhibits a gapless mode. In the case of the domain wall waves, there exists a translational symmetry, which ideally leads to a gapless mode. In the real microelement, the sample finiteness and the presence of defects, in principle, break the symmetry; however, the resulting gap is too small to play a significant role for the observed phenomena.

8.5 Summary

In Section 8.2 we reviewed present methods and processes for spin-wave excitation and propagation. Among these were for the first aspect thermal excitation, optical excitation, parametric excitation by spatially uniform fields of double frequency, local magnetic Oersted field excitation, spin-transfer torque and spin–orbit torque excitation, and utilization of sample internal fields in a Schlömann or Fano sense. With respect to spin-wave propagation and waveguiding we discussed the capabilities of chains of patterned structures, local field guiding by Oersted fields, and the use of spin textures such as domain walls as spin-wave channels.

In Section 8.3, we focused on the isotropic generation and 3D properties of spin waves with very short wavelengths using the driven cores of a layered pair of spin vortices [99]. In Section 8.4, we reviewed the extension of this concept to a system with anisotropy, essentially allowing for the excitation of directional 2D plane waves by domain walls as well as the channeling of 1D spin waves within such walls [71].

Acknowledgments

We would like to thank Tobias Schneider, Attila Kákay, Tobias Warnatz, Alina Deac, Jürgen Lindner, Artur Erbe and Jürgen Fassbender (Helmholtz-Zentrum Dresden-Rossendorf, Dresden, Germany), Vasil Tiberkevich, and Andrei Slavin (Oakland University, Rochester, USA); Markus Weigand and Gisela Schütz (Max-Planck-Institut für Intelligente Systeme, Stuttgart, Germany); Jörg Raabe (Paul Scherrer

Institut, Villigen PSI, Switzerland); Roland Mattheis (Leibniz Institut für Photonische Technologien, Jena, Germany); Rodolfo Gallardo and Pedro Landeros (Universidad Técnica Federico Santa María, Valparaíso, Chile); and Alejandro Roldán-Molina (Universidad de Aysén, Coyhaique, Chile) for their contributions to the original works [71, 99].

Most of the experiments described in Refs. [71, 99] were performed at the Maxymus endstation at BESSY2, HZB, Berlin, Germany. We thank HZB for the allocation of synchrotron radiation beam time. The Maxymus endstation is operated by MPI-IS, Stuttgart, Germany. Some experiments were performed at the PolLux endstation at SLS, PSI, Villigen PSI, Switzerland. PolLux is financed by BMBF via contracts 05KS4WE1/6 and 05KS7WE1. Support by the Nanofabrication Facilities Rossendorf at IBC, HZDR, Dresden, Germany is acknowledged.

References

1. Albisetti, E., Petti, D., Pancaldi, M., Madami, M., Tacchi, S., Curtis, J., King, W. P., Papp, A., Csaba, G., Porod, W., Vavassori, P., Riedo, E. and Bertacco, R. (2016). Nanopatterning reconfigurable magnetic landscapes via thermally assisted scanning probe lithography, *Nat. Nanotech.*, **11**, p. 545.
2. Albisetti, E., Petti, D., Sala, G., Silvani, R., Tacchi, S., Finizio, S., Wintz, S., Cal, A., Zheng, X., Raabe, J., Riedo, E. and R., B. (2018). Nanoscale spin-wave circuits based on engineered reconfigurable spin-textures, *Comms. Phys.*, **1**, p. 56.
3. Aliev, F. G., Awad, A. A., Dieleman, D., Lara, A., Metlushko, V. and Guslienko, K. Y. (2011). Localized domain-wall excitations in patterned magnetic dots probed by broadband ferromagnetic resonance, *Phys. Rev. B*, **84**, p. 144406.
4. Au, Y., Ahmad, E., Dmytriiev, O., Dvornik, M., Davison, T. and Kruglyak, V. V. (2012). Resonant microwave-to-spin-wave transducer, *Appl. Phys. Lett.*, **100**, p. 182404.
5. Behncke, C., Adolff, C. F., Lenzing, N., Hänze, M., Schulte, B., Weigand, M., Schütz, G. and Meier, G. (2018). Spin-wave interference in magnetic vortex stacks, *Comms. Phys.*, **1**, p. 50.

6. Berger, L. (1996). Emission of spin waves by a magnetic multilayer traversed by a current, *Phys. Rev. B*, **54**, p. 9353.
7. Bloch, F. (1930). Zur Theorie des Ferromagnetismus, *Z. Phys.*, **61**, p. 206.
8. Bonetti, S., Kukreja, R., Chen, Z., Macià, F., Hernàndez, J. M., Eklund, A., Backes, D., Frisch, J., Katine, J., Malm, G., Urazhdin, S., Kent, A. D., Stöhr, J., Ohldag, H. and Dürr, H. A. (2015). Direct observation and imaging of a spin-wave soliton with p-like symmetry, *Nat. Commun.*, **6**, p. 8889.
9. Bonetti, S., Tiberkevich, V., Consolo, G., Finocchio, G., Muduli, P., Mancoff, F., Slavin, A. and Åkerman, J. (2010). Experimental evidence of self-localized and propagating spin wave modes in obliquely magnetized current-driven nanocontacts, *Phys. Rev. Lett.*, **105**, p. 217204.
10. Buchanan, K. S., Guslienko, K. Y., Doran, A., Scholl, A., Bader, S. D. and Novosad, V. (2005). Magnetic remanent states and magnetization reversal in patterned trilayer nanodots, *Phys. Rev. B*, **72**, p. 134415.
11. Buess, M., Höllinger, R., Haug, T., Perzlmaier, K., Krey, U., Pescia, D., Scheinfein, M. R., Weiss, D. and Back, C. H. (2004). Fourier transform imaging of spin vortex eigenmodes, *Phys. Rev. Lett.*, **93**, p. 077207.
12. Cherepov, S. S., Koop, B. C., Galkin, A. Y., Khymyn, R. S., Ivanov, B. A., Worledge, D. C. and Korenivski, V. (2012). Core–core dynamics in spin vortex pairs, *Phys. Rev. Lett.*, **109**, p. 097204.
13. Choe, S.-B., Acremann, Y., Scholl, A., Bauer, A., Doran, A., Stöhr, J. and Padmore, H. A. (2004). Vortex core-driven magnetization dynamics, *Science*, **304**, p. 420.
14. Choi, S., Lee, K.-S., Guslienko, K. Y. and Kim, S.-K. (2007). Strong radiation of spin-waves by core reversal of a magnetic vortex and their wave behaviors in magnetic nanowire waveguides, *Phys. Rev. Lett.*, **98**, p. 087205.
15. Chumak, A. V., Serga, A. A. and Hillebrands, B. (2014). Magnon transistor for all-magnon data processing, *Nat. Comun.*, **5**, p. 4700.
16. Chumak, A. V., Vasyuchka, V. I., Serga, A. A. and Hillebrands, B. (2015). Magnon Spintronics, *Nat. Phys.*, **11**, p. 453.
17. Conca, A., Papaioannou, E., Klingler, S., Greser, J., Sebastian, T., Leven, B., Lösch, J. and Hillebrands, B. (2014). Annealing influence on the gilbert damping parameter and the exchange constant of CoFeB films, *Appl. Phys. Lett.*, **104**, p. 182407.
18. Damon, R. W. and Eshbach, J. R. (1961). Magnetostatic modes of a ferromagnetic slab, *J. Phys. Chem. Solids*, **19**, p. 308.

19. Davies, C. S., Francis, A., Sadovnikov, A. V., Chertopalov, S. V., Bryan, M. T., Grishin, S. V., Allwood, D. A., Sharaevskii, Y. P., Nikitov, S. A. and Kruglyak, V. V. (2015). Towards gradedindex magnonics: Steering spin-waves in magnonic networks, *Phys. Rev. B*, **92**, p. 020408(R).

20. Davies, C. S. and Kruglyak, V. V. (2016). Generation of propagating spin-waves from regions of increased dynamic demagnetising field near magnetic antidots, *IEEE Trans. Magn.*, **52**, p. 2300504.

21. Davies, C. S., Poimanov, V. D. and Kruglyak, V. V. (2017). Mapping the magnonic landscape in patterned magnetic structures, *Phys. Rev. B*, **96**, p. 094439.

22. Davies, C. S., Sadovnikov, A. V., Grishin, S. V., Sharaevskii, Y. P., Nikitov, S. A. and Kruglyak, V. V. (2015). Generation of propagating spin-waves from regions of increased dynamic demagnetising field near magnetic antidots, *Appl Phys. Lett.*, **107**, p. 162401.

23. Demidov, V. E., Kostylev, M. P., Rott, K., Münchenberger, J., Reiss, G. and Demokritov, S. O. (2011). Excitation of short-wavelength spin-waves in magnonic waveguides, *Appl. Phys. Lett.*, **99**, p. 082507.

24. Demidov, V. E., Urazhdin, S. and Demokritov, S. O. (2010). Direct observation and mapping of spin waves emitted by spin-torque nano-oscillators, *Nature Mater.*, **9**, p. 984.

25. Demidov, V. E., Urazhdin, S., Ulrichs, H., Tiberkevich, V., Slavin, A., Baither, D., Schmitz, G. and Demokritov, S. O. (2012). Magnetic nano-oscillator driven by pure spin current, *Nature Mater.*, **11**, p. 1028.

26. Demokritov, S. O., Demidov, V. W., Dzyapko, O., Melkov, G. A., Serga, A. A., Hillebrands, B. and Slavin, A. N. (2006). Bose-Einstein condensation of quasi-equilibirum magnons at room temperture under pumping, *Nature*, **443**, p. 430.

27. Dieterle, G., Förster, J., Stoll, H., Semisalova, A. S., Finizio, S., Gangwar, A., Weigand, M., Noske, M., Fähnle, I., Bykova, Gräfe, J., Bozhko, D. A., Musiienko-Shmarova, H. Y., Tiberkevich, V., Slavin, A. N., Back, C. H., Raabe, J., Schütz, G. and Wintz, S. (2017). Coherent excitation of heterosymmetric spin waves with ultrashort wavelengths, *arXiv* https: //arxiv.org/abs/1712.00681.

28. Ding, J., Kakazei, G. N., Liu, L., Guslienko, K. Y. and Adeyeye, A. O. (2014). Higher-order vortex gyrotropic modes in circular ferromagnetic nanodots, *Sci. Rep.*, **4**, p. 4796.

29. Dyson, F. (1956). General theory of general spin wave interactions, *Phys. Rev.*, **102**, p. 1217.

30. Fano, U. (1961). Effects of configuration interaction on intensities and phase shifts, *Phys. Rev.*, **124**, p. 1866.
31. Feldtkeller, E. and Thomas, H. (1965). Struktur und energie von blochlinien in dünnen ferromagnetischenen schichten, *Phys. Kondens. Materie*, **4**, p. 8.
32. Garcia-Sanchez, F., Borys, P., R., S., Adam, J.-P., Stamps, R. and Kim, J.-V. (2015). Narrow magnonic waveguides based on domain walls, *Phys. Rev. Lett.*, **114**, p. 247206.
33. Gilbert, T. L. (1955). A Lagrangian formulation of the gyromagnetic equation of the magnetization Field, *Phy. Rev.*, **100**, p. 1243.
34. Goldstone, J., Salam, A. and Weinberg, S. (1962). Broken symmetries, *Phys. Rev.*, **127**, p. 965.
35. Grünberg, P. (1981). Magnetostatic spinwave modes of a heterogeneous ferromagnetic double layer, *J. Appl. Phys.*, **52**, p. 6824.
36. Grünberg, P., Schreiber, R., Pang, Y., Brodsky, M. B. and Sowers, H. (1986). Layered magnetic structures: evidence for antiferromagnetic coupling of fe layers across cr interlayers, *Phys. Rev. Lett.*, **57**, p. 2442.
37. Gruszecki, P., Kasprzak, M., Serebryannikov, A., Krawczyk, M. and Smigaj, W. (2016). Microwave excitation of spin wave beams in thin ferromagnetic films, *Sci. Rep.*, **6**, p. 22367.
38. Gurevich, A. G. and Melkov, G. A. (1996). *Magnetization Oscillations and Waves* (CRC, New York).
39. Guslienko, K. Y., Buchanan, K. S., Bader, S. D. and Novosad, V. (2005). Dynamics of coupled vortices in layered magnetic nanodots, *Appl. Phys. Lett.*, **86**, p. 223112.
40. Haldar, A., Kumar, D. and Adeyeye, A. (2016). A reconfigurable waveguide for energy-efficient transmission and local manipulation of information in a nanomagnetic device, *Nat. Nanotech.*, **11**, p. 437.
41. Hämäläinen, S. J., Brandl, F., Franke, K. J. A., Grundler, D. and van Dijken, S. (2017). Tunable short-wavelength spin-wave emission and confinement in anisotropy-modulated multiferroic heterostructures, *Phys. Rev. Appl.*, **8**, p. 014020.
42. Hänze, M., Adolff, C. F., Velten, S., Weigand, M. and Meier, G. (2016). Two-body problem of core-region coupled magnetic vortex stacks, *Phys. Rev. B*, **93**, p. 054411.
43. Hermsdoerfer, S. J., Schultheiss, H., Rausch, C., Schäfer, S., Leven, B., Kim, S.-K. and Hillebrands, B. (2009). A spin-wave frequency doubler by domain wall oscillation, *Appl. Phys. Lett.*, **94**, p. 223510.

44. Hertel, R., Gliga, S., Fähnle, M. and Schneider, C. M. (2007). Ultrafast nanomagnetic toggle switching of vortex cores, *Phys. Rev. Lett.*, **98**, p. 117201.
45. Hoefer, M. A., Silva, T. J. and Stiles, M. D. (2008). Model for a collimated spin-wave beam generated by a single-layer spin torque nanocontact, *Phys. Rev. B*, **77**, p. 144401.
46. Holländer, R. B., Müller, C., Schmalz, J., Gerken, M. and McCord, J. (2018). Magnetic domain walls as broadband spin wave and elastic magnetisation wave emitters, *Sci. Rep.*, **8**, p. 13871.
47. Holstein, T. and Primakoff, H. (1940). Field dependence of the intrinsic domain magnetization of a ferromagnet, *Phys. Rev.*, **58**, p. 1098.
48. Kalinikos, B. A. and Slavin, A. N. (1986). Theory of dipole-exchange spin wave spectrum for ferromagnetic films with mixed exchange boundary conditions, *J. Phys. C: Solid State Phys.*, **19**, p. 7013.
49. Kittel, C. (1949). Physical theory of ferromagnetic domains, *Rev. Mod. Phys.*, **21**, p. 541.
50. Kruglyak, V. V., Demokritov, S. O. and Grundler, D. (2010). Magnonics, *J. Phys. D: Appl. Phys.*, **43**, p. 264001.
51. Labrune, M. and Miltat, J. (1995). Wall structures in ferro/antiferromagnetic exchange-coupled bilayers: a numerical micromagnetic approach, *J. Magn. Magn. Mater.*, **151**, p. 231.
52. Lan, J., Yu, W., Wu, R. and Xiao, J. (2015). Spin-wave diode, *Phys. Rev. X*, **5**, p. 041049.
53. Lee, K.-S., Choi, S. and Kim, S.-K. (2005). Radiation of spin waves from magnetic vortex cores by their dynamic motion and annihilation processes, *Appl. Phys. Lett.*, **87**, p. 192502.
54. M., K., Weigand, M., Curcic, M., Noske, M., Sproll, M., Vansteenkiste, A., Van Waeyenberge, B., Stoll, H., Woltersdorf, G., Back, C. H. and Schütz, G. (2011). Vortex core reversal by excitation of spin-waves, *Nat. Commun.*, **2**, p. 279.
55. Macià, F., Backes, D. and Kent, A. D. (2014). Stable magnetic droplet solitons in spintransfer nanocontacts, *Nat. Nanotech.*, **9**, p. 992.
56. Madami, M., Bonetti, S., Consolo, G., Tacchi, S., Carlotti, G., Gubbiotti, G., Mancoff, F. B., Yar, M. A. and Âkerman, J. (2011). Direct observation of a propagating spin wave induced by spin-transfer torque, *Nat. Nanotech.*, **6**, p. 635.
57. Mozooni, B. and McCord, J. (2015). Direct observation of closure domain wall mediated spin waves, *Appl. Phys. Lett.*, **107**, p. 042402.

58. Mushenok, F., Dost, R., Davies, C., Allwood, D., Inkson, B., Hrkac, G. and Kruglyak, V. (2017). Broadband conversion of microwaves into propagating spin-waves in patterned magnetic structures, *J. Appl. Phys.*, **111**, p. 042404.

59. Perzlmaier, K., Buess, M., Back, C. H., Demidov, V. E., Hillebrands, B. and Demokritov, S. O. (2005). Spin-wave eigenmodes of permalloy squares with a closure domain structure, *Phys. Rev. Lett.*, **94**, p. 057202.

60. Raabe, J., Tzvetkov, G., Flechsig, U., Bögel, M., Jaggi, A., Sarafimov, B., Vernooij, B. G. C., Huthwelker, T., Ade, H., Kilcoyne, D., Tyliszczak, T., Fink, R. H. and Quitmann, C. (2008). Pollux: a new facility for soft x-ray spectromicroscopy at the swiss light source, *Rev. Sci. Instrum.*, **79**, p. 113704.

61. Scheinfein, M. R. LLG micromagnetics simulator, http://planet-tech.com.hk/product/soft/llg.html.

62. Schlömann, E. (1964). Generation of spin-waves in nonuniform magnetic fields, I. Conversion of electromagnetic power into spin-wave power and vice versa. *J. Appl. Phys.*, **35**, p. 159.

63. Schütz, G., Wagner, W., Wilhelm, W., Kienle, P., Zeller, R., Frahm, R. and Materlik, G. (1987). Absorption of circularly polarized x rays in iron, *Phys. Rev. Lett.*, **58**, p. 737.

64. Shinjo, T., Okuno, T., Hassdorf, R., Shigeto, K. and Ono, T. (2000). Magnetic vortex core observation in circular dots of permalloy, *Science*, **289**, p. 930.

65. Slavin, A. and Tiberkevich, V. (2005). Spin wave mode excited by spin-polarized current in a magnetic nanocontact is a standing self-localized wave bullet, *Phys. Rev. Lett.*, **95**, p. 237201.

66. Slonczewski, J. C. (1996). Current-driven excitation of magnetic multilayers, *J. Magn. Magn. Mater.*, **159**, p. L1.

67. Slonczewski, J. C. (1999). Exitation of spin waves by an electric current, *J. Magn. Magn. Mater.*, **195**, p. L261.

68. Sluka, V., Kakay, A., Deac, A. M., Bürgler, D. E., Hertel, R. and Schneider, C. M. (2011). Spin-transfer torque induced vortex dynamics in fe/ag/fe nanopillars, *J. Phys. D: Appl. Phys.*, **44**, p. 384002.

69. Sluka, V., Kakay, A., Deac, A. M., Bürgler, D. E., Hertel, R. and Schneider, C. M. (2012). Quenched slonczewski windmill in spin-torque vortex oscillators, *Phys. Rev. B*, **86**, p. 214422.

70. Sluka, V., Kakay, A., Deac, A. M., Bürgler, D. E., Schneider, C. M. and Hertel, R. (2015). Spin-torque-induced dynamics at fine-split frequencies in nano-oscillators with two stacked vortices, *Nat. Commun.*, **6**, p. 6409.

71. Sluka, V., Schneider, T., Gallardo, R. A., Kakay, A., Weigand, W., Warnatz, T., Mattheis, R., Roldan-Molina, A., Landeros, P., Tiberkevich, A., V. Slavin, Schütz, G., Erbe, A., Deac, A., Lindner, J., Raabe, J., Fassbender, J. and Wintz, S. (2019). Emission and propagation of 1d and 2d spin-waves with nanoscale wavelengths in anisotropic spin textures, *Nat. Nanotech.*, **14**, 328–333.

72. Sluka, V., Weigand, M., Kakay, A., Erbe, A., Tyberkevych, V., Slavin, A., Deac, A., Lindner, J., Fassbender, J., Raabe, J. and Wintz, S. (2015). Stacked topological spin textures as emitters for multidimensional spin wave modes, IEEE International Magnetics Conference (INTER-MAG), Beijing, 11–15 May, 2015.

73. Stamps, R. L., Carrjco, A. S. and Wigen, P. E. (1997). Domain-wall resonance in exchange-coupled magnetic films, *Appl. Phys. Lett.*, **55**, p. 6473.

74. Stancil, D. D. (1993). *Theory of Magnetostatic Waves* (Springer).

75. Stigloher, J., Decker, M., Körner, H. S., Tanabe, K., Moriyama, T., Taniguchi, T., Hata, H., Madami, M., Gubbiotti, G., Kobayashi, K., Ono, T. and Back, C. H. (2016). Excitation of short-wavelength spin-waves in magnonic waveguides, *Phys. Rev. Lett.*, **117**, p. 037204.

76. Stoll, H., Puzic, A., van Wayenberge, B., Fischer, P., Raabe, J., Buess, M., Haug, T., Höllinger, R., Back, C., Weiss, D. and Denbeaux, G. (2004). High-resolution imaging of fast magnetization dynamics in magnetic nanostructures, *Appl. Phys. Lett.*, **84**, p. 3328.

77. Thiele, A. A. (1973). Steady-state of motion of magnetic domains, *Phys. Rev. Lett.*, **30**, p. 230.

78. Topp, J., Heitmann, D., Kostylev, M. P. and Grundler, D. (2010). Making a reconfigurable artificial crystal by ordering bistable magnetic nanowires, *Phys. Rev. Lett.*, **104**, p. 207205.

79. Tretiakov, O. A. and Tchernyshyov, O. (2007). Vortices in thin ferromagnetic films and the skyrmion number, *Phys. Rev. B*, **75**, p. 012408.

80. Tsoi, M., Jansen, A. G. M., Bass, J., Chiang, W.-C., Tsoi, V. and Wyder, P. (2000). Generation and detection of phase-coherent current-driven magnons in magnetic multilayers, *Nature*, **406**, p. 46.

81. Urazhdin, S., Demidov, V. E., Ulrichs, H., Kendziorczyck, T., Kuhn, T., Lethold, J., Wilde, G. and Demokritov, S. O. (2014). Nanomagnetic devices based on the spin-transfer torque, *Nat. Nanotech.*, **9**, p. 509.

82. Vansteenkiste, A., Chou, K. W., Weigand, M., Curcic, M., Sackmann, V., Stoll, H., Tyliszczak, T., Woltersdorf, G., Back, C. H. and Schütz, G. (2009).

X-ray imaging of the dynamic magnetic vortex core deformation, *Nat. Phys.*, **5**, p. 532.

83. Vansteenkiste, A., Leliaert, J., Dvornik, M., Helsen, M., Garcia-Sanchez, F. and Van Waeyenberge, B. (2014). The design and verification of mumax3, *AIP Advances*, **4**, p. 107133.
84. Verba, R. V., Hierro-Rodriguez, A., Navas, D., Ding, J., Liu, X. M., Adeyeye, A. O., Guslienko, K. Y. and Kakazei, G. N. (2016). Spin-wave excitation modes in thick vortex-state circular ferromagnetic nanodots, *Phys. Rev. B.*, **93**, p. 214437.
85. Vogt, K., Fradin, F. Y., Pearson, J. E., Sebastian, T., Bader, S. D., Hillebrands, B., Hoffmann, A. and Schultheiss, H. (2014). Realization of a spin-wave multiplexer, *Nat. Commun.*, **5**, p. 3727.
86. Vogt, K., Schultheiss, H., Jain, S., Pearson, J. E., Hoffmann, A., Bader, S. D. and Hillebrands, B. (2012). Spin waves turning a corner, *Applied Physics Letters*, **101**, p. 042410.
87. Völkel, A. R., Wysin, G. M., Mertens, F. G., Bishop, A. R. and Schnitzer, H. J. (1994). Collective-variable approach to the dynamics of nonlinear magnetic excitations with application to vortices, *Phys. Rev. B*, **50**, p. 12711.
88. Voto, M., Lopez-Diaz, L. and Martinez, E. (2017). Pinned domain wall oscillator as tunable direct current spin wave emitter, *Sci. Rep.*, **7**, p. 13559.
89. Wagner, K., Kákay, A., Schultheiss, K., Henschke, A., Sebastian, T. and Schultheiss, H. (2016). Magnetic domain walls as reconfigurable spin-wave nanochannels, *Nat. Nanotech.*, **11**, p. 432.
90. Wang, D., Nordman, C., Qian, Z., Daughton, J. M. and Myers, J. (2005). Magnetostriction effect of amorphous CoFeB thin films and application in spin-dependent tunnel junctions, *J. Appl. Phys.*, **97**, p. 10c906.
91. Wei, J., Zhu, Z., Song, C., Feng, H., Jing, P., Wang, x., Liu, Q. and Wang, J. (2016). Annealing influence on the exchange stiffness constant of permalloy films with stripe domains, *J. Phys. D: Appl. Phys.*, **49**, p. 265002.
92. Whitehead, N., Horsley, S., T.G., P., Kuchko, A. and Kruglyak, V. (2017). Theory of linear spin wave emission from a Bloch domain wall, *Phys. Rev. B*, **96**, p. 064415.
93. Wiele, B., Hämäläinen, S. J., Baláz, P., Montoncello, F. and Dijken, S. (2016). Tunable short-wavelength spin wave excitation from pinned magnetic domain walls, *Sci. Rep.*, **6**, p. 21330.

94. Winter, J. M. (1961). Bloch wall excitation. Application to nuclear resonance in a Bloch wall. *Phys. Rev.*, **124**, p. 452.

95. Wintz, S., Bunce, C., Banholzer, A., Körner, M., Strache, T., Mattheis, R., McCord, J., Raabe, J., Quitmann, C., Erbe, A. and Fassbender, J. (2012). Interlayer-coupled spin vortex pairs and their response to external magnetic fields, *Phys. Rev. B*, **85**, p. 224420.

96. Wintz, S., Bunce, C., Neudert, A., Körner, M., Strache, T., Buhl, M., Erbe, A., Gemming, S., Raabe, J., Quitmann, C. and Fassbender, J. (2013). Topology and origin of effective spin meron pairs in ferromagnetic multilayer elements, *Phys. Rev. Lett.*, **110**, p. 177201.

97. Wintz, S., Strache, T., Körner, M., Bunce, C., Banholzer, A., Mönch, I., Mattheis, R., Raabe, J., Quitmann, C., McCord, J., Erbe, A., Lenz, K. and Fassbender, J. (2012). Control of vortex pair states by post-deposition interlayer exchange coupling modification, *Phys. Rev. B*, **85**, p. 134417.

98. Wintz, S., Strache, T., Körner, M., Fritzsche, M., Marko, D., Mönch, I., Mattheis, R., Raabe, J., Quitmann, C., McCord, J., Erbe, A. and Fassbender, J. (2011). Direct observation of antiferromagnetically oriented spin vortex states in magnetic multilayer elements, *Appl. Phys. Lett.*, **99**, p. 149901.

99. Wintz, S., Tiberkevich, V., Weigand, M., Raabe, J., Lindner, J., Erbe, A., Slavin, A. and Fassbender, J. (2016). Magnetic vortex cores as tunable spin-wave emitters, *Nature Nanotech.*, **11**, p. 948.

100. Xing, X. and Zhou, Y. (2016). Fiber optics for spin waves, *NPG Asia Materials*, **8**, p. e246.

101. Yu, H., d'Allivy Kelly, O., Cros, V., Bernard, R., Bortolotti, P., Anane, A., Brandl, F., Heimbach, F. and Grundler, D. (2016). Approaching soft x-ray wavelengths in nanomagnet-based microwave technology, *Nat. Commun.*, **7**, p. 11255.

102. Yu, H., Duerr, G., Huber, R., Bahr, M., Schwarze, T., Brandl, F. and Grundler, D. (2013). Omnidirectional spin-wave nanograting coupler, *Nat. Commun.*, **4**, p. 2702.

Chapter 9

Precessional Magnetization Dynamics and Spin Waves in 3D Ferromagnetic Nanostructures

Sucheta Mondal, Sourav Sahoo, and Anjan Barman

Department of Condensed Matter Physics and Material Sciences, S. N. Bose National Centre for Basic Sciences, Block JD, Sector III, Salt Lake, Kolkata 700106, India

abarman@bose.res.in

The experimental and theoretical studies of patterned magnetic nanostructures have gained huge interest in the scientific community because of their potential applications in magnetic data storage, logic, sensors, microwave devices, and various biomedical applications. Their implementation in multidisciplinary technology requires exploitations of new phenomena ranging over different time frames, such as magnetic domain wall dynamics and magnetic vortex core motion to faster phenomena such as spin-wave dynamics, ultrafast demagnetization, and relaxation. The spin structures and magnetization dynamics of these nanostructures are distinct from their bulk counterparts and thin films. The magnetic ground state highly influences the magnetization dynamics. Not only their intrinsic material parameters like exchange stiffness constant, saturation magnetization, magnetocrystalline anisotropy

Three-Dimensional Magnonics: Layered, Micro- and Nanostructures
Edited by Gianluca Gubbiotti

ISBN 978-981-4800-73-0 (Hardcover), 978-0-429-29915-5 (eBook)
www.jennystanford.com

etc., but also patterning to submicron length scale can introduce unique magnetic properties that can modulate the magnetization dynamics over this broad timescale. Magnetic nanowires, waveguides, nanostripes, nanodots, antidots, and bicomponent or binary structures have been fabricated and studied over the years to explore the dependence of the collective magnetization dynamics of these systems upon various geometric or physical parameters such as shape, size, aspect ratio, and lattice arrangement. On the other hand, artificially patterned nanomagnets, which can also act as artificial crystals, emerged as magnonic crystals (MCs) where spin waves or magnons can carry and process information. External stimuli such as magnetic and electric fields can also modulate the dynamics of these systems. Here, we will present some remarkable versatility of these multifunctional magnetic nanostructures by exploring the generation, control, and detection of spin waves in such systems. The increasing demand of miniaturized devices has inclined the research toward energy-efficient storage, memory, signal processing, and neuromorphic computing using spin waves across all three dimensions in the nanostructures. Several state-of-the-art fabrication and characterization techniques have been developed to probe the magnetic properties in 3D nanostructures, which will be discussed to shed light on their potential for real-world application.

9.1 Introduction

Nanostructured magnetic materials offer several new functionalities that could not be achieved from their bulk counterparts [1, 2]. Currently a broad range of real-world applications of nanostructured materials exploiting many of these functionalities is available and emerging. Magnetic recording technology with magnetoresistive properties, such as giant magnetoresistance and tunneling magnetoresistance, is a successful example of magnetic multilayers (MLs) [3–5]. Over the last few decades, several prototype devices based on magnetic nanostructures have been proposed by the scientific community, which show great promises for next-generation miniaturized and energy-efficient devices. High-density media for data storing [6], spin-based transistors [7], spin-based logic [8, 9], reconfigurable waveguides for energy-efficient transmission of signals [10], switching of nanomagnets for nonvolatile memory application [11, 12], spin-torque nano-oscillators (STNOs) [13], spin-Hall nano-oscillators (SHNOs) [14], spin-based qubits [15], nanobiomedicines [16], and neuromorphic computation [17] are just a few examples of those devices that are proposed to be designed in submicron to atomic length scales.

Spin-wave (SW)-based systems can offer excellent scalability, which is not accessible in electromagnetic wave-based system (e.g., photonic crystal [18]). Additionally, it promises low energy consumption and a high data processing rate. These exciting opportunities fueled an upsurge in the research interest during the last two decades, and enormous efforts have been made to develop advanced fabrication procedures and new state-of-the-art characterization techniques for studying the static and dynamic properties of nanostructured magnetic materials. This journey began with patterning a thin ferromagnetic film with a certain periodicity or preparing a chemically grown assembly of micro- and nanostructures in the monodimension (1D). Stripes [19–21], waveguides [22, 23], nanowires [24, 25], chains of nanoparticles [26], and magnonic crystals (MCs) with unidirectional periodic modulation of magnetic properties [27, 28] are the members of this 1D nanostructure family. They exhibit some exciting phenomena such as SW propagation along with transmission and reflection from artificial boundaries, magnonic band formation with a partial bandgap, etc. However, researchers soon realized that an extension to another dimension will provide more control points for creation of stop bands for propagating SWs and processing and storing of information within lesser space. Most of these features originate from the competition between short-range exchange coupling, long-range dipolar coupling, and magnetic anisotropy. Initially a lattice of bubble domains inside a ferromagnetic film was considered as 2D structuring, which was also used in the primary stage of magnetic recording. Bit-patterned media (BPM) [29] was developed utilizing patterned 2D arrays of magnetic bits that rely on data storage by switching the magnetic field. Further progress in fabrication techniques facilitates the construction and investigation of magnetization dynamics of nanodot arrays [30], nanohole arrays (or antidot arrays) [31, 32], MCs [33–36], spin ice structures [37], magnetic solitons [38], and skyrmion lattices [39, 40]. However, along the thickness of these 2D arrays of nanostructures, the magnetization is homogeneous, and without introducing additional structuring along the third dimension, it is impossible to get any periodicity to further control the magnonic band structure. The idea of breaking this monodomain formalism in the vertical direction along the film thickness ignited the development of 3D nanostructures [41, 42].

During the last few years, 3D magnetism has become an emergent research topic due to its fascinating spin configuration and potential applications in storage and other magnetic devices. The discovery of "racetrack memory" [43] boosted the 3D nanostructures as a potential successor for ultrahigh-density data storage devices, as the conventional 2D magnetic storage devices have almost reached saturation of storage density due to various bottlenecks of confined 2D structures. On the contrary, 3D magnetic nanostructures offer several advantages due to more flexibility toward geometrical confinement, broad tunability of spin textures, and also the possibility of sustaining temperature differentials.

A system can be considered as a 3D magnetic nanostructure if it has the capability to grow nonplanar spin textures within it, induced by either pattering or topology. Advancement in the fabrication procedure has been made in the last few years, facilitating the formation of a high-quality interface in multilayerd heterostructures and the construction of 3D nano-objects. A variety of magnetic structures with 3D magnetic features such as topological magnetic solitons, domain walls in vertical nanowires, magnetic nanoparticles, 3D MCs, and several other 3D nanostructures have been realized. Additionally, while harvesting the properties of 3D nanostructures, some fascinating physics have emerged, such as curvature-induced effective anisotropy and Dzyaloshinskii–Moriya interaction (DMI) [44], existence of Bloch point [45] of 0D micromagnetic objects, magnetic charge transport in 3D artificial spin ice [46], etc. The 3D magnetic structures give a new wing to control and manipulate spin configuration and SW dynamics. The 3D nanostructures have a possibility to boost future applications in different aspects,

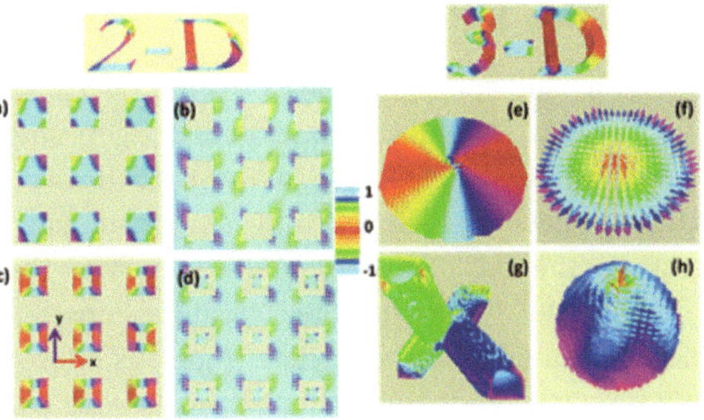

Figure 9.1 Ground-state magnetization configuration of 2D structures: (a) dot, (b) antidot, (c) nanoring, (d) antinanoring and 3D structures, (e) vortex, (f) skyrmion, (g) tetrapod, and (h) sphere.

such as sensors [47], actuators [48], 2.5D spintronic devices [42], neuromorphic computer architecture [17], and storage devices [43]. The researchers are dragged into the world of 3D nanomagnetism for exploring its rich physics and future applications.

In this chapter, we discuss the development of 3D magnetic nanostructures and the precessional magnetization dynamics and SWs in different classes of 3D magnetic nanostructures. The recent developments in the field of 3D micro- and nano-objects (see Fig. 9.1) have been discussed in detail. This is followed by a discussion of the advantages of topological magnetic objects with 3D spin texturing, patterned MLs, spin configuration in 3D magnetic curved surface, and spin ice over the 2D planar arrays. Finally, a brief and relevant discussion about the fabrication procedures and experimental and numerical techniques for studying spin texture and SW dynamics in 3D nanostructures is presented for convenience of the readers.

9.2 Experimental Methods

In this section we will discuss various fabrication and characterization methods generally used for the growth of the magnetic nanostructures and for investigation of the magnetization dynamics in the time domain, frequency domain, and wave vector domain.

9.2.1 Fabrication Techniques

Fabrication is an indispensable part of rapid development in any modern technology. The miniaturization of gadgets solely depends on the efficiency and sensitivity of on-chip device fabrication with ultrahigh packing density and choice of raw materials. To understand the basic physics and to explore the industrial application potential, researchers face great challenges to overcome various design and fabrication hurdles. Vigorous efforts made over the sample preparation down to the nanoscale with high precision led to the development of a number of state-of-the-art fabrication techniques based on bottom-up and top-down approaches. The former approach relies on chemical processes that involve nanometer-scale molecular self-organization and mesoscopic pattern formation [49–52]. Primarily by using the solution-phase colloidal chemistry and electrochemistry, a series of nanoparticles, nanowires, and hierarchically arranged biological organisms can be synthesized. Templates, that is, track-etched polymers, anodic alumina, and diblock copolymer membranes, are used for these processes. However, using this process it is often difficult to obtain highly ordered and monodispersed structures that are essential components for device fabrication. On the other hand, the top-down approach is primarily used for construction of nanodot arrays, antidot arrays, MCs, bi-MCs, nanostripes, quasicrystals, and several other micro- and nanostructures in 1D and 2D by pattering magnetic thin films and multilayers with high repeatability but with lower yield as compared to its bottom-up counterpart. It is extremely important to optimize the steps of fabrication to achieve high-quality interfaces, while involving different kinds of materials, and to get very sharp edges, while fabricating narrow-sized elements avoiding structural distortion. This approach deals with the physical processes, which include several kinds of lithography techniques, such as photolithography [53], electron beam lithography (EBL) [54], deep ultraviolet lithography (DUV) [55], ion beam lithography (IBL) [56], shadow masking [57], and laser or ion beam irradiation [58]. Besides these methods, X-ray lithography [59], scanning probe lithography [60], holographic or interface lithography (IL) [61], nanoimprint lithography (NIL) [62], and step growth methods [63] are also very popular methods. Recently a combination of both top-down and bottom-up techniques is being used as per convenience. Some of the most challenging procedures to fabricate high-quality 3D micro- and nanostructures are two-photon lithography (TPL) [64] combined with the electrodeposition technique, 3D printing [65], and advanced chemical synthesis. TPL is a powerful tool for fabrication of 3D structures. For TPL, a femtosecond pulsed laser beam is focused on a polymer, which only interacts with the focused area without disturbing the neighboring region. Its ability to write a pattern inside a photoresist leads to fabrication of a 3D structure. Focused electron-beam-induced deposition (FEBID) is a direct writing nanolithography technique called "3D nanoprinting" with the ability of fabrication of 3D structures with dimensions of tens of nanometers within a few seconds to a few minutes. Chemical synthesis techniques are commonly used in 3D nanostructure fabrication. The materials are deposited through the pores of nanotemplates such as anodic alumina templates, block copolymer templates, track-etched polycarbonate (PCTE), and other self-assembly methods. The 3D magnetic nanostructures fabricated using these techniques have been discussed briefly in Section 9.3.

9.2.2 Measurement Techniques

The next challenge is to probe the static or dynamic properties of spins inside the micro- and nanostructures. In the following two subsections we have discussed some of those popular experimental techniques.

9.2.2.1 Static characterization

Very early studies on the spin textures, magnetization, and stray field distribution, developed in 2D nanostructures, were conducted using magnetic force microscopy (MFM) [66], Lorentz transmission electron microscopy (LTEM) [67], magneto-optical Kerr effect (MOKE) [68], etc. MFM works based on the mechanism where magnetic field lines originated form the domains inside the sample interact with the highly localized field lines of the magnetic tip of the instrument. This induces repulsion in the tip and systematic deflection of the tip is sensed by an incident laser beam and a position sensitive detector providing information about the magnetic topography of the structure. The spatial resolution is limited by the radius of curvature of the tip. In the case of LTEM, the accelerated electrons across the magnetic material experience a Lorentz force and they map the spatial spin arrangement. MOKE is another very popular technique being used since the last century and is capable of imaging with submicron resolution. The working principle is based on the magneto-optical Kerr effect, where linearly polarized light converts to elliptically polarized light with the major axis rotated after interacting with the magnetization in the material. Various kinds of domain formation in the magnetic ground state or magnetic field hysteresis inside the nanostructures can be studied using this tool. Researchers have imaged magnetic states of magnetization reversal modes, magnetic vortex, domain walls, and skyrmions using the above-mentioned techniques. However, better spatial resolution requires further improvement. These techniques are also used to study magnetization of 3D magnetic nanostructures. Neutron scattering is a suitable technique for probing magnetization of a large-volume magnetic system. X-ray microscopies (XM) [69] are also useful for 3D magnetization imaging. To probe magnetization in 3D structures, different XM have been developed, such as X-ray photoemission electron microscopy (XPEEM) [70], ptychography [71], magnetic X-ray tomography (MXT) [72], etc. Photoemission electron microscopy (PEEM) [73], spin-polarized low-energy electron microscopy (SPLEEM) [74], scanning electron microscopy with polarization analysis (SEMPA) [75], spin-polarized scanning tunneling microscopy (SPSTM) [76], and ballistic electron magnetic microscopy (BEMM) [77] are the techniques where spatial resolution better than 10 nm can be achieved. Electron holography is another very strong imaging tool where amplitude and phase information of the spin distribution can be studied with resolution down to 2 nm [78].

9.2.2.2 Dynamic characterization

The magnetization dynamics generally take place in the subnanosecond timescale. There are various tools that have emerged with the requirements of quantitative and qualitative characterization of SW dynamics of the nanostructures. The well-known techniques in the time domain, frequency domain, and wave vector domain are time-resolved MOKE (TRMOKE) [79], ferromagnetic resonance (FMR) spectroscopy [80], and Brillouin light scattering (BLS) spectroscopy [81] (see Fig. 9.2).

There are a number of variants of the TRMOKE technique, which can be broadly categorized into field-pumped and optical-pumped techniques. In both techniques the magnetization of the sample is perturbed by either a femtosecond or a picosecond pump laser or with an ultrashort magnetic field, and spin dynamics is triggered due to the non-equilibrium configuration of the system. Explicitly, the precessional motion of the magnetic moments about an effective magnetic field becomes the matter of interest. A probe laser is used for the detection purpose using MOKE. Depending upon the spatiotemporal resolution offered by the system, direct time domain detection of the precession frequency and Gilbert damping of the SW modes is possible using this technique. This offers non-invasive and highly sensitive detection of ultrafast magnetization dynamics [82]. Time-resolved scanning Kerr microscopy (TRSKM) [83] can also be exploited to image the time evolution of magnetization in confined magnetic structures.

Broadband FMR is a global excitation method where external perturbation in the MHz to GHz frequency range is launched into the sample and the magnetization precession with certain resonant frequencies exhibits absorption from the external source in the frequency spectrum. This technique requires delicate microfabrication of waveguide structures during sample preparation. To measure the field dispersion of SW frequencies with high spectral resolution, FMR is found to be an extremely useful technique. There are other variants of these techniques, which include the conventional cavity-based FMR, spatially resolved FMR, spin-torque FMR [84–86], etc.

In the BLS technique, the photons interact with relatively low-frequency quasiparticles such as phonons, magnons, plasmons, etc., present in the system. In the case of magnetic materials, they scatter from the thermally excited SWs, which are both elastic and inelastic in nature. In the former case the frequency remains unchanged, whereas in the latter case, a shift in frequency is observed. The

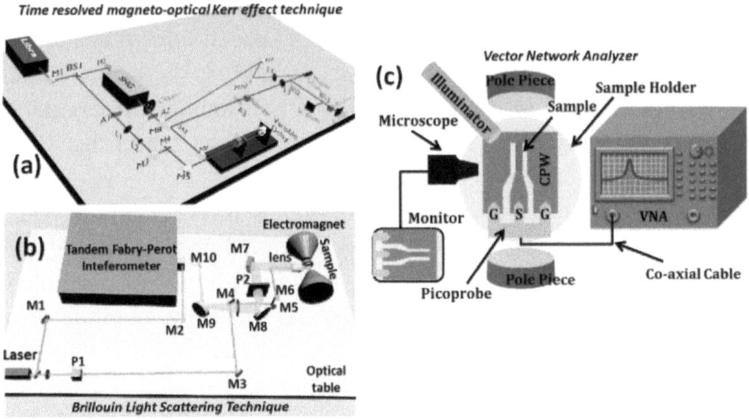

Figure 9.2 (a) Schematic diagram of the time-resolved magneto-optical Kerr effect (TRMOKE) technique. BS = beam splitter; M = mirror; A = attenuator; and L = lens. (b) Schematic diagram of a conventional Brillouin light scattering (BLS) setup. P = polarizer. (c) Schematic diagram of a broadband ferromagnetic resonance (FMR) spectrometer.

scattered light is collected in back-reflected geometry, and then the frequency is analyzed using a multipass tandem Fabry–Perot (FP) interferometer to extract the information about the bulk and surface magnons. It is a versatile tool for studying SW dynamics with flexibility in frequency, phase, time, and spatial domains. Pulsed inductive microwave magnetometry (PIMM) is an oscilloscope-based time domain detection tool to measure the dynamics with picosecond resolution [87]. The magnetoresistive methods and X-ray microscopy are also very powerful tools that are widely being used.

9.3 3D Magnetic Structures

The realization of a magnetic configuration of 3D nanostructure counterpart of a 2D nanostructure faces great challenges due to limitations of the fabrication technique. However, some of the 2D fabrication techniques have been promoted for 3D magnetic structure fabrication. In a broad sense, the fabrication techniques can be categorized into three different groups [42]: physical vapor deposition (PVD), chemical synthesis, and 3D nanoprinting. The PVD technique is used to fabricate most of the 2D nanopatterns and thin films. PVD can be used to fabricate 3D magnetic structures by depositing materials on a prepatterned 3D scaffold [71]. This frame can be fabricated by using TPL. The characterization of the Co layer on a buckyball [71] polymer 3D structure was studied by using ptychographic X-ray tomography. Helical 3D magnetic micromachines [88] were fabricated using TPL and electron-beam evaporation. These structures can be useful for localized drug delivery. Recently, May et al. demonstrated a 3D frustrated magnetic diamond lattice of Py nanowires by using TPL and electron beam evaporation [89]. Also, free-standing nanoscale Co tetrapod structures [64] have been fabricated on glass/indium tin oxide (ITO) by using a combination of TPL and the electrodeposition technique. The schematic of the TPL technique and a SEM image of a tetrapod structure are shown in Fig. 9.3. The curved magnetic structures can be fabricated by deposition of magnetic material [90, 91] on self-assembled spherical nonmagnetic nanoparticles. A multilayered curved surface such as nanotubes [41] can be fabricated by using rolled-up technology and self-assembled methods.

Chemical synthesis techniques are popular techniques for the fabrication of 3D magnetic nanostructures. A 3D interconnected NiCo nanowires structure [92] was fabricated by using PCTE membranes and the electrodeposition technique. Also, patterned MLs can be fabricated by using these techniques. However, this technique is not suitable for complex 3D nanostructure fabrication. Recent advancement of the FEBID technique makes it capable of growing polycrystalline magnetic metals and alloys [65] and promotes it as a powerful and flexible technique for free-form 3D complex magnetic structure fabrication. One of the major drawbacks of the FEBID technique is the contamination of deposited material with oxygen and carbon. However, MOKE loops obtained from FEBID-fabricated Co nanowires with diameter 100 ± 25 nm and length 4–12 μm [93] demonstrated high-quality samples obtained from this technique. Micromagnetic simulations revealed the Bloch-point vortex domain wall in a nanowire. Also, a 3D nanomagnetic logic device based on magnetic coupling between Fe nanopillars

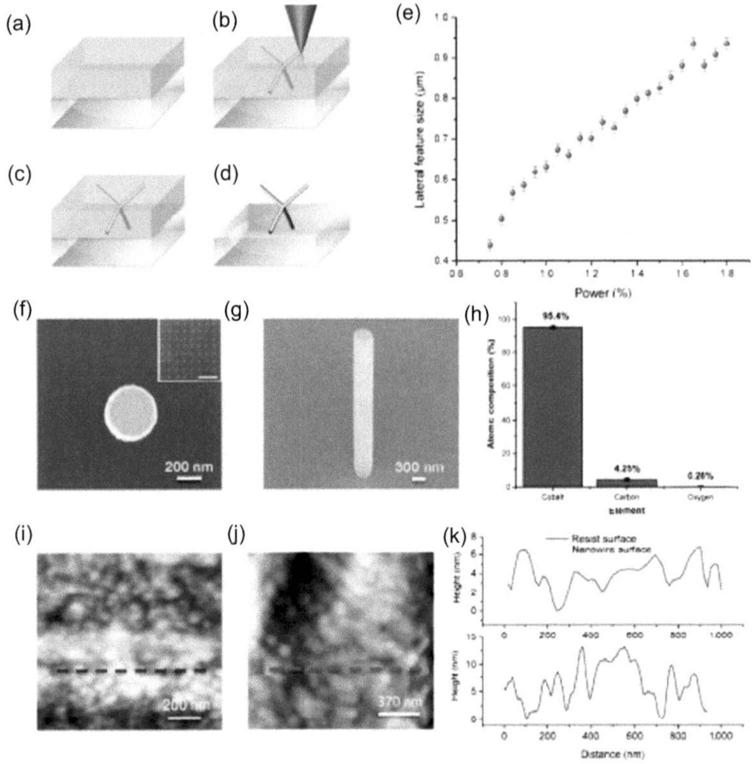

Figure 9.3 Fabrication and structural characterization of 3D magnetic nanostructures. (a) Spin coating of a positive resist onto a glass/indium tin oxide (ITO) substrate. (b) Two-photon lithography of a 3D structure into the positive resist. (c) Electrodeposition of Co into the channels. (d) Lift-off of the resist. (e) Lateral feature sizes obtained for pillars fabricated at different powers. (f) SEM micrograph of a single 435 nm Co nanowire. Inset: an array of sub-500-nm pillars. Scale bar = 8 μm. (g) SEM micrograph of a single 435 nm Co nanowire captured at a 60° angle. (h) Energy-dispersive X-ray analysis of Co pillars. (i) Atomic force microscopy (AFM) image of the resist channel surface. (j) AFM image of a nanowire sidewall. (k) Profiles obtained from the nanowire sidewall and AFM images of the resist channel (i) and (j) [64] (under the terms of the Creative Commons CC BY license).

and nanowires [94] and nanomagnetomehanical actuator-based Co nanorods [48] has been demonstrated in 3D magnetic structures.

Recently, FEBID-grown FeCo magnetic structures [95] have been proposed for 3D artificial spin ice. Simulation-guided fabrication of 3D nanostructures [96] can be used in desired 3D magnetic structure fabrication. However, large-area fabrication of magnetostatically coupled and uncoupled 3D nanostructures with better material purity [97] requires a new technique and strategy of fabrication for systematic understanding of magnetism in the third dimension.

The dynamics of 3D nano- and microstructures have not been explored so much in the literature. Recently, researchers have become interested in studying magnetization dynamics of 3D magnetic structures from fundamental and application points of view. Rana et al. studied magnetization dynamics of Ni nanoparticles assembled in various 3D structures and compared the ultrafast demagnetization, relaxation, and precessional dynamics in various types of assemblies [98]. In Ref. [99], the microwave response of self-assembled magnetoferritin nanoparticles arranged in 3D face-centered cubic (FCC) arrays is shown. Below the blocking temperature of nanoparticles, the linewidth of ferromagnetic resonance was broadened due to the misalignment of crystal lattices of individual nanoparticles with respect to each other. A large number of works on dynamics of magnetic nanowires can also be found in the literature [25, 100, 101]. However, the study of magnetization dynamics of lithographically patterned high-quality 3D nanostructures has been missing in the literature until recently, when Sahoo et al. [102] reported for the first time the study of magnetization dynamics from a 3D nanoscale complex tetrapod structure using the all-optical TRMOKE microscopy technique.

The magnetic tetrapod structures were fabricated by using a combination of TPL and electrodeposition. The details of sample fabrication can be found in Ref. [58]. The time-resolved magnetization dynamics were measured from the junction of a tetrapod by using a focused submicron probe laser after excitation of another pump laser of similar spot size. Three well-separated SW modes of frequencies between 1 and 30 GHz have been observed. The nature of the SW modes was studied by using micromagnetic

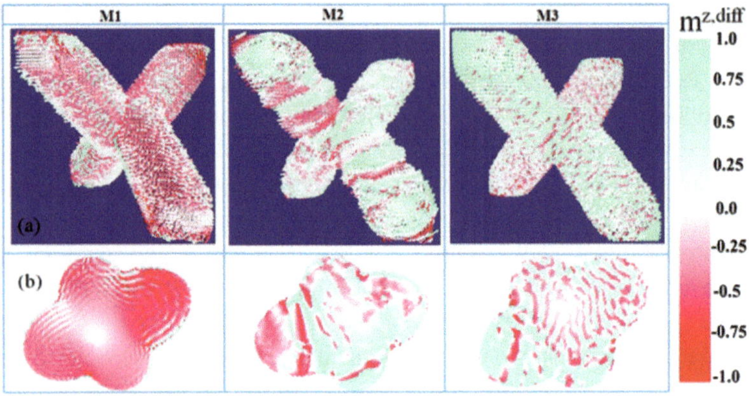

Figure 9.4 Magnified view of the mode profile and (b) sliced view of the tetrapod junction for different modes at $H = 4.50$ kOe. From Ref. [102]. Published by The Royal Society of Chemistry (https://creativecommons.org/licenses/by/3.0/).

simulations. The simulated mode profiles are shown in Fig. 9.4. The highest frequency mode showed uniform precession, while the two lower frequency modes showed quantized behavior where the quantization number increased with a decrease in frequency.

Numerically calculated magnonic band structures of 3D MCs were first reported in Ref. [103]. In the literature, SW properties in 3D magnetic structures have been studied by using the plane wave method (PWM) [103–108], the linear combination of atomic orbitals (LCAO) [109] method, and micromagnetic simulations [108, 110]. The 3D structures have been studied in systems of ferromagnetic spheres arranged in sites of a simple cubic (SC) [103, 104, 107, 109], body-centered cubic (BCC) [104], FCC [104, 106], and hexagonal close-packed (HCP) [105, 109] crystallographic structures embedded in a matrix [103–107] of ferromagnetic materials. The dipolar interaction affects the band width [104] and spin relaxation [109] in these structures. The magnonic band gaps were found to be sensitive to exchange and magnetization contrast in between magnetic materials of sphere and matrix. The bandgap opening occurred above a critical value of magnetization contrast (even without exchange contrast) depending on the lattice type. The exchange contrast also played an important role on magnonic gap opening, but it needed very large values for opening of the bandgap in the absence of magnetization contrast. The normalized bandgap varied almost linearly with both the contrasts. The bandgaps also depended on lattice symmetry and scattering center shape. Among the cubic lattices, the FCC structure showed the highest bandgap. A systematic study of ferromagnetic material combination effect on the bandgap in HCP lattices was reported in Ref. [105]. The highest bandgap was found for Ni spheres embedded in an Fe matrix. The PWM calculations of 3D MCs have been extended for magnetoferritin-based 3D MCs. In the calculations, the interparticle space was filled by ferromagnetic material. The introduction of ferromagnetic material ignited the magnonic bandgap opening in these systems. The FCC lattice produced by self-assembled magneto-ferrotin nanoparticles is suitable for a magnonic bandgap system. The damping of SW in a homogeneous medium is independent of the direction of SW propagation. This scenario changed for MCs. The damping has been considered in PWM calculations for SW in 3D MCs in Ref. [107]. The anisotropic nature and wave vector dependence of damping can be tuned by changing the magnetic material and filling fraction of the MC structure. Recently, the meander-like structures of yttrium iron garnet (YIG) fabricated on a prepatterned substrate have been proposed for 3D MCs [108, 110]. SW propagation and dispersion for isolated and coupled 3D MCs were studied by using the finite element method, micromagnetic simulation, and transfer matrix method. The bandgap of the Damon–Eshbach (DE) mode increased with the increase in the width of the groove. For the coupled case, the dispersion curve splits into two, symmetric and antisymmetric modes. The calculation of SW dispersion was extended for the volume magnetostatic surface wave [108]. The dispersion curves were getting narrower with the increase in the groove width, which means the volume wave can propagate with small wavenumbers. Both the volume modes can coexist in this MC system.

9.3.1 Topological Magnetic Solitons

Solitons are mathematical solutions of a broad class of weakly nonlinear dispersive partial differential equations describing a physical system. They were first theorized in 1970s and were found to occur due to a delicate balance of magnetic forces. There is an inherent link between topology and dynamics in topological magnetic solitons, as found in ferromagnetic bubbles. They can potentially be harnessed for transmitting and storing data in all-magnetic circuits in a more energy-efficient manner than the present technology. Here, in this section, we will discuss the static and dynamic magnetic properties of two special types of topological magnetic solitons, magnetic vortex and skyrmions, which exhibit spin reorientation from planar to nonplanar configuration.

9.3.1.1 Magnetic vortex

A magnetic vortex is a class of topological defect that mostly stabilizes within submicron magnets following a specific range of thickness-to-diameter ratio. Being confined by the geometrical restriction, spins inside cylindrical magnetic platelets form a curling in-plane magnetization with an out-of-plane core [111]. This is characterized by two binary properties: one is polarity and another is chirality. Polarity of a vortex can be denoted by $p = 1$ or $p = -1$, depending upon the direction of magnetization of the core pointing upward or downward, respectively. This is also denoted as the topological charge (Q) of the vortex. Chirality is the sense of in-plane rotation of magnetization in counterclockwise (CCW) or clockwise (CW) direction and is also denoted by $C = 1$ (CCW) and $C = -1$ (CW). Without losing much of the exchange energy, spins align themselves in the plane of the structure to cancel the total dipole energy. However, near the center, the angle between neighboring spins becomes increasingly larger if the spins try to remain in the sample plane. As a result, at the core of the vortex, magnetization turns out of plane within a small area with a Gaussian shape. The relatively isotropic shape (i.e., circle, square, ellipse, ring) of a submicron dot facilitates the formation of a vortex. A magnetization reversal study of cylindrical nanomagnets by varying their aspect ratio (thickness and diameter) revealed a phase transition with the lateral confinement from the vortex to single-domain states [111]. Depending upon the order of the exchange length and aspect ratio $\beta = d/R = 0.1 - 0.5$ of the magnetic disks (d is the thickness, and R

is the radius) the vortex can be stabilized. In this phase, nucleation and annihilation of a vortex is observed from the squeezed loop near the origin. In the monodomain phase, reversal progresses by a coherent rotation of the spins, forming a square-type loop. The most interesting feature that makes this topological soliton the center of huge attraction in the last two decades is the exchange-coupled core, which points along the third dimension. This can store information in the form of a binary state within itself. However, it adds much complexity to the dynamics of the system.

The vortex core was first observed from MFM imaging in a 50-nm-thick circular Py dot with varying diameter from 1 μm to 100 nm, though the core size could not be quantified [112]. Subsequently, spin-polarized scanning tunneling microscopy was exploited to map the internal spin configuration on nanoscale iron islands [113]. It was found that the size and shape of the vortex core depend on the exchange stiffness constant and saturation magnetization of the material. In a patterned thin Py film, a cross-tie-wall can be probed, which evidences the existence of perpendicularly magnetized cores, even for an antivortex texture with diverging spin configuration at the vicinity of the core [114]. Sometimes, the relatively larger dot can accommodate two vortex cores with four different possible combination of polarities ($p1p2 = 1$ or $p1p2 = -1$) [115]. This coupled vortex dynamics can be solved by following Thiele's equations:

$$\mathbf{G}_j \times \frac{dX_j}{dt} - \frac{\partial W(X_1, X_2)}{\partial X_j} = 0, \tag{9.1}$$

where W is the energy function for restoring force, $\mathbf{G}$ is the gyrovector for gyroforce, and X_j is the position of the jth core. If external perturbation is applied to the core in the presence of another bias magnetic field, the frequency of excitations scales with the field value. It is also found that the vortex state can still exist without having an out-of-plane core if the "singularity" can be physically removed. Inside an exchange-biased elliptical nanoring in the absence of any core, the chirality (CW or CCW) of the vortex varies as a function of the external magnetic field and exchange bias angle [116]. The core of the vortex can be reversed by an application of a short burst of magnetic field [117], from current-driven resonant dynamics of the vortex [118], spin current [119], or spin-polarized current [120]. The switching of the vortex core can be useful for efficient data writing in a storage device having data stored inside a nanometric core.

The time domain study of vortex dynamics was first reported by Park et al. using TRSKM in polar geometry. Magnetic disks with $\beta = d/R = 0.03, 0.06$, and 0.12 were subjected to the excitation near the core and damped sinusoidal oscillations were obtained with a couple of high-frequency as well as low-frequency mode superposed in it [121]. The high-frequency mode was found to be originated from the precession of the magnetization about the internal field. The low-frequency mode, which increased with increasing β (in the order of 1–2 GHz), corresponded to the spiral gyrotropic motion of the entire vortex in the presence of a magnus force. According to the theoretical prediction made by Guslienko et al., the frequency of this gyrotropic motion is

$$\omega_0 = \gamma M_s \xi^2 / 2\chi(0), \tag{9.2}$$

where γ is the gyromagnetic ratio and M_s is the saturation magnetization. The parameter ξ and the susceptibility $\chi(0)$ depend on the magnetization distribution of the vortex. The nature of this motion can be portrayed from two possible models according to them: the entire vortex can move rigidly in the presence of edge poles of the disk (rigid vortex model) and a model avoiding edge poles. The experimental situation agrees mostly with the second model [122, 123]. Along with the low-frequency radial mode, other azimuthal modes were also experimentally observed in the magnetic vortex. The degeneracy of those modes was found to be broken depending upon the internal domain structure with decreasing radius of the magnetic cylinder [124]. The length scale and timescale of the vortex core deviation from its equilibrium are evidenced from the measurements using broad-band FMR with both the real and the imaginary impedance measured [125] and very sensitive X-ray photoemission electron microscopy [126]. During this radial motion of the core, the streamlines of the stray field outside the vortex can experience a dynamic motion, too. It is pertinent to mention here that magnetic vortices cannot, in general, interact among themselves with dipolar coupling due to their flux closure structures in equilibrium. However, the collective gyrotropic motion is affected by this dynamic coupling between the interacting stray fields in an array of magnetic vortices [127, 128]. Barman et al. proposed a unique energy transfer mechanism based on the dynamics of coupled vortices [129]. In a 1D chain of magnetic vortices without having any physical connections, they numerically demonstrated that local excitation given to a single vortex can travel across the chain via dynamic dipolar coupling. The polarization, chirality, and shape of the disks were optimized to achieve reasonable transmittance at the output end. They have clearly lineated the role of sign, magnitude, and separation of the magnetic side charges on determining the timescale and length scale of this energy propagation. Later, more detailed experimental and numerical studies on the anisotropic energy transfer were pursued on coupled magnetic vortices [130, 131]. The amplitude of gyration was detected, and mode splitting was observed due to a chirality sensitive phase difference in gyrations of the cores. The magnitude of the splitting was determined by their polarity configuration.

The exciting properties of the magnetic vortex make it a possible candidate for highly compact spin-torque vortex-oscillator devices, which are important in microwave signal-processing applications [132]. Kim et al. demonstrated logic operations using magnetic vortices based on the idea proposed by Barman et al. They utilized a simple chain structure consisting of three physically separated but dipolar-coupled Py disks having a vortex state as well as two electrodes for launching the logical inputs and directly monitored the vortex gyrations in the middle disk, as the logical output, by time-resolved full-field soft X-ray microscopy measurements. By manipulating the relative polarization configurations of both end disks, two different logic operations are programmable: the XOR operation for the parallel polarization and the OR operation for the antiparallel polarization. The advantages are as follows: (i) a low-power input signal by means of resonant vortex excitation, (ii) low-energy dissipation during signal transportation by selection of low-damping materials, and (iii) a simple patterned-array structure [133].

Recently, the idea of a magnetic vortex-based transistor (MVT) was proposed by Kumar et al. There, a three-vortex sequence (material Py: diameter = 200 nm; thickness = 40 nm; separation

= 50 nm) with polarity combinations (1, −1, −1) was proposed to work as an MVT, providing a gain (B, the amplification of core gyration amplitude) of ~15 dB at the output end, similar to an electronic bipolar junction transistor (BJT) [134].

By the switching the polarity of the middle vortex the transistor can be operated between the ON and OFF states, similar to the BJT driven by a base current (see Fig. 9.5a,b). Further in a similar system, a giant amplification MVT was achieved by introducing geometrical asymmetry: tuning the intervortex separation $S1$ and $S2$ in the three-vortex network. The gain B for an asymmetric MVT (AMVT) started to increase with the increase of $S2$. A maximum of $B = 28.28$ dB was observed at $S2 = 100$ nm and it decreased with a further increase of $S2$. Temporal evolution of the stray field showed that stray-field antivortices are created, which move along nearly elliptical paths, and the length of the major axis of this ellipse increased with an increase in $S2$ and reached a maximum at $S2 = 100$ nm, beyond which it dropped. Furthermore, the output from this AMVT [135] was fed into the input of another two AMVTs (placed parallel with the vertical shift) to perform a successful fan-out operation and nearly equal gains were observed at both the outputs (see Fig. 9.5c–g).

Exploiting the asymmetric nature of the energy transfer mechanism by antivortex solitons, a successful fan-in operation can be shown. A tristate buffer gate with fan-in of two AMVTs can be formed. The loss (0) and gain (1) of energy at the output vortex can be controlled only by manipulating the polarities of the middle vortices in four different combinations (1, −1), (−1, −1), (−1, 1), and (1, 1) placed at input branches. This gate can be used as a "switch" to the logic circuit, and it has technological importance for energy transfer to large-scale vortex networks. With the rapid advancement of memory devices, this finding will accelerate the development of all-magnetic computation.

9.3.1.2 Magnetic skyrmions

Magnetic skyrmions are topologically protected magnetic solitons with a whirling spin texture [136]. This texture allows spins to align along all three dimensions. The topology is described by a winding number W:

$$W = \frac{1}{4\pi}\int \mathbf{m}\cdot(\partial_x\mathbf{m}\times\partial_y\mathbf{m})dx\,dy = \pm 1, \qquad (9.3)$$

where $\mathbf{m}$ is the normalized magnetization vector, which covers a sphere, and hence the winding number turns out to be unity for skyrmions. This exhibits a particle-type nature. Primarily there are two prominent types of skyrmions, Néel (hedgehog) and Bloch skyrmions. Experimentally, skyrmions were first observed in itinerant-electron magnet MnSi with a noncentrosymmteric lattice [137] well below the Curie temperature. There, the symmetry breaking originates from bulk DMI. At the ferromagnet and heavy metal interface, broken structure-inversion symmetry appears from interfacial DMI (IDMI). It is due to spin–orbit coupling, which links the spins to the real space. SPSTM is exploited to image the skyrmion lattice in the magnetic ground state of such an interface between hexagonal monolayer Fe and Ir at low temperature. Since then, due to the evergrowing demand for data storage, the quest for generation of skyrmions at room temperature started growing. In both the magnetic systems, in the presence and absence of DMI, stable skyrmion phases were detected at room temperature [138–142].

The additive nature of IDMI in the presence of two interfaces at either side of a ferromagnetic layer was found to be more suitable for the generation of skyrmions [143]. Tuning the size of the skyrmion became a challenging task, as it depends on many external factors such as DMI energy, saturation magnetization, strength of the external magnetic field, and current. Skyrmions with sub-100-nm size can evolve from the micron-sized labyrinth, stripe, or bubble domain. It has been shown by using the XMCD technique that sub-100-nm skyrmions can be stabilized by designing multiple repeats of the Pt/Co/Ir trilayer. The higher value of DMI energy facilitates skyrmion formation, and with increasing magnitude of the external magnetic field, the sizes of these skyrmions were found to be decreased. A very interesting idea was surfaced by various groups to create artificial skyrmionic crystals within 2D nanodot arrays [144, 145]. On top of a perpendicularly magnetized thin film, arrays of D-shaped nanodots can be fabricated, where the aspect ratio of the dots prefers the stabilization of

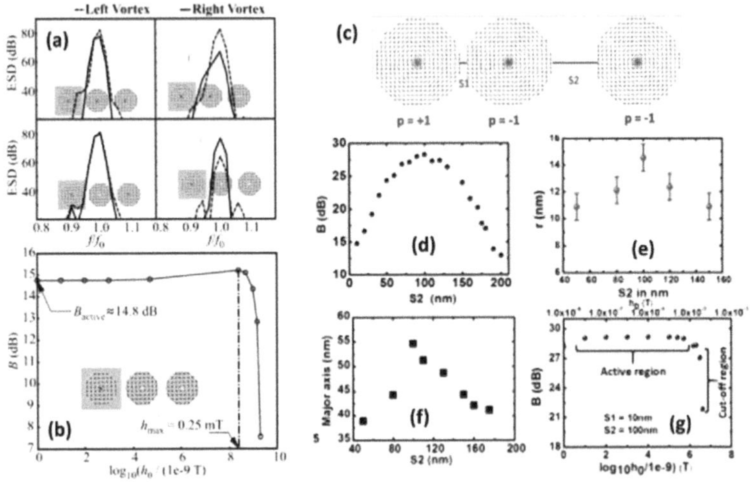

Figure 9.5 (a) ESDs of left and right magnetic vortices with ($p1$, $p2$, $p3$) equaling (1, 1, 1), (1, 1, −1), (1, −1, 1), and (1, −1, −1), as shown in the respective insets. A 1.5 mT signal rotating CCW at frequency f_0 is applied only to the left (shaded) vortex. (b) Gain B, versus logarithm of signal amplitude h_0. (c) The ground-state spin configuration of a magnetic disk of diameter 200 nm and thickness 40 nm arranged in a three-vortex system with a separation between left and center vortices ($S1$) of 10 nm and a separation between center and right vortices ($S2$) of 100 nm. (d) Variation of gain (B) with $S2$ for AMVT. (e) Variation of relative core displacement (r) of the right vortex with respect to the left vortex with $S2$ for AMVT. (f) Variation of length of the major axis of the ellipse with $S2$ for AMVT. (g) Variation of gain (B) with strength of the applied magnetic field. Reprinted with permission from Ref. [134]. Copyright © 2014, Springer Nature. From Ref. [135] (https://creativecommons.org/licenses/by/4.0/).

the vortex state with an out-of-plane core [144]. Gilbert et al. demonstrated room-temperature realization of large-area artificial Bloch skyrmion lattices by using asymmetric magnetic nanodots with controlled circularity fabricated on a Co/Pd multilayer with perpendicular magnetic anisotropy (PMA). By using a critical ion irradiation step they suppressed the PMA and thereby imprinted the vortex structure into the interfacial region of the dot and the underlayer. The polarity was controlled by a tailored magnetic field sequence. The imprinted skyrmion lattices were probed directly with polarized neutron reflectometry, which was further confirmed by magnetoresistance measurements (see Fig. 9.6) [39]. More recently search for antiskyrmions has started, and at room temperature using the LTEM technique an antiskyrmion lattice was experimentally imaged in a $Mn_{1.4}Pt_{0.9}Pd_{0.1}Sn$ Heusler alloy by Nayak et al. [146]. These antiskyrmions have cylindrical symmetry and a quadrupolar moment of magnetic charges, and their motion can be controlled by applying a magnetic field. Skyrmions have been proposed to be extremely potential candidates for many of the new-age magnetic technologies, such as skyrmion racetrack memory, skyrmion logic devices, skyrmion MCs, skyrmion-based radiofrequency devices, etc. [136].

These possibilities can be implemented as real-world applications only if the individual skyrmion can be efficiently manipulated and its dynamics controlled [147]. Researchers attempted to simulate the skyrmion dynamics early in this decade. Ma et al. proposed a unique design for skyrmion-based dynamic magnonic crystals (SDMCs) [148]. A 3200-nm-long and 40-nm-wide waveguide with a 1-nm-thick Co and Pt bilayer was chosen to be a convenient system for this study where IDMI can be added in terms of a magnetic field. On application and withdrawal of a spin-polarized current via nanocontacts, the linear skyrmion lattice appeared (ON) and disappeared (OFF) in the waveguide. The size of individual skyrmions was controlled by an external field or amplitude of current. The magnonic band structure and corresponding SW mode profiles were calculated for the SDMCs, which revealed that during transition from OFF to ON state, the allowed SW modes became forbidden. Experimentally, the skyrmion lattice was stabilized in many other heterostructures along with the Pt/Co/Ta film, and by using scanning transmission X-ray microscopy (STXM) their velocity of about 50 m/s was recorded in a waveguide on application of a current (see Fig. 9.7) [149]. Skyrmion velocity increased with diameter. Interestingly, in a system with a ferromagnetic layer attached to a heavy metal layer with high spin–orbit coupling, the spin Hall effect–generated spin-orbit torque could control the velocity more efficiently [150]. However, in the ferromagnet-based system, a transverse movement of the skyrmions was observed in the track and the straight-line motion of the skyrmions was hindered. This phenomenon is known as the skyrmion Hall effect,

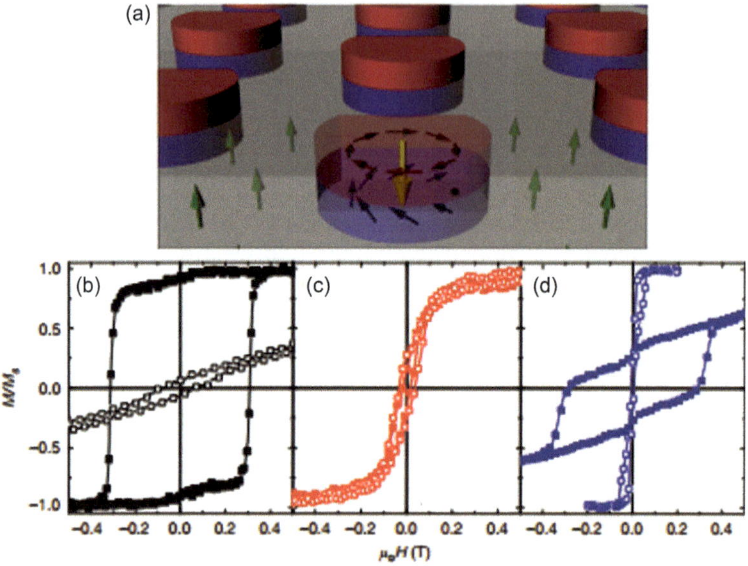

Figure 9.6 Schematic diagram of artificial skyrmion lattices and measured magnetic hysteresis loops. (a) The hybrid structure consists of Co dots (red) on top of a Co/Pd PMA underlayer (gray), where the in-plane spin texture of the Co dots (purple arrows) is imprinted into an irradiated Co/Pd region (light blue) underneath the dots (tilted blue arrows). Green and yellow arrows indicate the moments in the Co/Pd underlayer and the core region of the (imprinted) vortex, respectively. Major in-plane (open symbols) and perpendicular (solid symbols) hysteresis loops are shown for (b) the Co/Pd underlayer as grown, (c) the irradiated Co/Pd witness sample, and (d) the hybrid Co + Co/Pd sample. From Ref. [39] (https://creativecommons.org/licenses/by/4.0/).

which originates from magnus force [151, 152]. This effect increases with increasing velocity, and it is intriguing to examine how can it be suppressed. Later, it was found that for antiferromagnetically exchange-coupled skyrmions, this effect does not arise due to the absence of a net topological charge [153]. The SWs associated with the skyrmion dynamics were studied in great detail by Montoya et al. by using the FMR technique. In a Gd/Fe multilayer, the dipole skyrmions lattice was stabilized on application of a magnetic field perpendicular to the substrate when temperature was changed from 180 K to 100 K via the formation of several domains [154].

It was identified that five distinct magnetic textures appear in the field dispersion curve of the precessional frequency with decreasing temperature: i) disordered stripes, ii) coexisting stripes and dipole skyrmions, iii) skyrmion lattices, iv) disordered isolated skyrmions, and v) magnetic saturation. Except in the saturation state, spin dynamics do not exactly follow the Kittel formula but frequency increases with decreasing field in the lower field regime. They also observed the existence of a breathing mode proposed earlier by Mochizuki [155], which originated due to expansion and contraction of a skyrmion under the sinusoidal field in the out-of-plane direction. With the help of micromagnetic simulations they have explained the nature of the SW modes with more clarity. A great deal of further work on skyrmion dynamics is under progress at present.

9.3.2 Patterned Magnetic Multilayer Structure

Patterned ML structures with PMA have potential applications in high-density magnetic storage devices, magnetic sensors, and magnonic devices. Patterned ML structures possess periodic modulation of magnetic properties along all three spatial directions. Magnetic properties of ML dot and antidot lattices with PMA, mostly based on Co/Pd and Co/Pt MLs, have been studied in detail for their possible applications in storage devices. As an outcome, it was found that PMA in these structures originates from *d* shell electron hybridization [156]. Magnetic properties of patterned structures strongly depend on pattern size and spacing between them. In the literature, the reported coercivity values of patterned MLs are higher (by almost 1 order) than continuous MLs film due to additional pinning center at their edges and shape anisotropy of the patterned structure [157, 158]. Ion irradiation was exploited to tune the coercive field [159] and anisotropy [160] of these systems. It was observed that PMA of the ion-irradiated areas decreases and may switch to the in-plane direction [160]. Lithographically patterned discrete ML dots showed higher coercivity than the exchange-coupled ML dot patterns prepared by focused ion beam irradiation, via the in-plane anisotropic magnetic regions. Nanomagnetic logic devices have been demonstrated based

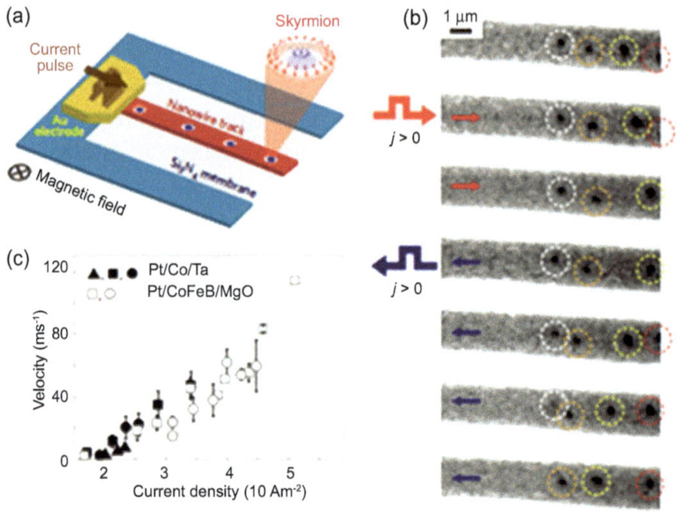

Figure 9.7 Current-driven skyrmion motion. (a) Schematic of a magnetic track on a Si_3N_4 membrane with current contacts and skyrmions stabilized by a down-directed applied magnetic field. (b) Sequential STXM images showing skyrmion displacement in Pt/Co/Ta after injecting 20 unipolar current pulses along the track, with amplitude $j_a = 2.2 \times 10^{11}$ A m^{-2} and polarity, as indicated. Individual skyrmions are outlined in colored circles for clarity. (c) Average skyrmion velocity of Pt/Co/Ta (filled symbols) and Pt/CoFeB/MgO (open symbols) versus current density. Different-shaped symbols correspond to different devices. Error bars denote standard deviation of multiple measurements. Reprinted with permission from Springer Customer Service Centre GmbH: Springer Nature, *Nature Materials*, Ref. [149], Copyright (2016).

upon these properties of MLs [161]. For PMA antidot systems, magnetic properties strongly depend on antidot size and separation. For self-assembled structures, the coercive field and pinning field increase with increasing pore density [162]. Also, it can be tuned by varying the pore diameter and edge-to-edge separation [158]. In nanorings sculpted from the ML structure [163], it was observed that with increasing interelement separation, the magnetostatic interaction decreases. As a result, the coercivity increased. In patterned ML structures the coercive field and domain wall pinning field can be predicted by the Kondrosky model [164, 165]. The coercivity variation for transition from the dot to the antidot regime was studied in Ref. [166]. It reached a maximum value for an array of antidots with a separation close to the domain wall width for which the pinning was strongest. Switching field distribution (SFD) is an important parameter for stability of storage devices. Grains with the in-plane vector are correlated to dots with small switching fields. The switching field suppression is linked to the size of these grains [167]. SFD can be controlled by varying the antidot diameter [168] and spacing between nanoring patterns [163].

Theoretically, SW dispersion in patterned MLs with repetition of two layers of different magnetization was studied in Ref. [169] using discrete dipolar approximation (DDA). The effects of exchange and dipolar interactions on SW spectra were reported. For dipolar interaction, the SW spectra showed a quasi-continuous dispersion band for uniform ML magnetization. For non-uniform magnetization the band splits into two and the bandgap increases with increasing inhomogeneity. The quasi-continuous dispersion band splits into six bands when exchange interaction is included in the calculations. The high frequencies of exchange spectra were in the order of tens of THz. Only the lowest two modes were modified by dipolar interaction and other modes were controlled by exchange interaction between the layers. The experimentally observed SW modes in patterned ML structures have been demonstrated recently [170, 171]. In Ref. [170], the magnetization dynamics in Py dots arranged in a square lattice underneath the Py thin film has been studied using the FMR technique. The patterned structure periodically modulated the potential of the film surface, and consequently the SW spectra got modified. Two SW modes were observed, of which the lower-frequency mode originated from the dot, whereas higher-frequency mode originated from the thin film. The mode intensities got modified with varying film thickness. On the other hand, the magnetization dynamics in Co/Pd ML PMA antidot lattices were studied using a TRMOKE microscope in Ref. [171]. The SW mode frequency was found to decrease with decreasing antidot lattice constant for a fixed antidot diameter. Considering modified magnetic properties around the shell of the antidot, created due to ion irradiation during the FIB milling procedure of the ML, in PWM calculation, they deciphered the origin of SW modes (see Fig. 9.8).

The dynamical coupling between these shell modes modified the frequencies in the SW spectra.

Further works on the dynamics of patterned MLs with the aim of understanding various multiphysics phenomena such as IDMI, spin–orbit torque, and voltage-controlled magnetic anisotropy (VCMA)-based SW generation and control are under progress but beyond the scope of this chapter.

9.3.3 3D Spin Configurations in the Curved Surface

Geometrical curvature in a planar system can induce spin reorientation from tangential to normal direction of the plane, giving rise to a huge anisotropy in the nanostructure. The curvature can be induced by patterning, chemical treatment, and reshaping of the geometry after patterning or modifying the topology. The types of curved nano-objects gaining recent scientific interest are curved wire, curved film, rolled-up surface, Möbius band, helices, and magnetic cap [44]. There are lots of physical and chemical methods available for preparation of 3D magnetic objects with different sizes and shapes. Mainly the nanotube and nanowire type of structures are grown by electrochemical deposition techniques within corresponding templates. Nanospheres, nanocaps, and nanocylinders are fabricated using physical deposition techniques, which provide solid, smooth interfaces. Explicitly, glancing angle deposition is required for helices, springs, etc. Another technique combining chemical deposition with strain engineering is exploited to prepare a swiss roll with different helicities. The ground states of these 3D magnetic structures exhibit peculiar spin textures, which are not observed usually in other nanostructures. Specially, the formation of a domain wall in the third dimension is a matter of huge interest due to its already proposed manifestation as "racetrack memory." To study those features, extremely sensitive and high-resolution characterization techniques are used, such as MFM, scanning magnetoresistive microscopy, TEM, XPEEM, X-ray holography, neutron tomography, X-ray tomography, 3D imaging, etc. However, to probe the feature size in the nanoscale, XMCD is a better tool. The basic understating of curvature-induced effects can be built up by studying the geometry-induced potential in the system. Magnetic energy can be expressed as a combined effect from isotropic Heisenberg exchange and two magnetostatic interactions: i) effective anisotropy and ii) effective DMI. In the case of nanowires or nanoribbons, effective anisotropy creates tilted spin configuration, which can be tuned by changing the curvature and torsion of the structure. A ring-shaped wire has a finite constant curvature and zero torsion originated from shape-induced pattering, and depending upon that, the magnetization forms a flux-closure vortex-like distribution with large anisotropy. A Möbius ring develops topologically induced domain walls within it due to a discontinuity of any normal vector field arising from its non-orientable surface. In the case of nanoshells, the curvature-induced DMI introduces broken inversion symmetry in the system and gives rise to magnetochiral effects. A common example is the magnetic vortex-like state in 3D nanostructures where the degeneracy in the energy with respect to polarities and chiralities is lifted. This type of symmetry breaking associated with magnetic vortices can be realized in square-shaped platelets, magnetic cap structures, spherical magnetic shells, etc. There are two categories: i) patterned induced and ii) geometry induced. The former is common in nanotube structures. The nanotubes having two energetically equivalent vortex domain walls with different chiralities show that having highly symmetric energy of the magnetic states is degenerate. Also, Walker breakdown (an abrupt decrease of DW velocity) in 3D nanotubes can be avoided by increasing the tube radius and chirality, leading to Cherenkov radiation of magnons for fast domain walls [172]. Geometry-induced symmetry breaking can be observed in nanowires or nanotubes with a curvature, resulting in domain wall pinning [173]. Similar pinning was observed earlier in nanowires with notches, roughness, or magnetic inhomogeneity [174]. Interestingly, in the presence of an external magnetic field, the wall oscillates in the pinning potential with certain frequency. In the case of magnetic helices, curvature- and torsion-induced DMI can break the symmetry of SWs by inducing an effective magnetic field. Some of these structures are found to have application potentials in modern technology. A magnetic nanocap fabricated using Co/Pt multilayer stacks with out-of-plane anisotropy deposited on top of monolayers of silica microbeads resembles magnetic Janus particles,

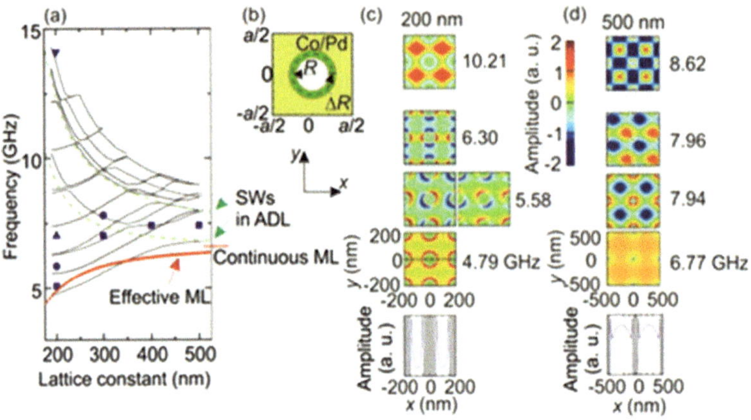

Figure 9.8 (a) Dependence of the frequency of SW excitations in ADL with shells (ADLS) on the lattice constant at $\mu_0 H = 0.084$ T. The full blue symbols represent frequencies extracted from the TRMOKE data. The frequency of the SW in the continuous ML is marked by the red line, and the two dashed green lines point at the SW in the ADL without shells. The dotted red line shows the frequency of the SW in the film of the effective parameters. (b) The unit cell of the ADLS. (c, d) Amplitude of the dynamical component of the magnetization vector in ADLS. The profiles of five modes with the lowest frequencies for $a = 200$ nm and 500 nm are shown. At the bottom, the amplitude of the SW excitation with the lowest frequency (the first mode) in ADLS across the center of the unit cell along the x axis. Reprinted from Ref. [171], with the permission of AIP Publishing.

which can receive, transport, and deliver drugs. Nanocaps can be of the following types with different spin textures: out-of-plane cap, soft cap, single cap, and cap array.

A magnetic tubular structure also can be categorized into several types: cylinders, hollow tubes, rolled-up magnetic membranes, etc. These can be used as magnetic field sensors, giant magnetoresistive sensorics, etc. The dynamics of these structures was revealed early in this decade. Microwave absorption spectroscopy was exploited to probe the SW excitations in rolled-up Py microtubes (RUPTs) (see Fig. 9.9) [41]. A 20-nm-thick Py layer was first deposited on top of an MBE-grown heterostructure of 15 nm (10 nm) GaAs, 15 nm pseudomorphically strained $In_{20}Ga_{80}As$, and a 40 nm AlAs sacrificial layer on a GaAs substrate. After etching out the AlAs sacrificial layer,

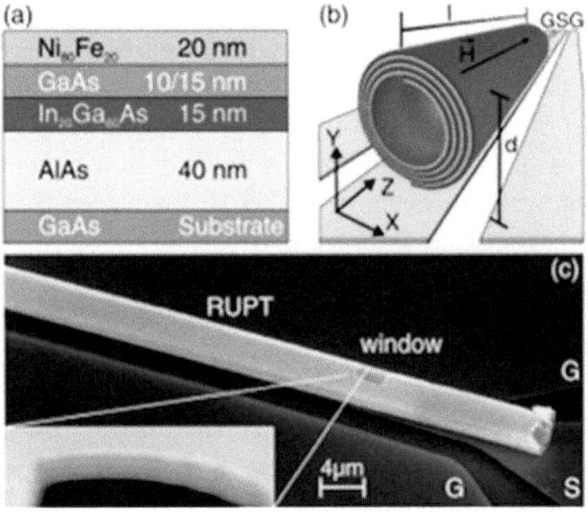

Figure 9.9 (a) Layer sequence used to roll up a strained $Ni_{80}Fe_{20}$/ GaAs/$In_{20}Ga_{80}As$ layer system into a Permalloy microtube (RUPT). (b) Sketch of a RUPT positioned on the signal line (S) between the two ground lines (G) of a coplanar waveguide to perform microwave absorption measurements. (c) SEM image of RUPT A corresponding to the scheme in (b). The inset shows 3.5 tight windings of the strained layer system in a tube cross section drilled by focused ion beams after the measurements were performed. Reprinted (figure) with permission from Ref. [41]. Copyright (2010) by the American Physical Society.

the rest was pulled off the substrate, and it formed a nanoroll by minimizing the surface energy. The absorption frequency of the quantized SW modes appearing due to constructive interference on the circumference of the tube was found to increase with increasing magnetic field and can be tuned by varying the radius and winding number.

9.3.4 Artificial Spin Ice

The frustration occurs in a physical system where all the interaction terms are unable to minimize simultaneously. In an extreme situation, frustration can give rise to some novel and complex phenomena. In 1935, Pauling [175] described the origin of observed [176] frustration in water ice. The frustration occurs due to the ambiguity of forming hydrogen bonds between H_2O molecules of water ice. It leads to a number of energy degenerate states, resulting in zero-point entropy in water ice. In the early 1990s, researchers observed the frustrated state in disorder magnetic materials [177, 178]. Most of the materials are antiferromagnetically coupled. The presence of the frustrated state is unlikely in a ferromagnetically interacting system. Harris et al. [179] discovered the frustration in a ferromagnetically interacting pyrochlore material ($Ho_2Ti_2O_7$). They named this frustrated state "spin ice." The frustration occurs because of the local magnetic moment of pyrochlore material inhabiting a tetrahedra lattice, same as the hydrogen ion positions in water ice. Magnetic monopoles were observed in these systems [180]. This topological defect exists in monopole-antimonopole pairs. They are connected via a Dirac string and can move back and forth. This movement of monopoles gives rise to a magnetic current and is promising for magnetronic devices [181]. These geometrical frustrated structures were mimicked in ferromagnetic nanopatterned structures to probe the topological defects locally and exploring possible applications in magnetic devices. These tailored structures are known as artificial spin ice structures. The magnetic defects and their roles in magnetic properties have been observed in square lattices and kagome lattices [182, 183]. The interaction and frustration in artificial spin ice can be tuned by varying the material parameters, lattice types, and nanomagnet separation [182, 183]. The defects strongly influence the quasistatic and equilibrium response of artificial spin ice lattices. The artificial spin ice structure of Py nanobars arranged in a square geometry was studied experimentally [37] for the first time. The researchers observed water ice–like frustration in this artificial magnetic system. Later, artificial spin ice structures of square [184–186], kagome [186–188], and other geometry [189–191] made of Py or Co were studied experimentally to reveal the role of defects in their static and dynamic magnetic properties. The topological defects can modify SW mode frequencies and behavior. SW mode spectra of stadium-like islands arranged in a square lattice made of Py were studied [192] theoretically. The mode nature depends on coupling between islands and local magnetic configuration. The defects introduce new modes and modify the mode behavior. The monopoles modify the edge mode frequency and a splitting of the mode related to monopole resonance. The SW spectral amplitude of theses modes is proportional to the number of defects present in the system. The length of the Dirac string can affect the uniform mode frequency and split the mode because of the different magnetostatic interactions at the string extremities. The appearance of localized modes at topological defects opens up a new way of manipulating magnetization modes in spin-ice lattices, which can be useful for magnonic devices.

The microwave response of artificial spin ice structures shows that the SW mode spectra highly depend on the applied field history. The resonance spectra of Py square artificial spin ice [185] revealed the influence of local magnetization of neighboring islands during field hysteresis. The static and dynamic properties of spin ice and anti–spin ice Py structures with two different lattices of six different configurations were investigated [186] to explore systematic variation of the interesting static and dynamic behaviors. This may open a new possibility of application in the design of magnetically controlled tunable microwave filters. Recently, a systematic study of field- [187] and angular-dependent hysteresis [188] of SW modes in interconnected kagome lattices was reported. The resonance modes are affected by the spin solid state and monopole pairs on Dirac strings (see Fig. 9.10). The vertices in interconnected artificial spin ice are key to observing significant resonance frequency modifications induced by topological defects and can manipulate exchange-dominated SWs. These findings can be useful for designing reconfigurable magnonic devices. For realization of magnetic monopoles in 3D, the layered structure of 2D artificial spin ice has been proposed [193, 194]. Spin ice behavior, such as an effective Coulomb interaction between magnetic monopoles interacting via Coulomb law, has been demonstrated. The topological defect states in 3D inverse opal-like nanostructure (IOLS) of cobalt have been studied by small-angle neutron scattering [46]. The geometry of the opal structure consists of two quasi-tetrahedra separated by a quasi-cube (see Fig. 9.11). This geometry manifests the defects in the structure. The local magnetization of this ferromagnetic 3D lattice follows the "ice rule." Recently, magnetization reversal from Ni IOLS has been studied [195]. The researchers found a drastic change in irreversibility and

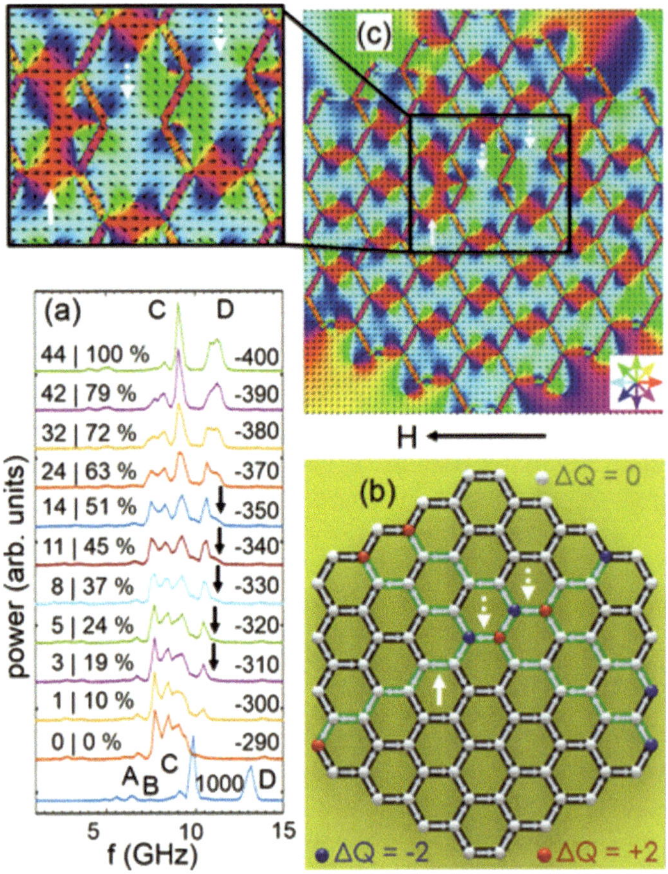

Figure 9.10 (a) Simulated power spectra for the saturated state at $H = 1$ kOe, -290 Oe, and successively decreased H (from bottom to top). Numbers on the left (right) indicate the number n of monopole-antimonopole pairs and percentage of nanobars switched, respectively (field values in Oe). (b) Charge model representation of Dirac strings with monopole-antimonopole pairs. **M** of green (black) segments points to the left (right). Red, blue, and gray spheres represent monopoles, antimonopoles, and reference configuration, respectively. (c) Directions of the local demagnetization field $\mathbf{H_d}(x,y)$ (black arrows and color wheel) at $H = -320$ Oe ($\phi = 180°$) for the artificial spin ice state shown in panel (b). The inset highlights the region where three Dirac strings cross. Between monopole-antimonopole pairs (downward dashed arrows) $H_d < 0$. $H_d > 0$ in remaining segments (upward solid arrow) on the same Dirac strings. Reprinted (figure) with permission from Ref. [187]. Copyright (2016) by the American Physical Society.

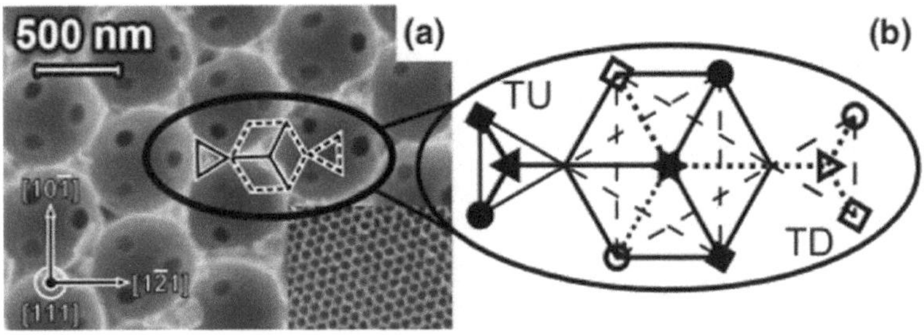

Figure 9.11 Top view of the Co IOLS sample representing the (111) plane of the FCC structure (a). The unit element of IOLS consists of two tetrahedra connected by a cube (b). Reprinted (figure) with permission from Ref. [46]. Copyright (2013) by the American Physical Society.

saturation fields in between the transition from the antidot array to the 3D IOLS. This behavior was explained on the basis of the ice rule. Different types of 3D frustrated artificial lattices [89, 95] are being fabricated to explore topological defect behaviors in the third dimension.

9.4 Conclusion

In this chapter, we discussed magnetic properties of 3D magnetic nanostructures possessing fascinating ground-state spin configurations, which are manifested in the dynamic magnetic properties of the magnetic systems. The 3D magnetic structures have been emerging as an interesting research topic because of their potential applications in ultrahigh-density storage, memory devices, and other applications. We discussed 3D magnetic spin structures, including patterned 3D elements, magnetic vortices and skyrmions, patterned MLs, and artificial spin ice systems. The fabrication of 3D magnetic structures and their properties have also been discussed in this chapter. For the fabrication of complex 3D structures FEBID and TPL are suitable techniques. FEBID can produce any arbitrary nanoscale 3D structure, and the simulation-guided fabrication makes this a mighty candidate for nanoscale 3D fabrication. However, the purity of the material needs to be improved for FEBID. On the other hand, the chemical route does not have purity issues, but fabrication is limited to the prepatterned template. Some exotic spin textures such as skyrmions, Bloch points, monopoles, and Möbius rings in 3D magnetic structures have been introduced from the recent literature. X-ray-based techniques are suitable to study the spin configuration of 3D magnetic structures due to element-specific mapping. However, conventional techniques are also used to study the static and dynamic properties of magnetic structures. Recently, the magnetization dynamics and SW mode profiles of a nanoscale 3D tetrapod structure were studied using the TRMOKE technique and micromagnetic simulations, as presented here. However, the study of 3D magnetic nanostructures is restricted by the limitations of fabrication and measurement techniques. The large-area fabrication of magnetostatically coupled and uncoupled 3D nanostructures with better material purity requires a new fabrication technique. Also, modified measurement and analyses tools need to be engaged to study spin configuration and SWs in large-area 3D magnetic nanostructures.

Acknowledgments

The authors gratefully acknowledge financial assistance from the S. N. Bose National Centre for Basic Sciences, Kolkata, India, under project number SNB/AB/18-19/211. S.M. acknowledges the Department of Science and Technology India for the INSPIRE fellowship (no. IF140998), while S.S. acknowledges the S. N. Bose National Centre for Basic Sciences for the senior research fellowship.

References

1. Bader, S. D. (2006). Colloquium: opportunities in nanomagnetism, *Rev. Mod. Phys.*, **78**, 1–15.
2. Skomski, R. (2003). Nanomagnetics, *J. Phys.: Condens. Matter*, **15**, R841–R896.
3. Binasch, G., Grünberg, P., Saurenbach, F. and Zinn, W. (1989). Enhanced magnetoresistance in layered magnetic structures with antiferromagnetic interlayer exchange, *Phys. Rev. B*, **39**, 4828–4830.

4. Baibich, M. N., Broto, J. M., Fert, A., Van Dau, F. N., Petroff, F., Etienne, P., Creuzet, G., Friederich, A. and Chazelas, J. (1988). Giant magnetoresistance of (001)Fe/(001)Cr magnetic superlattices, *Phys. Rev. Lett.*, **61**, 2472–2475.
5. Moodera, J. S., Kinder, L. R., Wong, T. M. and Meservey, R. (1995). Large magnetoresistance at room temperature in ferromagnetic thin film tunnel junctions, *Phys. Rev. Lett.*, **74**, 3273–3276.
6. Thomson, T., Hu, G. and Terris, B. D. (2006). Intrinsic distribution of magnetic anisotropy in thin films probed by patterned nanostructures, *Phys. Rev. Lett.*, **96**, 257204.
7. Nikonov, D. E. and Bourianoff, G. I. (2005). Spin gain transistor in ferromagnetic semiconductors-the semiconductor Bloch-equations approach, *IEEE Trans. Nanotechnol.*, **4**, 206–214.
8. Bandyopadhyay, S. and Cahay, M. (2009). Electron spin for classical information processing: a brief survey of spin-based logic devices, gates and circuits, *Nanotechnology*, **20**, 412001.
9. Imre, A., Csaba, G., Ji, L., Orlov, A., Bernstein, G. H. and Porod, W. (2006). Majority logic gate for magnetic quantum-dot cellular automata, *Science*, **311**, 205.
10. Haldar, A., Kumar, D. and Adeyeye, A. O. (2016). A reconfigurable waveguide for energy-efficient transmission and local manipulation of information in a nanomagnetic device, *Nat. Nanotechnol.*, **11**, 437.
11. Biswas, A. K., Ahmad, H., Atulasimha, J. and Bandyopadhyay, S. (2017). Experimental demonstration of complete 180° reversal of magnetization in isolated Co nanomagnets on a PMN–PT substrate with voltage generated strain, *Nano Lett.*, **17**, 3478–3484.
12. Bhowmik, D., You, L. and Salahuddin, S. (2013). Spin Hall effect clocking of nanomagnetic logic without a magnetic field, *Nat. Nanotechnol.*, **9**, 59.
13. Kaka, S., Pufall, M. R., Rippard, W. H., Silva, T. J., Russek, S. E. and Katine, J. A. (2005). Mutual phase-locking of microwave spin torque nano-oscillators, *Nature*, **437**, 389.
14. Demidov, V. E., Urazhdin, S., Ulrichs, H., Tiberkevich, V., Slavin, A., Baither, D., Schmitz, G. and Demokritov, S. O. (2012). Magnetic nano-oscillator driven by pure spin current, *Nat. Mater.*, **11**, 1028.
15. Wernsdorfer, W., Aliaga-Alcalde, N., Hendrickson, D. N. and Christou, G. (2002). Exchange-biased quantum tunnelling in a supramolecular dimer of single-molecule magnets, *Nature*, **416**, 406.

16. Arruebo, M., Fernández-Pacheco, R., Ibarra, M. R. and Santamaría, J. (2007). Magnetic nanoparticles for drug delivery, *Nano Today*, **2**, 22–32.
17. Lequeux, S., Sampaio, J., Cros, V., Yakushiji, K., Fukushima, A., Matsumoto, R., Kubota, H., Yuasa, S. and Grollier, J. (2016). A magnetic synapse: multilevel spin-torque memristor with perpendicular anisotropy, *Sci. Rep.*, **6**, 31510.
18. Yablonovitch, E. (1987). Inhibited spontaneous emission in solid-state physics and electronics, *Phys. Rev. Lett.*, **58**, 2059–2062.
19. Tacchi, S., Madami, M., Gubbiotti, G., Carlotti, G., Goolaup, S., Adeyeye, A. O., Singh, N. and Kostylev, M. P. (2010). Analysis of collective spin-wave modes at different points within the hysteresis loop of a one-dimensional magnonic crystal comprising alternative-width nanostripes, *Phys. Rev. B*, **82**, 184408.
20. Wang, Z. K., Zhang, V. L., Lim, H. S., Ng, S. C., Kuok, M. H., Jain, S. and Adeyeye, A. O. (2010). Nanostructured magnonic crystals with size-tunable bandgaps, *ACS Nano*, **4**, 643–648.
21. Saha, S., Barman, S., Otani, Y. and Barman, A. (2015). All-optical investigation of tunable picosecond magnetization dynamics in ferromagnetic nanostripes with a width down to 50 nm, *Nanoscale*, **7**, 18312–18319.
22. Demidov, V. E., Kostylev, M. P., Rott, K., Krzysteczko, P., Reiss, G. and Demokritov, S. O. (2009). Excitation of microwaveguide modes by a stripe antenna, *Appl. Phys. Lett.*, **95**, 112509.
23. Chumak, A. V., Pirro, P., Serga, A. A., Kostylev, M. P., Stamps, R. L., Schultheiss, H., Vogt, K., Hermsdoerfer, S. J., Laegel, B., Beck, P. A. and Hillebrands, B. (2009). Spin-wave propagation in a microstructured magnonic crystal, *Appl. Phys. Lett.*, **95**, 262508.
24. Gubbiotti, G., Tacchi, S., Carlotti, G., Vavassori, P., Singh, N., Goolaup, S., Adeyeye, A. O., Stashkevich, A. and Kostylev, M. (2005). Magnetostatic interaction in arrays of nanometric permalloy wires: a magneto-optic Kerr effect and a Brillouin light scattering study, *Phys. Rev. B*, **72**, 224413.
25. Dmytriiev, O., Al-Jarah, U. A. S., Gangmei, P., Kruglyak, V. V., Hicken, R. J., Mahato, B. K., Rana, B., Agrawal, M., Barman, A., Mátéfi-Tempfli, M., Piraux, L. and Mátéfi-Tempfli, S. (2013). Static and dynamic magnetic properties of densely packed magnetic nanowire arrays, *Phys. Rev. B*, **87**, 174429.

26. Agrawal, M., Rana, B. and Barman, A. (2010). Magnetization reversal in chains and clusters of exchange-coupled Nickel nanoparticles, *J. Phys. Chem. C*, **114**, 11115–11118.

27. Kostylev, M., Schrader, P., Stamps, R. L., Gubbiotti, G., Carlotti, G., Adeyeye, A. O., Goolaup, S. and Singh, N. (2008). Partial frequency band gap in one-dimensional magnonic crystals, *Appl. Phys. Lett.*, **92**, 132504.

28. Nikitov, S. A., Tailhades, P. and Tsai, C. S. (2001). Spin waves in periodic magnetic structures—magnonic crystals, *J. Magn. Magn. Mater.*, **236**, 320–330.

29. Albrecht, T. R., Arora, H., Ayanoor-Vitikkate, V., Beaujour, J., Bedau, D., Berman, D., Bogdanov, A. L., Chapuis, Y., Cushen, J., Dobisz, E. E., Doerk, G., Gao, H., Grobis, M., Gurney, B., Hanson, W., Hellwig, O., Hirano, T., Jubert, P., Kercher, D., Lille, J., Liu, Z., Mate, C. M., Obukhov, Y., Patel, K. C., Rubin, K., Ruiz, R., Schabes, M., Wan, L., Weller, D., Wu, T. and Yang, E. (2015). Bit-patterned magnetic recording: Theory, media fabrication, and recording performance, *IEEE Trans. Magn.*, **51**, 1–42.

30. Jung, S., Watkins, B., DeLong, L., Ketterson, J. B. and Chandrasekhar, V. (2002). Ferromagnetic resonance in periodic particle arrays, *Phys. Rev. B*, **66**, 132401.

31. Neusser, S., Botters, B. and Grundler, D. (2008). Localization, confinement, and field-controlled propagation of spin waves in $Ni_{80}Fe_{20}$ antidot lattices, *Phys. Rev. B*, **78**, 054406.

32. Neusser, S., Botters, B., Becherer, M., Schmitt-Landsiedel, D. and Grundler, D. (2008). Spin-wave localization between nearest and next-nearest neighboring holes in an antidot lattice, *Appl. Phys. Lett.*, **93**, 122501.

33. Lenk, B., Ulrichs, H., Garbs, F. and Münzenberg, M. (2011). The building blocks of magnonics, *Phys. Rep.*, **507**, 107–136.

34. Kruglyak, V. V., Demokritov, S. O. and Grundler, D. (2010). Magnonics, *J. Phys. D: Appl. Phys.*, **43**, 264001.

35. Neusser, S. and Grundler, D. (2009). Magnonics: Spin waves on the nanoscale, *Adv. Mater.*, **21**, 2927–2932.

36. Krawczyk, M. and Grundler, D. (2014). Review and prospects of magnonic crystals and devices with reprogrammable band structure, *J. Phys.: Condens. Matter*, **26**, 123202.

37. Wang, R. F., Nisoli, C., Freitas, R. S., Li, J., McConville, W., Cooley, B. J., Lund, M. S., Samarth, N., Leighton, C., Crespi, V. H. and Schiffer, P. (2006).

Artificial 'spin ice' in a geometrically frustrated lattice of nanoscale ferromagnetic islands, *Nature*, **439**, 303.

38. Fernández-Pacheco, A., Petit, D., Mansell, R., Lavrijsen, R., Lee, J. H. and Cowburn, R. P. (2012). Controllable nucleation and propagation of topological magnetic solitons in CoFeB/Ru ferrimagnetic superlattices, *Phys. Rev. B*, **86**, 104422.
39. Gilbert, D. A., Maranville, B. B., Balk, A. L., Kirby, B. J., Fischer, P., Pierce, D. T., Unguris, J., Borchers, J. A. and Liu, K. (2015). Realization of ground-state artificial skyrmion lattices at room temperature, *Nat. Commun.*, **6**, 8462.
40. Heinze, S., von Bergmann, K., Menzel, M., Brede, J., Kubetzka, A., Wiesendanger, R., Bihlmayer, G. and Blügel, S. (2011). Spontaneous atomic-scale magnetic skyrmion lattice in two dimensions, *Nat. Phys.*, **7**, 713.
41. Balhorn, F., Mansfeld, S., Krohn, A., Topp, J., Hansen, W., Heitmann, D. and Mendach, S. (2010). Spin-wave interference in three-dimensional rolled-up ferromagnetic microtubes, *Phys. Rev. Lett.*, **104**, 037205.
42. Fernández-Pacheco, A., Streubel, R., Fruchart, O., Hertel, R., Fischer, P. and Cowburn, R. P. (2017). Three-dimensional nanomagnetism, *Nat. Commun.*, **8**, 15756.
43. Parkin, S. S. P., Hayashi, M. and Thomas, L. (2008). Magnetic domain-wall racetrack memory, *Science*, **320**, 190.
44. Streubel, R., Fischer, P., Kronast, F., Kravchuk, V. P., Sheka, D. D., Gaididei, Y., Schmidt, O. G. and Makarov, D. (2016). Magnetism in curved geometries, *J. Phys. D: Appl. Phys.*, **49**, 363001.
45. Hertel, R. (2002). Computational micromagnetism of magnetization processes in Nickel nanowires, *J. Magn. Magn. Mater.*, **249**, 251–256.
46. Mistonov, A. A., Grigoryeva, N. A., Chumakova, A. V., Eckerlebe, H., Sapoletova, N. A., Napolskii, K. S., Eliseev, A. A., Menzel, D. and Grigoriev, S. V. (2013). Three-dimensional artificial spin ice in nanostructured Co on an inverse opal-like lattice, *Phys. Rev. B*, **87**, 220408.
47. Ernst, D., Melzer, M., Makarov, D., Bahr, F., Hofmann, W., Schmidt, O. G. and Zerna, T. (2014). Packaging technologies for ultra-thin sensor applications in active magnetic bearings. In *Proceedings of the 2014 37th International Spring Seminar on Electronics Technology*, 125–129.
48. Vavassori, P., Pancaldi, M., Perez-Roldan, M. J., Chuvilin, A. and Berger, A. (2016). Remote magnetomechanical nanoactuation, *Small*, **12**, 1013–1023.

49. Rogach, A. L., Talapin, D. V., Shevchenko, E. V., Kornowski, A., Haase, M. and Weller, H. (2002). Organization of matter on different size scales: monodisperse nanocrystals and their superstructures, *Adv. Funct. Mater.*, **12**, 653–664.
50. Williams, W. D. and Giordano, N. (1986). Experimental study of localization and electron-electron interaction effects in thin Au wires, *Phys. Rev. B*, **33**, 8146–8154.
51. Masuda, H., Yamada, H., Satoh, M., Asoh, H., Nakao, M. and Tamamura, T. (1997). Highly ordered nanochannel-array architecture in anodic alumina, *Appl. Phys. Lett.*, **71**, 2770–2772.
52. Shimomura, M. and Sawadaishi, T. (2001). Bottom-up strategy of materials fabrication: a new trend in nanotechnology of soft materials, *Curr. Opin. Colloid Interface Sci.*, **6**, 11–16.
53. Aizenberg, J., Rogers, J. A., Paul, K. E. and Whitesides, G. M. (1997). Imaging the irradiance distribution in the optical near field, *Appl. Phys. Lett.*, **71**, 3773–3775.
54. Fischer, P. B. and Chou, S. Y. (1993). 10 nm electron beam lithography and sub-50 nm overlay using a modified scanning electron microscope, *Appl. Phys. Lett.*, **62**, 2989–2991.
55. Singh, N., Goolaup, S. and Adeyeye, A. O. (2004). Fabrication of large area nanomagnets, *Nanotechnology*, **15**, 1539–1544.
56. Parikh, D., Craver, B., Nounu, H. N., Fong, F. and Wolfe, J. C. (2008). Nanoscale pattern definition on nonplanar surfaces using ion beam proximity lithography and conformal plasma-deposited resist, *J. Microelectomech. Syst.*, **17**, 735–740.
57. Marty, F., Vaterlaus, A., Weich, V., Stamm, C., Maier, U. and Pescia, D. (1999). Ultrathin magnetic particles, *J. Appl. Phys.*, **85**, 6166–6168.
58. Devolder, T., Chappert, C., Chen, Y., Cambril, E., Bernas, H., Jamet, J. P. and Ferré, J. (1999). Sub-50 nm planar magnetic nanostructures fabricated by ion irradiation, *Appl. Phys. Lett.*, **74**, 3383–3385.
59. Rousseaux, F., Decanini, D., Carcenac, F., Cambril, E., Ravet, M. F., Chappert, C., Bardou, N., Bartenlian, B. and Veillet, P. (1995). Study of large area high density magnetic dot arrays fabricated using synchrotron radiation based x-ray lithography, *J. Vac. Sci. Technol.*, **13**, 2787–2791.
60. Bessho, K., Iwasaki, Y. and Hashimoto, S. (1996). Fabricating nanoscale magnetic mounds using a scanning probe microscope, *J. Appl. Phys.*, **79**, 5057–5059.

61. Wassermann, E. F., Thielen, M., Kirsch, S., Pollmann, A., Weinforth, H. and Carl, A. (1998). Fabrication of large scale periodic magnetic nanostructures, *J. Appl. Phys.*, **83**, 1753–1757.

62. Moritz, J., Landis, S., Toussaint, J. C., Bayle-Guillemaud, R., Rodmacq, B., Casali, G., Lebib, A., Chen, Y., Nozires, J. P. and Dieny, B. (2002). Patterned media made from pre-etched wafers: a promising route toward ultrahigh-density magnetic recording, *IEEE Trans. Magn.*, **38**, 1731–1736.

63. Dallmeyer, A., Carbone, C., Eberhardt, W., Pampuch, C., Rader, O., Gudat, W., Gambardella, P. and Kern, K. (2000). Electronic states and magnetism of monatomic Co and Cu wires, *Phys. Rev. B*, **61**, R5133–R5136.

64. Williams, G., Hunt, M., Boehm, B., May, A., Taverne, M., Ho, D., Giblin, S., Read, D., Rarity, J., Allenspach, R. and Ladak, S. (2018). Two-photon lithography for 3D magnetic nanostructure fabrication, *Nano Res.*, **11**, 845–854.

65. De Teresa, J. M., Fernández-Pacheco, A., Córdoba, R., Serrano-Ramón, L., Sangiao, S. and Ibarra, M. R. (2016). Review of magnetic nanostructures grown by focused electron beam induced deposition (FEBID), *J. Phys. D: Appl. Phys.*, **49**, 243003.

66. Martin, Y. and Wickramasinghe, H. K. (1987). Magnetic imaging by "force microscopy" with 1000 Å resolution, *Appl. Phys. Lett.*, **50**, 1455–1457.

67. Chapman, J. N. (1984). The investigation of magnetic domain structures in thin foils by electron microscopy, *J. Phys. D: Appl. Phys.*, **17**, 623–647.

68. Argyle, B. E. and McCord, J. G. (2000). New laser illumination method for Kerr microscopy, *J. Appl. Phys.*, **87**, 6487–6489.

69. Fischer, P. (2015). X-Ray imaging of magnetic structures, *IEEE Trans. Magn.*, **51**, 1–31.

70. Kimling, J., Kronast, F., Martens, S., Böhnert, T., Martens, M., Herrero-Albillos, J., Tati-Bismaths, L., Merkt, U., Nielsch, K. and Meier, G. (2011). Photoemission electron microscopy of three-dimensional magnetization configurations in core-shell nanostructures, *Phys. Rev. B*, **84**, 174406.

71. Donnelly, C., Guizar-Sicairos, M., Scagnoli, V., Holler, M., Huthwelker, T., Menzel, A., Vartiainen, I., Müller, E., Kirk, E., Gliga, S., Raabe, J. and Heyderman, L. J. (2015). Element-specific x-ray phase tomography of 3D structures at the nanoscale, *Phys. Rev. Lett.*, **114**, 115501.

72. Streubel, R., Kronast, F., Fischer, P., Parkinson, D., Schmidt, O. G. and Makarov, D. (2015). Retrieving spin textures on curved magnetic thin films with full-field soft X-ray microscopies, *Nat. Commun.*, **6**, 7612.

73. Stohr, J., Wu, Y., Hermsmeier, B. D., Samant, M. G., Harp, G. R., Koranda, S., Dunham, D. and Tonner, B. P. (1993). Element-specific magnetic microscopy with circularly polarized x-rays, *Science*, **259**, 658–661.

74. Pinkvos, H., Poppa, H., Bauer, E. and Hurst, J. (1992). Spin-polarized low-energy electron microscopy study of the magnetic microstructure of ultra-thin epitaxial Cobalt films on W(110), *Ultramicroscopy*, **47**, 339–345.

75. Koike, K. and Hayakawa, K. (1984). Scanning electron microscope observation of magnetic domains using spin-polarized secondary electrons, *Jpn. J. Appl. Phys.*, **23**, L187–L188.

76. Johnson, M. and Clarke, J. (1990). Spin-polarized scanning tunneling microscope: concept, design, and preliminary results from a prototype operated in air, *J. Appl. Phys.*, **67**, 6141–6152.

77. Rippard, W. H. and Buhrman, R. A. (1999). Ballistic electron magnetic microscopy: imaging magnetic domains with nanometer resolution, *Appl. Phys. Lett.*, **75**, 1001–1003.

78. Cohen, M. S. (1967). Wave-optical aspects of lorentz microscopy, *J. Appl. Phys.*, **38**, 4966–4976.

79. Barman, A., Wang, S., Maas, J. D., Hawkins, A. R., Kwon, S., Liddle, A., Bokor, J. and Schmidt, H. (2006). Magneto-optical observation of picosecond dynamics of single nanomagnets, *Nano Lett.*, **6**, 2939–2944.

80. Vonsovskii, S. V. (1966). *Ferromagnetic Resonance: The Phenomenon of Resonant Absorption of a High-Frequency Magnetic Field in Ferromagnetic Substances* (Pergamon Press, Oxford; New York).

81. Demokritov, S. O., Hillebrands, B. and Slavin, A. N. (2001). Brillouin light scattering studies of confined spin waves: linear and nonlinear confinement, *Phys. Rep.*, **348**, 441–489.

82. Barman, A. and Sinha, J. (2018) *Spin Dynamics and Damping in Ferromagnetic Thin Films and Nanostructures* (Springer Publishing AG; Switzerland).

83. Hiebert, W. K., Stankiewicz, A. and Freeman, M. R. (1997). Direct observation of magnetic relaxation in a small Permalloy disk by time-resolved scanning Kerr microscopy, *Phys. Rev. Lett.*, **79**, 1134–1137.

84. Denysenkov, V. P. and Grishin, A. M. (2003). Broadband ferromagnetic resonance spectrometer, *Rev. Sci. Instrum.*, **74**, 3400–3405.

85. Tamaru, S., Bain, J. A., van de Veerdonk, R. J. M., Crawford, T. M., Covington, M. and Kryder, M. H. (2002). Imaging of quantized magnetostatic modes using spatially resolved ferromagnetic resonance, *J. Appl. Phys.*, **91**, 8034–8036.
86. Liu, L., Moriyama, T., Ralph, D. C. and Buhrman, R. A. (2011). Spin-torque ferromagnetic resonance induced by the spin Hall effect, *Phys. Rev. Lett.*, **106**, 036601.
87. Silva, T. J., Lee, C. S., Crawford, T. M. and Rogers, C. T. (1999). Inductive measurement of ultrafast magnetization dynamics in thin-film permalloy, *J. Appl. Phys.*, **85**, 7849–7862.
88. Tottori, S., Zhang, L., Qiu, F., Krawczyk, K. K., Franco-Obregón, A. and Nelson, B. J. (2012). Magnetic helical micromachines: Fabrication, controlled swimming, and cargo transport, *Adv. Mater.*, **24**, 811–816.
89. May, A., Hunt, M., Van Den Berg, A., Hejazi, A. and Ladak, S. (2018). Realisation of a frustrated 3D magnetic nanowire lattice, arXiv:1812.02442.
90. Albrecht, M., Hu, G., Guhr, I. L., Ulbrich, T. C., Boneberg, J., Leiderer, P. and Schatz, G. (2005). Magnetic multilayers on nanospheres, *Nat. Mater.*, **4**, 203.
91. Streubel, R., Kravchuk, V. P., Sheka, D. D., Makarov, D., Kronast, F., Schmidt, O. G. and Gaididei, Y. (2012). Equilibrium magnetic states in individual hemispherical permalloy caps, *Appl. Phys. Lett.*, **101**, 132419.
92. da Câmara Santa Clara Gomes, T., De La Torre Medina, J., Lemaitre, M. and Piraux, L. (2016). Magnetic and magnetoresistive properties of 3D interconnected NiCo nanowire networks, *Nanoscale Res. Lett.*, **11**, 466.
93. Fernández-Pacheco, A., Serrano-Ramón, L., Michalik, J. M., Ibarra, M. R., De Teresa, J. M., O'Brien, L., Petit, D., Lee, J. and Cowburn, R. P. (2013). Three dimensional magnetic nanowires grown by focused electron-beam induced deposition, *Sci. Rep.*, **3**, 1492.
94. Gavagnin, M., Wanzenboeck, H. D., Wachter, S., Shawrav, M. M., Persson, A., Gunnarsson, K., Svedlindh, P., Stöger-Pollach, M. and Bertagnolli, E. (2014). Free-standing magnetic nanopillars for 3D nanomagnet logic, *ACS Appl. Mater. Interfaces*, **6**, 20254–20260.
95. Keller, L., Al Mamoori, M. K. I., Pieper, J., Gspan, C., Stockem, I., Schröder, C., Barth, S., Winkler, R., Plank, H., Pohlit, M., Müller, J. and Huth, M. (2018). Direct-write of free-form building blocks for artificial magnetic 3D lattices, *Sci. Rep.*, **8**, 6160.

96. Fowlkes, J. D., Winkler, R., Lewis, B. B., Stanford, M. G., Plank, H. and Rack, P. D. (2016). Simulation-guided 3D nanomanufacturing via focused electron beam induced deposition, *ACS Nano*, **10**, 6163–6172.

97. Botman, A., Mulders, J. J. L. and Hagen, C. W. (2009). Creating pure nanostructures from electron-beam-induced deposition using purification techniques: a technology perspective, *Nanotechnology*, **20**, 372001.

98. Rana, B. and Barman, A. (2015). Ultrafast magnetization dynamics of chemically synthesized Ni nanoparticles, *J. Phys. Chem. C*, **119**, 17444–17449.

99. Okuda, M., Schwarze, T., Eloi, J. C., Ward Jones, S. E., Heard, P. J., Sarua, A., Ahmad, E., Kruglyak, V. V., Grundler, D. and Schwarzacher, W. (2017). Top-down design of magnonic crystals from bottom-up magnetic nanoparticles through protein arrays, *Nanotechnology*, **28**, 155301.

100. Pal, S., Saha, S., Kamalakar, M. V. and Barman, A. (2016). Field-dependent spin waves in high-aspect-ratio single-crystal ferromagnetic nanowires, *Nano Res.*, **9**, 1426–1433.

101. Xiong, L. L., Kostylev, M. and Adeyeye, A. O. (2017). Magnetization dynamics of $Ni_{80}Fe_{20}$ nanowires with continuous width modulation, *Phys. Rev. B*, **95**, 224426.

102. Sahoo, S., Mondal, S., Williams, G., May, A., Ladak, S. and Barman, A. (2018). Ultrafast magnetization dynamics in a nanoscale three-dimensional cobalt tetrapod structure, *Nanoscale*, **10**, 9981–9986.

103. Krawczyk, M. and Puszkarski, H. (2006). Magnonic excitations versus three-dimensional structural periodicity in magnetic composites, *Cryst. Res. Technol.*, **41**, 547–552.

104. Krawczyk, M. and Puszkarski, H. (2008). Plane-wave theory of three-dimensional magnonic crystals, *Phys. Rev. B*, **77**, 054437.

105. Krawczyk, M., Kłos, J., Sokolovskyy, M. L. and Mamica, S. (2010). Materials optimization of the magnonic gap in three-dimensional magnonic crystals with spheres in hexagonal structure, *J. Appl. Phys.*, **108**, 093909.

106. Mamica, S., Krawczyk, M., Sokolovskyy, M. L. and Romero-Vivas, J. (2012). Large magnonic band gaps and spectra evolution in three-dimensional magnonic crystals based on magnetoferritin nanoparticles, *Phys. Rev. B*, **86**, 144402.

107. Romero Vivas, J., Mamica, S., Krawczyk, M. and Kruglyak, V. V. (2012). Investigation of spin wave damping in three-dimensional magnonic crystals using the plane wave method, *Phys. Rev. B*, **86**, 144417.
108. Popov, P. A., Sharaevskaya, A. Y., Beginin, E. N., Sadovnikov, A. V., Stognij, A. I., Kalyabin, D. V. and Nikitov, S. A. (2019). Spin wave propagation in three-dimensional magnonic crystals and coupled structures, *J. Magn. Magn. Mater.*, **476**, 423–427.
109. Tartakovskaya, E., Kreuzpaintner, W. and Schreyer, A. (2008). Spin wave dynamics in two- and three-dimensional superlattices of nanosized ferromagnetic spheres, *J. Appl. Phys.*, **103**, 023913.
110. Beginin, E. N., Sadovnikov, A. V., Sharaevskaya, A. Y., Stognij, A. I. and Nikitov, S. A. (2018). Spin wave steering in three-dimensional magnonic networks, *Appl. Phys. Lett.*, **112**, 122404.
111. Cowburn, R. P., Koltsov, D. K., Adeyeye, A. O., Welland, M. E. and Tricker, D. M. (1999). Single-domain circular nanomagnets, *Phys. Rev. Lett.*, **83**, 1042–1045.
112. Shinjo, T., Okuno, T., Hassdorf, R., Shigeto, K. and Ono, T. (2000). Magnetic vortex core observation in circular dots of permalloy, *Science*, **289**, 930.
113. Wachowiak, A., Wiebe, J., Bode, M., Pietzsch, O., Morgenstern, M. and Wiesendanger, R. (2002). Direct observation of internal spin structure of magnetic vortex cores, *Science*, **298**, 577.
114. Shigeto, K., Okuno, T., Mibu, K., Shinjo, T. and Ono, T. (2002). Magnetic force microscopy observation of antivortex core with perpendicular magnetization in patterned thin film of permalloy, *Appl. Phys. Lett.*, **80**, 4190–4192.
115. Buchanan, K. S., Roy, P. E., Grimsditch, M., Fradin, F. Y., Guslienko, K. Y., Bader, S. D. and Novosad, V. (2005). Soliton-pair dynamics in patterned ferromagnetic ellipses, *Nat. Phys.*, **1**, 172.
116. Jung, W., Castaño, F. J. and Ross, C. A. (2006). Vortex chirality in exchange-biased elliptical magnetic rings, *Phys. Rev. Lett.*, **97**, 247209.
117. Van Waeyenberge, B., Puzic, A., Stoll, H., Chou, K. W., Tyliszczak, T., Hertel, R., Fähnle, M., Brückl, H., Rott, K., Reiss, G., Neudecker, I., Weiss, D., Back, C. H. and Schütz, G. (2006). Magnetic vortex core reversal by excitation with short bursts of an alternating field, *Nature*, **444**, 461.
118. Yamada, K., Kasai, S., Nakatani, Y., Kobayashi, K., Kohno, H., Thiaville, A. and Ono, T. (2007). Electrical switching of the vortex core in a magnetic disk, *Nat. Mater.*, **6**, 270.

119. Shibata, J., Nakatani, Y., Tatara, G., Kohno, H. and Otani, Y. (2006). Current-induced magnetic vortex motion by spin-transfer torque, *Phys. Rev. B*, **73**, 020403.

120. Caputo, J.-G., Gaididei, Y., Mertens, F. G. and Sheka, D. D. (2007). Vortex polarity switching by a spin-polarized current, *Phys. Rev. Lett.*, **98**, 056604.

121. Park, J. P., Eames, P., Engebretson, D. M., Berezovsky, J. and Crowell, P. A. (2003). Imaging of spin dynamics in closure domain and vortex structures, *Phys. Rev. B*, **67**, 020403.

122. Guslienko, K. Y., Ivanov, B. A., Novosad, V., Otani, Y., Shima, H. and Fukamichi, K. (2002). Eigenfrequencies of vortex state excitations in magnetic submicron-size disks, *J. Appl. Phys.*, **91**, 8037–8039.

123. Guslienko, K. Y., Scholz, W., Chantrell, R. W. and Novosad, V. (2005). Vortex-state oscillations in soft magnetic cylindrical dots, *Phys. Rev. B*, **71**, 144407.

124. Park, J. P. and Crowell, P. A. (2005). Interactions of spin waves with a magnetic vortex, *Phys. Rev. Lett.*, **95**, 167201.

125. Novosad, V., Fradin, F. Y., Roy, P. E., Buchanan, K. S., Guslienko, K. Y. and Bader, S. D. (2005). Magnetic vortex resonance in patterned ferromagnetic dots, *Phys. Rev. B*, **72**, 024455.

126. Guslienko, K. Y., Han, X. F., Keavney, D. J., Divan, R. and Bader, S. D. (2006). Magnetic vortex core dynamics in cylindrical ferromagnetic dots, *Phys. Rev. Lett.*, **96**, 067205.

127. Vogel, A., Drews, A., Kamionka, T., Bolte, M. and Meier, G. (2010). Influence of dipolar interaction on vortex dynamics in arrays of ferromagnetic disks, *Phys. Rev. Lett.*, **105**, 037201.

128. Barman, A., Barman, S., Kimura, T., Fukuma, Y. and Otani, Y. (2010). Gyration mode splitting in magnetostatically coupled magnetic vortices in an array, *J. Phys. D: Appl. Phys.*, **43**, 422001.

129. Barman, S., Barman, A. and Otani, Y. (2010). Dynamics of 1-D chains of magnetic vortices in response to local and global excitations, *IEEE Trans. Magn.*, **46**, 1342–1345.

130. Sugimoto, S., Fukuma, Y., Kasai, S., Kimura, T., Barman, A. and Otani, Y. (2011). Dynamics of coupled vortices in a pair of ferromagnetic disks, *Phys. Rev. Lett.*, **106**, 197203.

131. Hasegawa, N., Sugimoto, S., Kumar, D., Barman, S., Barman, A., Kondou, K. and Otani, Y. (2016). Observation of anisotropic energy transfer in magnetically coupled magnetic vortex pair, *Appl. Phys. Lett.*, **108**, 242402.

132. Pribiag, V. S., Krivorotov, I. N., Fuchs, G. D., Braganca, P. M., Ozatay, O., Sankey, J. C., Ralph, D. C. and Buhrman, R. A. (2007). Magnetic vortex oscillator driven by d.c. spin-polarized current, *Nat. Phys.*, **3**, 498.

133. Jung, H., Choi, Y.-S., Lee, K.-S., Han, D.-S., Yu, Y.-S., Im, M.-Y., Fischer, P. and Kim, S.-K. (2012). Logic operations based on magnetic-vortex-state networks, *ACS Nano*, **6**, 3712–3717.

134. Kumar, D., Barman, S. and Barman, A. (2014). Magnetic vortex based transistor operations, *Sci. Rep.*, **4**, 4108.

135. Barman, S., Saha, S., Mondal, S., Kumar, D. and Barman, A. (2016). Enhanced amplification and fan-out operation in an all-magnetic transistor, *Sci. Rep.*, **6**, 33360.

136. Fert, A., Reyren, N. and Cros, V. (2017). Magnetic skyrmions: advances in physics and potential applications, *Nat. Rev. Mater.*, **2**, 17031.

137. Mühlbauer, S., Binz, B., Jonietz, F., Pfleiderer, C., Rosch, A., Neubauer, A., Georgii, R. and Böni, P. (2009). Skyrmion lattice in a chiral magnet, *Science*, **323**, 915.

138. Yu, X. Z., Kanazawa, N., Onose, Y., Kimoto, K., Zhang, W. Z., Ishiwata, S., Matsui, Y. and Tokura, Y. (2010). Near room-temperature formation of a skyrmion crystal in thin-films of the helimagnet FeGe, *Nat. Mater.*, **10**, 106.

139. Yu, G., Upadhyaya, P., Li, X., Li, W., Kim, S. K., Fan, Y., Wong, K. L., Tserkovnyak, Y., Amiri, P. K. and Wang, K. L. (2016). Room-temperature creation and spin–orbit torque manipulation of skyrmions in thin films with engineered asymmetry, *Nano Lett.*, **16**, 1981–1988.

140. Makhfudz, I., Krüger, B. and Tchernyshyov, O. (2012). Inertia and chiral edge modes of a skyrmion magnetic bubble, *Phys. Rev. Lett.*, **109**, 217201.

141. Zhou, Y., Iacocca, E., Awad, A. A., Dumas, R. K., Zhang, F. C., Braun, H. B. and Åkerman, J. (2015). Dynamically stabilized magnetic skyrmions, *Nat. Commun.*, **6**, 8193.

142. Husain, S., Sisodia, N., Chaurasiya, A. K., Kumar, A., Singh, J. P., Yadav, B. S., Akansel, S., Chae, K. H., Barman, A., Muduli, P. K., Svedlindh, P. and Chaudhary, S. (2019). Observation of skyrmions at room temperature in Co2FeAl Heusler alloy ultrathin film heterostructures, *Sci. Rep.*, **9**, 1085.

143. Moreau-Luchaire, C., Moutafis, C., Reyren, N., Sampaio, J., Vaz, C. A. F., Van Horne, N., Bouzehouane, K., Garcia, K., Deranlot, C., Warnicke, P., Wohlhüter, P., George, J. M., Weigand, M., Raabe, J., Cros, V. and Fert, A. (2016). Additive interfacial chiral interaction in multilayers for

stabilization of small individual skyrmions at room temperature, *Nat. Nanotechnol.*, **11**, 444.

144. Sun, L., Cao, R. X., Miao, B. F., Feng, Z., You, B., Wu, D., Zhang, W., Hu, A. and Ding, H. F. (2013). Creating an artificial two-dimensional skyrmion crystal by nanopatterning, *Phys. Rev. Lett.*, **110**, 167201.
145. Guslienko, K. Y. (2015). Skyrmion state stability in magnetic nanodots with perpendicular anisotropy, *IEEE Magn. Lett.*, **6**, 1–4.
146. Nayak, A. K., Kumar, V., Ma, T., Werner, P., Pippel, E., Sahoo, R., Damay, F., Rößler, U. K., Felser, C. and Parkin, S. S. P. (2017). Magnetic antiskyrmions above room temperature in tetragonal Heusler materials, *Nature*, **548**, 561.
147. Romming, N., Hanneken, C., Menzel, M., Bickel, J. E., Wolter, B., von Bergmann, K., Kubetzka, A. and Wiesendanger, R. (2013). Writing and deleting single magnetic skyrmions, *Science*, **341**, 636.
148. Ma, F., Zhou, Y., Braun, H. B. and Lew, W. S. (2015). Skyrmion-based dynamic magnonic crystal, *Nano Lett.*, **15**, 4029–4036.
149. Woo, S., Litzius, K., Krüger, B., Im, M.-Y., Caretta, L., Richter, K., Mann, M., Krone, A., Reeve, R. M., Weigand, M., Agrawal, P., Lemesh, I., Mawass, M.-A., Fischer, P., Kläui, M. and Beach, G. S. D. (2016). Observation of room-temperature magnetic skyrmions and their current-driven dynamics in ultrathin metallic ferromagnets, *Nat. Mater.*, **15**, 501.
150. Legrand, W., Maccariello, D., Reyren, N., Garcia, K., Moutafis, C., Moreau-Luchaire, C., Collin, S., Bouzehouane, K., Cros, V. and Fert, A. (2017). Room-temperature current-induced generation and motion of sub-100 nm skyrmions, *Nano Lett.*, **17**, 2703–2712.
151. Jiang, W., Zhang, X., Yu, G., Zhang, W., Wang, X., Benjamin Jungfleisch, M., Pearson, John E., Cheng, X., Heinonen, O., Wang, K. L., Zhou, Y., Hoffmann, A. and te Velthuis, Suzanne G. E. (2016). Direct observation of the skyrmion Hall effect, *Nat. Phys.*, **13**, 162.
152. Litzius, K., Lemesh, I., Krüger, B., Bassirian, P., Caretta, L., Richter, K., Büttner, F., Sato, K., Tretiakov, O. A., Förster, J., Reeve, R. M., Weigand, M., Bykova, I., Stoll, H., Schütz, G., Beach, G. S. D. and Kläui, M. (2016). Skyrmion Hall effect revealed by direct time-resolved X-ray microscopy, *Nat. Phys.*, **13**, 170.
153. Woo, S., Song, K. M., Zhang, X., Zhou, Y., Ezawa, M., Liu, X., Finizio, S., Raabe, J., Lee, N. J., Kim, S.-I., Park, S.-Y., Kim, Y., Kim, J.-Y., Lee, D., Lee, O., Choi, J. W., Min, B.-C., Koo, H. C. and Chang, J. (2018). Current-driven dynamics and inhibition of the skyrmion Hall effect of ferrimagnetic skyrmions in GdFeCo films, *Nat. Commun.*, **9**, 959.

154. Montoya, S. A., Couture, S., Chess, J. J., Lee, J. C. T., Kent, N., Henze, D., Sinha, S. K., Im, M. Y., Kevan, S. D., Fischer, P., McMorran, B. J., Lomakin, V., Roy, S. and Fullerton, E. E. (2017). Tailoring magnetic energies to form dipole skyrmions and skyrmion lattices, *Phys. Rev. B*, **95**, 024415.

155. Mochizuki, M. (2012). Spin-wave modes and their intense excitation effects in skyrmion crystals, *Phys. Rev. Lett.*, **108**, 017601.

156. Smith, D., Parekh, V., E, C., Zhang, S., Donner, W., Lee, T. R., Khizroev, S. and Litvinov, D. (2008). Magnetization reversal and magnetic anisotropy in patterned Co/Pd multilayer thin films, *J. Appl. Phys.*, **103**, 023920.

157. Parekh, V., Chunsheng, E., Smith, D., Ruiz, A., Wolfe, J. C., Ruchhoeft, P., Svedberg, E., Khizroev, S. and Litvinov, D. (2006). Fabrication of a high anisotropy nanoscale patterned magnetic recording medium for data storage applications, *Nanotechnology*, **17**, 2079–2082.

158. Rahman, M. T., Shams, N. N., Lai, C. H., Fidler, J. and Suess, D. (2010). Co/Pt perpendicular antidot arrays with engineered feature size and magnetic properties fabricated on anodic aluminum oxide templates, *Phys. Rev. B*, **81**, 014418.

159. Suharyadi, E., Kato, T., Tsunashima, S. and Iwata, S. (2006). Magnetic properties of patterned Co/Pd nanostructures by e-beam lithography and Ga ion irradiation, *IEEE Trans. Magn.*, **42**, 2972–2974.

160. Hauet, T., Hellwig, O., Park, S. H., Beigné, C., Dobisz, E., Terris, B. D. and Ravelosona, D. (2011). Influence of ion irradiation on switching field and switching field distribution in arrays of Co/Pd-based bit pattern media, *Appl. Phys. Lett.*, **98**, 172506.

161. Breitkreutz, S., Kiermaier, J., Vijay Karthik, S., Csaba, G., Schmitt-Landsiedel, D. and Becherer, M. (2012). Controlled reversal of Co/Pt Dots for nanomagnetic logic applications, *J. Appl. Phys.*, **111**, 07A715.

162. Rahman, M. T., Shams, N. N., Wu, Y.-C., Lai, C.-H. and Suess, D. (2007). Magnetic multilayers on porous anodized alumina for percolated perpendicular media, *Appl. Phys. Lett.*, **91**, 132505.

163. Ren, Y., Liu, X. M., Singh, N. and Adeyeye, A. O. (2013). Influence of magnetostatic interaction on the magnetization reversal of patterned Co/Pd multilayers nanorings, *IEEE Trans. Magn.*, **49**, 3620–3623.

164. Lau, J. W., Liu, X., Boling, R. C. and Shaw, J. M. (2011). Decoupling nucleation and domain-wall propagation regimes in $(Co/Pd)_n$ multilayer nanostructures, *Phys. Rev. B*, **84**, 214427.

165. Delalande, M., de Vries, J., Abelmann, L. and Lodder, J. C. (2012). Measurement of the nucleation and domain depinning field in a single

Co/Pt multilayer dot by Anomalous Hall effect, *J. Magn. Magn. Mater.*, **324**, 1277–1280.

166. Krupinski, M., Mitin, D., Zarzycki, A., Szkudlarek, A., Giersig, M., Albrecht, M. and Marszałek, M. (2017). Magnetic transition from dot to antidot regime in large area Co/Pd nanopatterned arrays with perpendicular magnetization, *Nanotechnology*, **28**, 085302.
167. Lau, J. W., McMichael, R. D., Chung, S. H., Rantschler, J. O., Parekh, V. and Litvinov, D. (2008). Microstructural origin of switching field distribution in patterned Co/Pd multilayer nanodots, *Appl. Phys. Lett.*, **92**, 012506.
168. Chuang, V. P., Jung, W., Ross, C. A., Cheng, J. Y., Park, O.-H. and Kim, H.-C. (2008). Multilayer magnetic antidot arrays from block copolymer templates, *J. Appl. Phys.*, **103**, 074307.
169. Krawczyk, M., Mamica, S., Kłos, J. W., Romero-Vivas, J., Mruczkiewicz, M. and Barman, A. (2011). Calculation of spin wave spectra in magnetic nanograins and patterned multilayers with perpendicular anisotropy, *J. Appl. Phys.*, **109**, 113903.
170. Liu, X. M., Ding, J., Kakazei, G. N. and Adeyeye, A. O. (2013). Magnonic crystals composed of $Ni_{80}Fe_{20}$ film on top of $Ni_{80}Fe_{20}$ two-dimensional dot array, *Appl. Phys. Lett.*, **103**, 062401.
171. Pal, S., Kłos, J. W., Das, K., Hellwig, O., Gruszecki, P., Krawczyk, M. and Barman, A. (2014). Optically induced spin wave dynamics in $[Co/Pd]_8$ antidot lattices with perpendicular magnetic anisotropy, *Appl. Phys. Lett.*, **105**, 162408.
172. Yan, M., Kákay, A., Andreas, C. and Hertel, R. (2013). Spin-Cherenkov effect and magnonic Mach cones, *Phys. Rev. B*, **88**, 220412.
173. Hayward, T. J., Bryan, M. T., Fry, P. W., Fundi, P. M., Gibbs, M. R. J., Im, M. Y., Fischer, P. and Allwood, D. A. (2010). Pinning induced by inter-domain wall interactions in planar magnetic nanowires, *Appl. Phys. Lett.*, **96**, 052502.
174. Kim, J.-S., Mawass, M.-A., Bisig, A., Krüger, B., Reeve, R. M., Schulz, T., Büttner, F., Yoon, J., You, C.-Y., Weigand, M., Stoll, H., Schütz, G., Swagten, H. J. M., Koopmans, B., Eisebitt, S. and Kläui, M. (2014). Synchronous precessional motion of multiple domain walls in a ferromagnetic nanowire by perpendicular field pulses, *Nat. Commun.*, **5**, 3429.
175. Pauling, L. (1935). The structure and entropy of ice and of other crystals with some randomness of atomic arrangement, *J. Am. Chem. Soc.*, **57**, 2680–2684.

176. Giauque, W. F. and Ashley, M. F. (1933). Molecular rotation in ice at 10° K. free energy of formation and entropy of water, *Phys. Rev.*, **43**, 81–82.
177. Ramirez, A. P. (1994). Strongly geometrically frustrated magnets, *Annu. Rev. Mater. Sci.*, **24**, 453–480.
178. Balents, L. (2010). Spin liquids in frustrated magnets, *Nature*, **464**, 199.
179. Harris, M. J., Bramwell, S. T., McMorrow, D. F., Zeiske, T. and Godfrey, K. W. (1997). Geometrical frustration in the ferromagnetic pyrochlore $Ho_2Ti_2O_7$, *Phys. Rev. Lett.*, **79**, 2554–2557.
180. Castelnovo, C., Moessner, R. and Sondhi, S. L. (2008). Magnetic monopoles in spin ice, *Nature*, **451**, 42.
181. Bramwell, S. T., Giblin, S. R., Calder, S., Aldus, R., Prabhakaran, D. and Fennell, T. (2009). Measurement of the charge and current of magnetic monopoles in spin ice, *Nature*, **461**, 956.
182. Nisoli, C., Moessner, R. and Schiffer, P. (2013). Colloquium: artificial spin ice: designing and imaging magnetic frustration, *Rev. Mod. Phys.*, **85**, 1473–1490.
183. Marrows, C. (2016). Experimental studies of artificial spin ice, arXiv:1611.00744.
184. Jungfleisch, M. B., Zhang, W., Iacocca, E., Sklenar, J., Ding, J., Jiang, W., Zhang, S., Pearson, J. E., Novosad, V., Ketterson, J. B., Heinonen, O. and Hoffmann, A. (2016). Dynamic response of an artificial square spin ice, *Phys. Rev. B*, **93**, 100401.
185. Sklenar, J., Bhat, V. S., DeLong, L. E. and Ketterson, J. B. (2013). Broadband ferromagnetic resonance studies on an artificial square spin-ice island array, *J. Appl. Phys.*, **113**, 17B530.
186. Zhou, X., Chua, G.-L., Singh, N. and Adeyeye, A. O. (2016). Large area artificial spin ice and anti-spin ice $Ni_{80}Fe_{20}$ structures: static and dynamic behavior, *Adv. Funct. Mater.*, **26**, 1437–1444.
187. Bhat, V. S., Heimbach, F., Stasinopoulos, I. and Grundler, D. (2016). Magnetization dynamics of topological defects and the spin solid in a kagome artificial spin ice, *Phys. Rev. B*, **93**, 140401.
188. Bhat, V. S., Heimbach, F., Stasinopoulos, I. and Grundler, D. (2017). Angular-dependent magnetization dynamics of kagome artificial spin ice incorporating topological defects, *Phys. Rev. B*, **96**, 014426.
189. Bhat, V. S., Sklenar, J., Farmer, B., Woods, J., Hastings, J. T., Lee, S. J., Ketterson, J. B. and De Long, L. E. (2013). Controlled magnetic reversal in permalloy films patterned into artificial quasicrystals, *Phys. Rev. Lett.*, **111**, 077201.

190. Mironov, V. L., Skorohodov, E. V. and Blackman, J. A. (2014). Magnetic states and ferromagnetic resonance in geometrically frustrated arrays of multilayer ferromagnetic nanoparticles ordered on triangular lattices, *J. Appl. Phys.*, **115**, 184301.

191. Lao, Y., Caravelli, F., Sheikh, M., Sklenar, J., Gardeazabal, D., Watts, J. D., Albrecht, A. M., Scholl, A., Dahmen, K., Nisoli, C. and Schiffer, P. (2018). Classical topological order in the kinetics of artificial spin ice, *Nat. Phys.*, **14**, 723–727.

192. Gliga, S., Kákay, A., Hertel, R. and Heinonen, O. G. (2013). Spectral analysis of topological defects in an artificial spin-ice lattice, *Phys. Rev. Lett.*, **110**, 117205.

193. Chern, G.-W., Reichhardt, C. and Nisoli, C. (2014). Realizing three-dimensional artificial spin ice by stacking planar nano-arrays, *Appl. Phys. Lett.*, **104**, 013101.

194. Perrin, Y., Canals, B. and Rougemaille, N. (2016). Extensive degeneracy, Coulomb phase and magnetic monopoles in artificial square ice, *Nature*, **540**, 410.

195. Shishkin, I. S., Mistonov, A. A., Dubitskiy, I. S., Grigoryeva, N. A., Menzel, D. and Grigoriev, S. V. (2016). Nonlinear geometric scaling of coercivity in a three-dimensional nanoscale analog of spin ice, *Phys. Rev. B*, **94**, 064424.

Chapter 10

Spin Waves in Nanotubes: Impact of Curvature on Transport Properties

Jorge A. Otálora,[a] Helmut Schultheiss,[b] and Attila Kákay[b]

[a]*Institute of Metallic Materials at the Leibniz Institute for Solid State and Materials Research, IFW, 01069 Dresden, Germany*
[b]*Helmholtz-Zentrum Dresden - Rossendorf, Institute of Ion Beam Physics and Materials Research, Bautzner Landstr. 400, 01328 Dresden, Germany*
j.otalora@ifw-dresden.de, h.schultheiss@hzdr.de, a.kakay@hzdr.de

10.1 Introduction

Already in planar, thin magnetic films, spin-wave transport shows a rich spectrum of phenomena originating from the interplay of bulk and interface properties and magnetic dipolar interaction. One of the most prominent examples is the existence of the Damon–Eshbach (DE) magnetostatic surface waves [11]. These special type of spin waves shows a nonreciprocal propagation on the surface of an in-plane magnetized film, that is, these modes exist only in a narrow angular range of wave vectors, which is not symmetric around the center of the Brillouin zone and oriented perpendicular to the equilibrium direction of the magnetization. A reversal of the wave vector of DE surface modes is only possible if the direction

Three-Dimensional Magnonics: Layered, Micro- and Nanostructures
Edited by Gianluca Gubbiotti

ISBN 978-981-4800-73-0 (Hardcover), 978-0-429-29915-5 (eBook)
www.jennystanford.com

of the magnetization is inverted. A comprehensive review of these spin-wave properties, which are solely induced by magnetic dipolar interaction, is given in Refs. [18, 19]. In this sense, spin waves behave very similar to spin currents in topological insulators, and indeed, it could be shown in several material systems and geometries that nonreciprocal magnon properties can be traced back to the chiral nature of spin waves and the topology of the magnonic band structure [26, 37, 38].

When a magnetic thin film is bended into the third dimension, the resulting curvature is introducing an additional symmetry-breaking vector. As it turned out and as we will discuss in this chapter, this curvature is breaking the inversion symmetry of the spin-wave dispersion and, thus, introducing an additional degree of freedom for controlling the nonreciprocity of spin-wave transport and designing the spin-wave dispersion. We explore the origin and consequences of these curvature-induced effects in the simplest curved magnetic element, which is a magnetic nanotube with a constant, nonzero curvature. For this geometry an analytic solution of the spin-wave dispersion is calculated, which allows to compare the additional terms in the dispersion to other phenomena, which cause a nonreciprocal spin-wave dispersion, such as the intrinsic Dzyaloshinskii–Moriya interaction (DMI) [13, 27].

Due to the relevance of engineering the spin-wave dispersion for technological applications, we explicitly do the comparison of curvature and DMI-induced nonreciprocities of the spin-wave dispersion in Section 10.3. However, we deem it worthy to mention the advantages of curvature already at this point: Typically, in material systems with large interfacial DMI the magnetization is aligned perpendicular to the film plane due to a large perpendicular magnetic anisotropy (PMA). In order to observe nonreciprocities, the magnetization needs to be forced into the film plane by applying an external magnetic field large enough to overcome the PMA, which is impractical for applications. Moreover, a large PMA is linked to a large magnetic damping, hampering spin-wave transport over longer distances. In contrast to interfacial DMI, the curvature-induced nonreciprocity is not an interface effect, which means that the magnitude of the effect will not scale inversely proportional to the thickness of the magnetic material. This is a strong advantage for spin-wave applications because the group velocity of spin waves, which ultimately determines the signal speed and throughput in data processing applications, increases with increasing thickness of the magnetic material. Our analysis at the end of Section 10.3 shows that dipolar interaction in curved geometries can even overcome the impact of exchange interaction on the spin-wave transport properties once approaching spin-wave wavelength comparable to the exchange length.

The first pioneering studies of spin waves in waveguides with curved geometries are from the turn of the millennium. These works are focused on the analytical development of the dispersion relation of magnetic nanowires, nanowire arrays, and spheres [1, 3–5]. About 10 years after these works numerous exciting theoretical studies showed that bending a thin film into a curved surface will lead to interesting physical effects, especially if the curvature radius is comparable to the intrinsic magnetic length scales [17]. Often such a bending is enough to result in chiral domain walls [29, 43] or exceptionally high domain wall velocities that can lead to the Cherenkov-like spin-wave emission [42]. In the static regime curvature can stabilize chiral noncollinear magnetic textures, such as skyrmions [14, 24, 34], but the curvatures influence also extends into the dynamic regime, resulting in changes of the spin-wave transport [16, 28, 30, 35, 36]. These effects are a consequence of the break of the local inversion symmetry in curvilinear structures, which appear in conventional ferromagnetic materials and here are referenced as magnetochiral effects. Although the curvature-induced magnetochiral effects display phenomena similar to those of intrinsic DMI on the static and dynamic properties of magnetic materials and noncollinear textures, they originate from the symmetric exchange or magnetostatic interactions.

The physics of curved magnetic thin shells is driven by the interplay between local (exchange, anisotropy) and nonlocal (magnetostatic) interactions. The energy functionals of these interactions depend on the spatial derivatives of the magnetization distribution, which make the interactions sensitive to bends and twists in the physical space. So far both the exchange- and magnetostatic-induced chiral effects in ferromagnetic systems and their influence on the static magnetic configurations, on noncollinear magnetic textures, on their dynamics, as well as on the spin-wave transport were studied only theoretically. In this chapter we summarize the magnetostatic-induced chiral effects on the spin-wave transport in magnetic nanotubes and compare them to their thin-film counterparts. Since the full understanding of the spin-wave propagation in arbitrarily magnetized nanotubes is still under development, we focus to the particular case, when the magnetization in the ground state is a vortex-like state or, in other words, a circular magnetic state along the azimuthal direction.

10.2 Circular Magnetic State

The vortex-like magnetic ground state is illustrated in Fig. 10.1 together with the cylindrical coordinate system we use throughout this chapter. In the beginning, for simplicity we assume that the nanotube is fabricated from permalloy; therefore the magnetization dynamics is governed by the competition of the exchange, magnetostatic, anisotropy, and Zeeman interactions. We also assume that the nanotube shell thickness is in the range of the exchange length of permalloy; hence, along the thickness the magnetization profile is assumed to be constant in both static and dynamic scenarios.

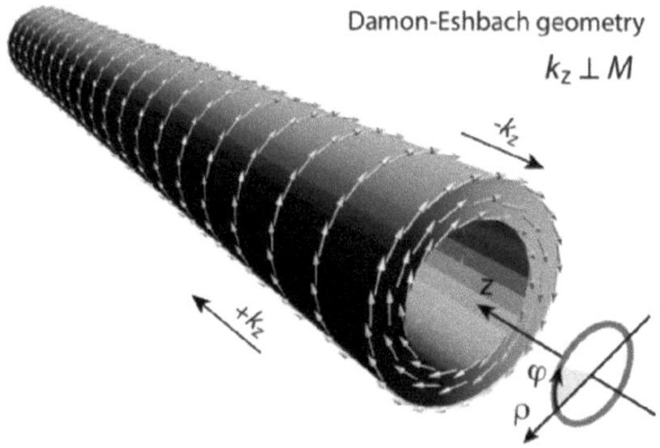

Figure 10.1 Schematic illustration of a nanotube in a vortex magnetic state and the cylindrical coordinate system. The $+k_z$ and $-k_z$ indicate the right and left propagation directions, respectively. Reprinted (figure) with permission from Ref. [31]. Copyright (2017) by the American Physical Society.

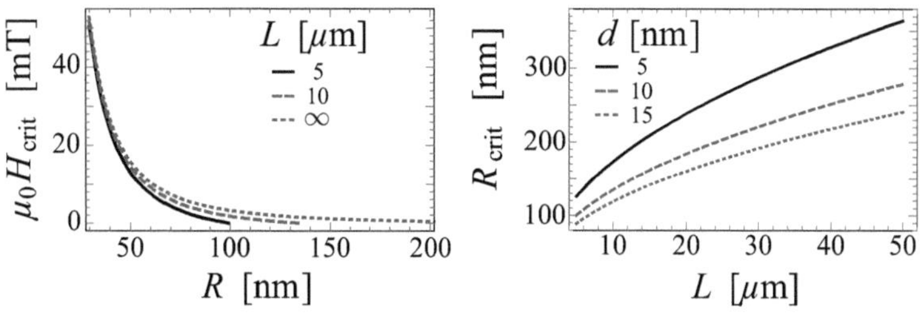

Figure 10.2 (Critical field H_{crit} above which the circular magnetic field H_0 stabilizes the circular magnetic configuration for different permalloy nanotubes of radius R and lengths L. The radius at which the $H_{crit} = 0$ is named the critical radius R_{crit} and is presented here as a function of L for a nanotube with different thicknesses d.

For material parameters of permalloy the ground magnetization state strongly depends on the geometrical dimensions. In the absence of any external fields for small nanotube radii the ground state would be a homogeneously magnetized state along the long axis with two flux closure caps of opposite vorticity at the ends of the nanotube. Above a critical radius, however, the ground state is a homogeneous vortex state, at least theoretically. The critical radius for a fixed nanotube length at which the ground state is a vortex state can be read from Fig. 10.2. It can be seen that for nanotubes with several tens of micrometer length this critical radius is around 200 nm. The vortex ground state for all radii can be stabilized, for example, by the Oersted field (or circular field) of an electric current inside the nanotube. The magnitude of the circular field can be calculated analytically [28] and depends on the thickness (d) as well as on the length (L) of the nanotube due to the magnetostatic interaction, as shown in Fig. 10.2. According to the analytical calculations the magnitude of the critical field increases proportional to $1/R$ with the decrease of the nanotube radius. Therefore, for nanotubes with radii comparable to the wavelength of the exchange spin waves, the critical fields cannot be reached without destroying the tube.

The large currents required to create Oersted fields that are able to stabilize the vortex state in tubes with small radii can be replaced by a uniaxial anisotropy with a hard axis along the tube's long axis. As shown in Ref. [45], this can be achieved experimentally with

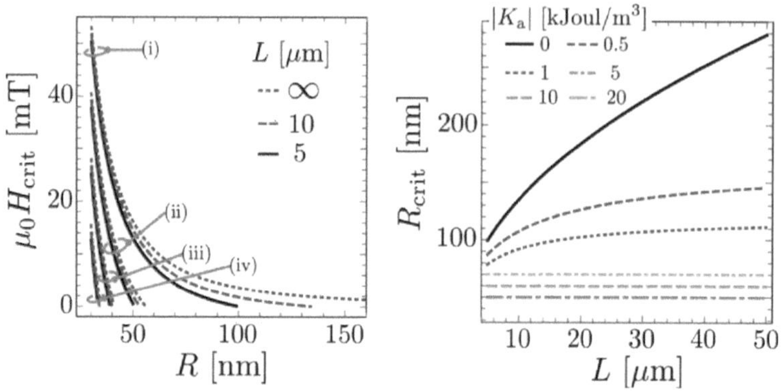

Figure 10.3 Critical field H_{crit} above which the circular magnetic field H_0 stabilizes the circular magnetic configuration for different permalloy nanotubes of radius R and lengths L. The radius at which the $H_{\text{crit}} = 0$ is named the critical radius R_{crit} and is presented here as a function of L for a nanotube with thicknesses $d = 10$ nm and different hard-axis anisotropies: (i) $K_a = 0$, (ii) $K_a = -5$ kJ/m^3, (iii) $K_a = -10$ kJ/m^3, and (iv) $K_a = -20$ kJ/m^3.

special growth conditions and the obtained anisotropy values can be so large (up to about $K_a = -20$ kJoul/m^3) that the vortex state will be the spontaneously forming ground state. The evolution of the critical field with the anisotropy constant is summarized in Fig. 10.3. The anisotropy significantly reduces the nanotube radii above which the vortex state is a ground state. As seen in the figure, a small induced anisotropy $K_a = -5$ kJ/m^3 can already lead to a vortex ground state for nanotubes above 50 nm and only requires couple of mT fields for radii below.

10.3 Curvature-Induced Nonreciprocities in Spin-Wave Dispersion Relation

The analytical description presented here is under the frame of the micromagnetic continuum theory. Accordingly, the dispersion relation is calculated by linearizing the equation of motion, that is, the Landau–Lifshitz–Gilbert (LLG) equation, given as

$$\dot{\vec{\Omega}} = -\gamma_0\, \vec{\Omega} \times \vec{H}_{\text{eff}} + \alpha_G \left(\vec{\Omega} \times \dot{\vec{\Omega}}\right), \tag{10.1}$$

where $\vec{\Omega} = \vec{M}/M_s$ is the unit magnetization vector, M_s is the saturation magnetization, $\gamma_0 = g\mu_B/\hbar$ is the gyromagnetic ratio, g is the electron's Landé factor, μ_B is the Bohr magneton, $\hbar$ is the Planck constant, $\vec{H}_{\text{eff}}$ is the effective magnetic field, and α_G is the phenomenological Gilbert damping parameter.

The analytical calculations include dipolar, exchange, and anisotropy interactions, as well as the Zeeman interaction related to an applied circular field $\vec{H}_0 = H_0\hat{\varphi}$. Hence, the effective field can be written as

$$\vec{H}_{\text{eff}} = \vec{H}_{\text{ex}} + \vec{H}_d + \vec{H}_{K_a} + \vec{H}_0, \tag{10.2}$$

where $\vec{H}_{\text{ex}} = D\nabla^2\vec{\Omega}$ is the Heisenberg exchange field, $D = l_{\text{ex}}^2 M_s$, $l_{\text{ex}} = \sqrt{A/K_d}$ the exchange length, A the stiffness constant, $K_d = (1/2)\mu_0 M_s^2$ a measure of the shape anisotropy, $H_d = -\nabla\Phi$ is the dipolar field arising from the scalar magnetostatic potential Φ, $\vec{H}_{K_a} = 2K_a/(\mu_0 M_s)$ is a uniaxial anisotropy field, and H_0 is the circular field strength that could be induced by the injection of an electric current I that flows along the nanotube length and through the magnetic material and/or the nanotube core when it is filled with a conductive wire.

To obtain the spin-wave dispersion one needs to consider small perturbations in the equilibrium magnetization, which are either generated by applying a low-amplitude radiofrequency (RF) field or generated by maintaining the system at finite temperatures. Our equilibrium assumption is the vortex state ($\vec{\Omega}_0 = \hat{\varphi}$), and any small deviation from it results in a magnetization configuration that can be expressed as $\vec{\Omega} \approx \vec{\Omega}_0 + \vec{m}$, where the dynamic part of the magnetization $\vec{m}$ is assumed to be (i) in a perturbative range $\|\vec{m}\| \ll 1$ and, consequently, (ii) perpendicular to the equilibrium magnetization $\vec{m} \cdot \vec{\Omega}_0 = 0$; hence, $\vec{m} = \hat{\rho}\, m_\rho + \hat{z}\, m_z$. Furthermore, given the small nanotube thickness considered here (not larger than 10 nm), it is reasonable to think that the excitation profile does not depend on the ρ coordinate, that is, $m_x = m_x(\varphi, z)$ with $x = \rho, z$. The dynamical equations governing the behavior of the perturbation $\vec{m}$ are obtained by linearizing the LLG equation up to the first order in $\vec{m}$. This can be done by expanding the interactions in terms of the excitation and keeping only the remaining linear terms in $\vec{m}$ in the LLG equation. Averaging along the nanotube cross section and considering only the linear terms in $\vec{m}$, the dynamic equation for the excitation components m_ρ and m_z have the form

$$\dot{m}_\rho = \omega_{M_s}\left\{-l_{\rm ex}^2\left(\frac{1}{b^2}+\nabla^2\right)m_z - h_{K_a}m_z + h_0 m_z + \frac{1}{M_s}\langle h_d\rangle_z\right\},$$

$$\dot{m}_z = \omega_{M_s}\left\{l_{\rm ex}^2\nabla^2 m_\rho - h_0 m_\rho + \frac{1}{M_s}\langle h_d\rangle_\rho\right\}, \tag{10.3}$$

where $h_{K_a} = \frac{2K_a}{\mu_0 M_s^2}$, $h_0 = \bar{H}_0/M_s$, $\omega_{M_s} = \gamma_0\mu_0 M_s$, $\langle h_d\rangle_x = (2\pi/S)\int_r^R d\rho (h_d)_x$ with x $= \rho$, z, and $S = \pi(R^2 - r^2)$ being the cross section area of the nanotube. The Gilbert damping is assumed to be zero since the focus here is finding an analytic expression for the dispersion relation.

Equation 10.3 is an integrodifferential relation and any attempt to extract from it an analytical expression for the spin-wave dispersion requires previous knowledge of the dynamic excitation $\vec{m}$. Accordingly, one can argue that any perturbation can be thought as a superposition of individual oscillations or Fourier decomposition with well-defined oscillatory numbers, like wave vectors, and frequency. In our case, the azimuthal symmetry and elongated tube suggest that any of such individual oscillations can be described with the wave vector along the $\hat{z}$ direction, k_z, the integer wavenumber, n, to account for the azimuthal oscillation phase along $\hat{\varphi}$ direction, and the angular frequency ω. In this way, the excitations can be written as

$$m_z(\varphi, z, t) = \sum_{n=-\infty}^{\infty}\int_{-\infty}^{\infty}\mathcal{Z}_n(k_z)\, e^{ik_z z}e^{in\varphi}e^{-i\omega_n(k_z)t}dk_z,$$

$$m_\rho(\varphi, z, t) = \sum_{n=-\infty}^{\infty}\int_{-\infty}^{\infty}\mathcal{R}_n(k_z)\, e^{ik_z z}e^{in\varphi}e^{-i\omega_n(k_z)t}dk_z, \tag{10.4}$$

where $\mathcal{R}_n(k_z)$ and $\mathcal{Z}_n(k_z)$ are the amplitude projections of an oscillation featured by k_z and n, along $\hat{\rho}$ and $\hat{\varphi}$ directions, respectively. It is worth mentioning that once the dispersion relation is calculated, it can be realized that a given oscillation defined with k_z, n and ω corresponds to an eigen-excitation of the thin and elongated nanotube when it is in the equilibrium vortex state, wherein the magnon occupancy is expressed by $|\mathcal{R}_n(k_z)|^2 + |\mathcal{Z}_n(k_z)|^2$. According to these assumptions, even in the presence of the dynamic dipolar fields the eigenmodes with well-defined sets of k_z and n are orthogonal.

Taking into account the excitation expansion in Eq. 10.4, the Fourier coefficients of the dynamic dipolar fields $\langle h_d \rangle_\rho$ and $\langle h_d \rangle_z$ (which are the only components that participate in the dynamic Eq. 10.3) can be calculated. Using the averaged dynamic dipolar fields h_ρ and h_z for the excitation $\vec{m}$, the dispersion relation takes the form

$$\frac{\omega_n(k_z)}{\omega_{M_s}} = \mathcal{K}_n(k_z) + \sqrt{\mathcal{A}_n(k_z)\mathcal{B}_n(k_z)}, \tag{10.5}$$

where the quantities $\mathcal{A}$ and $\mathcal{B}$ are defined as

$$\begin{aligned} \mathcal{A}_n(k_z) &= l_{\text{ex}}^2 k_z^2 + (n^2 - 1)h_u - h_{K_a} + h_0 + \mathcal{L}_n(k_z), \\ \mathcal{B}_n(k_z) &= l_{\text{ex}}^2 k_z^2 + n^2 h_u + h_0 + \mathcal{J}_n(k_z), \end{aligned} \tag{10.6}$$

with the functions $\mathcal{J}$, $\mathcal{K}$, and $\mathcal{L}$ being given as

$$\begin{aligned} \mathcal{J}_n(k_z) &= \frac{\pi}{S}\int_0^\infty dq \frac{q^3}{2\left(q^2 + k_z^2\right)} \left(\Gamma_n(q)\right)^2 \\ \mathcal{K}_n(k_z) &= \frac{\pi}{S}\int_0^\infty dq \frac{q^2 k_z}{q^2 + k_z^2} \Gamma_n(q)\Lambda_n(q) \\ \mathcal{L}_n(k_z) &= \frac{\pi}{S}\int_0^\infty dq \frac{2qk_z^2}{q^2 + k_z^2} \left(\Lambda_n(q)\right)^2, \end{aligned} \tag{10.7}$$

with $\Lambda_n(q) = \int_r^R d\rho\, \rho J_n(q\rho)$, $\Gamma_n(q) = \Lambda_{n-1}(q) - \Lambda_{n+1}(q)$, $J_n(x)$ Bessel functions of the first kind of order n, $S = \pi(R^2 - r^2)$ the nanotube cross section area, and the term

$$h_\text{u} = \frac{H_\text{u}}{M_s} = \left(\frac{l_\text{ex}}{b}\right)^2 \geq 0 \tag{10.8}$$

being an exchange field arising from the vortex state of the magnetization with $b = \sqrt{S/(2\pi \log(R/r))}$. This field term is equivalent to an anisotropy field with an easy axis along the long axis of the tube and vanishes in the thin-film limiting case, when the radius of the nanotube is infinity. Such field prefers the magnetization oriented along the nanotube axis. For infinite nanotubes, H_u defines the lower field limit (also called the critical field) of the applied circular field $H_0 = M_s h_0$ required to stabilize the magnetization into the vortex state [28]. The lack of mixing between different values of k_z and n in the dispersion relation reveals that an oscillation defined with the set $(k_z, n, \omega_n(k_z))$ corresponds to an eigenmode of the thin and elongated nanotube when it is in the equilibrium vortex state, as mentioned previously. The detailed derivation of the dispersion relation can be found in Ref. [31].

The dispersion relation in Eq. 10.5 results in an asymmetric dispersion regarding the propagation vectors for the DE spin waves, similar to that reported in Ref. [44] for thin films with interfacial DMI. Moreover, Eq. 10.5 has the same mathematical form as the dispersion relation known for thin films with interfacial DMI or for crystals with a special symmetry (C_{nv}) and bulk DMI (see Refs. [12] and [10]). The $\mathcal{K}_n(k_z)$ plays the same role in nanotubes as the well-known asymmetrical terms in thin films (crystals) with interfacial (bulk) DMI (i.e., the term $\frac{2\gamma_0}{M_s} Dk$ in the dispersion of thin films with interfacial DMI [12], where D is the DMI constant) but with the difference that $\mathcal{K}_n(k_z)$ originates from the dynamic volume charges created by the spin waves as a result of the tubular curvature.

The dispersion relation in Eq. 10.5 is plotted in Fig. 10.4 for different nanotube radii and as a density plot. The dispersion is shown for the first three azimuthal spin-wave modes, that is, $n = 0$, $n = \pm 1$, and $n = \pm 2$. Note, due to the periodic boundary condition along the circumference, the $n = \pm 1$ modes have two nodal lines and the $n = \pm 2$ modes four nodal lines. It can be seen from Fig. 10.4 that the $n = 0$ modes show the largest asymmetry. To give a flavor for the magnitude of the asymmetry in wavelength for the counterpropagating spin waves, the wavelengths for two exemplarily chosen frequencies are given for the zero- and higher-order modes. The difference in wavelength can vary from several tens of nanometers to several hundreds of nanometers, depending on the nanotube radius.

The origin of the asymmetry is the dipole–dipole interaction. The term $\mathcal{K}_n(k_z)$, which can only be calculated by numerical integration of the corresponding Bessel functions, comprises the dynamic dipolar energy arising from the surface as well as from the volume charges [7] $\rho_V \equiv -(M_s/4\pi)\nabla \cdot \vec{M}$. If the spin waves are considered in the linear excitation regime, with the magnetization is in the vortex state for a spin wave with wave vector k_z, wavenumber $n = 0$, and eigenfrequency ω, the volume charges averaged over the nanotube

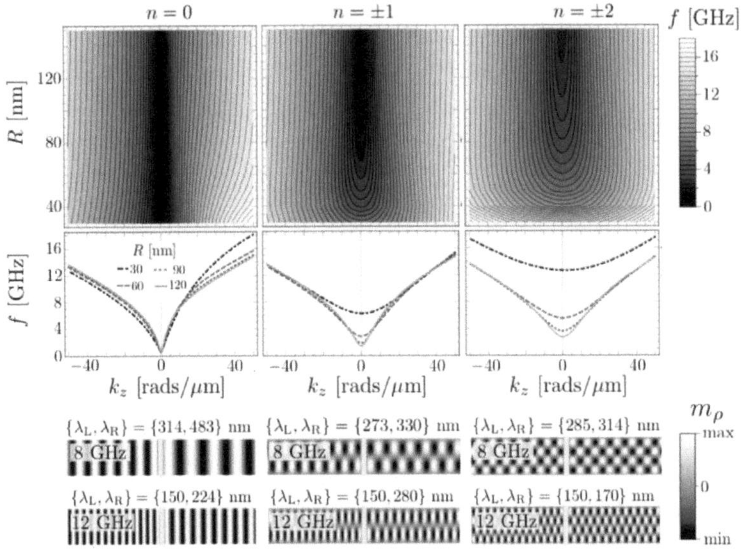

Figure 10.4 Density plot of the dispersion relation $f = \omega/2\pi$ as a function of the nanotube radius R and wave vector k_z for the zero-order $n = 0$, first-order $n = \pm 1$, and second-order $n = \pm 2$ azimuthal modes. The dispersion relations for four different radius $R = (30, 60, 90, 120)$ nm are shown, as well as the phase distribution of spin waves at two different frequencies $f = (8,12)$ GHz. For these frequencies the corresponding wavelengths of the counterpropagating spin waves are given above the phase plots. λ_L and λ_R represent the left- and right-propagating spin waves, respectively. The nanotube is made of permalloy, infinitely long and with thickness $d = 10$ nm.

radius are $\langle \rho_V \rangle = \langle \rho_V \rangle_0 \, e^{i(k_z z - \omega t + \xi)}$, with

$$\langle \rho_V \rangle_0 = -\frac{M_s^2}{4\pi}\left(\frac{1}{\bar{\rho}} + k_z \sqrt{\frac{\mathcal{B}_0(k_z)}{\mathcal{A}_0(k_z)}}\right)\left(1 + \frac{\mathcal{B}_0(k_z)}{\mathcal{A}_0(k_z)}\right)^{-\frac{1}{2}} \tag{10.9}$$

where $\mathcal{A}_0(k_z)$ and $\mathcal{B}_0(k_z)$ are defined in Eq. 10.6; ξ is the phase constant of the radial and axial spin-wave components. The amplitude of the volume charges is proportional to two terms: The first term $1/\bar{\rho}$ is the inverse of the nanotube average radius, which is proportional to the mean nanotube curvature [6]. The second term $k_z \sqrt{\frac{\mathcal{B}_0(k_z)}{\mathcal{A}_0(k_z)}}$ depends on the propagation vector k_z. This term is equivalent to the divergence term in thin films. Hence, the sum of the two terms depends on the sign of k_z. Therefore, for

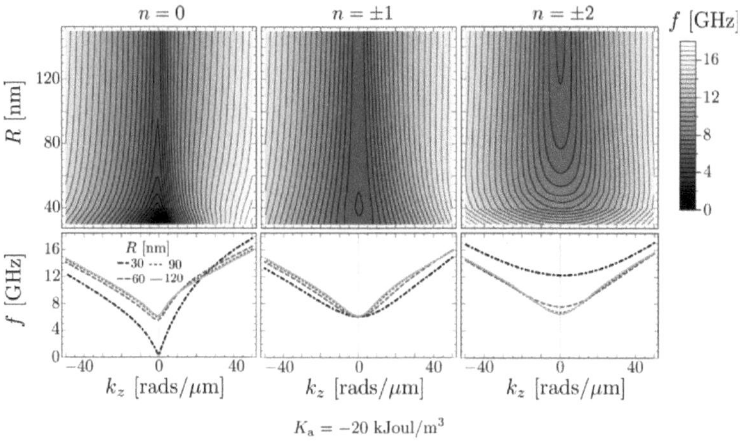

Figure 10.5 Density plot of the dispersion relation $f = \omega/2\pi$ as a function of the nanotube radius R and wave vector k_z for the zero-order $n = 0$, first-order $n = \pm 1$, and second-order $n = \pm 2$ azimuthal modes with hard-axis anisotropy along the nanotube long axis and strength $K_a = -20$ kJ/m^3.

opposite propagation directions the dynamic volume charges are different. Moreover, for small wave vectors a negative dispersion is obtained for one of the travel directions. The negative group velocity is the consequence of the volume charges close to zero and thus of the energy reduction due to the surface charges. Later in the Section 10.5 we show that for small k_z values the integrals in Eq. 10.6 can be approximated with simple functions directly proportional to the mean curvature of the nanotube and depending linearly on k_z. This linear term in the dispersion relation is similar to that reported for thin films with interfacial DMI.

The influence of the hard-axis anisotropy on the dispersion is plotted in Fig. 10.5. Due to the anisotropy the frequency of the modes will be shifted to larger values. While the anisotropy influences the frequency only, the Oersted field has an effect also on the wave vector at which the dispersion reaches its minimum value. In Fig. 10.6 the dispersion relation is shown for nanotubes with 30 nm and 150 nm diameters. As seen, the minima of the dispersion are shifted toward larger wave vector values with an increasing Oersted field. From the application point of view, both the anisotropy and

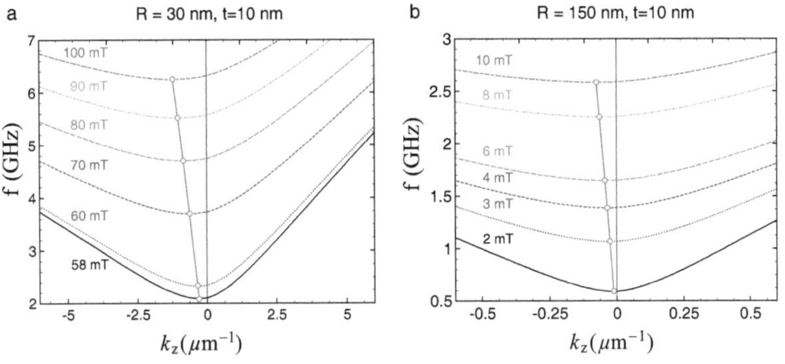

Figure 10.6 The dispersion of spin waves is summarized for several circular fields as a function of wavenumber for nanotubes with (a) 30 nm and (b) 150 nm outer radius and a 10 nm film thickness. The minima of the dispersion are shifted toward larger k_z values with increasing circular field for both diameters. The open dots represent the minima for each circular field and the solid line connecting them is a guide to the eye only. Reprinted (figure) with permission from Ref. [30]. Copyright (2016) by the American Physical Society.

the Oersted field can be used to tune the dispersion relation and its degree of asymmetry.

The asymmetry of the dispersion relation can be quantified by introducing the spin-wave asymmetry Δf_n, defined as the difference of the frequencies between the left- and right-propagating spin waves with the same wave vector k_z. This can be calculated analytically:

$$\Delta f_n = \frac{1}{2\pi}\left|\omega_n(k_z) - \omega_n(-k_z)\right| = \frac{\omega_{M_s}}{\pi}|\mathcal{K}_n(k_z)|. \tag{10.10}$$

The spin-wave asymmetry as a function of the wave vector k_z is shown for nanotubes with four different radii in Fig. 10.7. Clearly, the highest asymmetry is obtained for nanotubes with the smallest outer radii, which can be approximately as high as 5 GHz for the lowest-order spin-wave mode. However, for nanotubes with diameters that can be produced nowadays (the state of the art in the current materials science [39, 41] is around 200 nm), the frequency asymmetry is still as large as 1 GHz. It should be noted that above a certain radius the asymmetry remains constant up to very large wave vectors. Unfortunately, the asymmetry is strongly reduced for

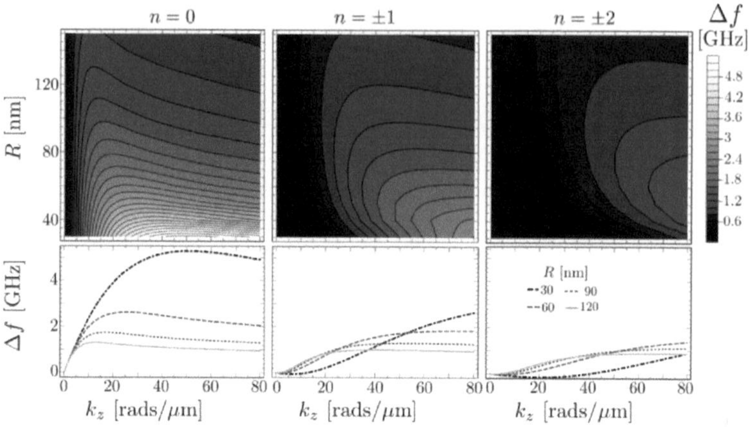

Figure 10.7 Density plot of the frequency asymmetry Δf as a function of the nanotube radius R and wave vector k_z for the zero-order $n = 0$, first-order $n = \pm 1$, and second-order $n = \pm 2$ azimuthal modes. The frequency asymmetries for four different radii $R = (30, 60, 90, 120)$ nm are shown. The nanotube is considered to be made of permalloy, infinitely long with thickness $d = 10$ nm.

the higher-order spin-wave modes for small wave vectors but still reaches values of about 1 GHz for large wave vectors.

In flat thin films the DE modes show a nonreciprocity regarding the inversion of the wave vector caused by dipolar interaction. When the propagation direction is reversed, these spin waves switch from the top to the bottom surface of the thin film. This effect is the result of the surface and volume charge distributions caused by the dynamic magnetization components. The dispersion relation of these modes is very typical: It has a steep slope for small wave vectors, thus resulting in high group velocities. In the literature one will find a clear distinction between the magnetostatic spin waves and the exchange spin-wave modes (k^2 dependence of the dispersion). In flat thin films, therefore, it is mostly claimed that the dipole–dipole interaction will not have influence on the exchange spin-wave modes. On the contrary, for nanotubes, after a short analysis we can conclude that for wave vectors for which the exchange interaction is not anymore neglected, especially for nanotubes with small radii, the effect of the dipole–dipole interaction on the dispersion relation is the largest. Decreasing

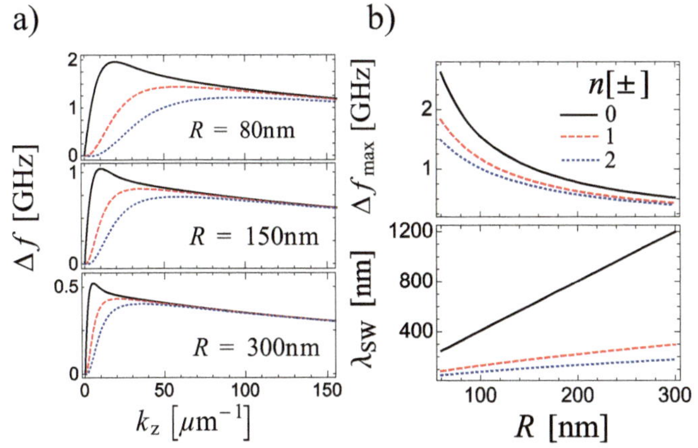

Figure 10.8 (a) Spin-wave asymmetry of the fundamental $n = 0$ (black lines) and the first two higher azimuthal modes ($n = \pm1, \pm2$, red dashed and blue dotted lines, respectively) in a function of the wave vector k_z for nanotubes with 80 nm, 150 nm, and 300 nm outer radii. (b) The maximum asymmetry Δf_{max} and the wavelength λ_{SW} of the excited spin waves at which the asymmetry reaches its maximum in a function of the nanotube radius. All cases presented here are calculated for a nanotube with $d = 10$ nm thickness and using a circular field of $\mu_0 H_0 = 6$ mT. Reprinted (figure) with permission from Ref. [31]. Copyright (2017) by the American Physical Society.

the nanotube radius the position of the largest asymmetry is clearly moving into the exchange spin-wave modes' region. Thus for nanotubes, in contrast to the flat thin films, the dipole–dipole interaction will also influence the exchange spin-wave modes, which one should name as the magnetostatic exchange spin-wave modes. In Fig. 10.8a it can be seen that even for very large nanotubes ($R >$ 100 nm) for spin waves with wavelengths smaller than 125 nm the asymmetry arising from the dipole–dipole interaction is present for the zero-, first-, and second-order modes; however, they converge to a linear dependence on the wave vector.

Assuming that during the spin-wave propagation the local spins precess on a circular trajectory (ferromagnetic resonance [FMR] precession), with a simple analysis of the dynamic volume charge distribution shown in Eq. 10.9 we can already make an assumption on the wave vector for which the maximum asymmetry should arise for a nanotube with a well-defined radius. According to Eq. 10.9, the magnitude of the volume charge distribution of the counterpropagating waves should show a maximum when the wavelength of the spin wave (in one of the directions) is comparable to the nanotube radius. In Fig. 10.8b we show the maximum frequency asymmetry as well as the wavelength for which the maximum frequency asymmetry is obtained in a function of the nanotube radius. It can be seen that indeed the wavelength is comparable to the radius of the nanotubes for which the frequency difference between the counterpropagating spin waves has the maximum. Since the local precession of the spins is rather elliptical than circular, the radial component of the magnetization dynamics is much smaller than the in-plane component of the magnetization. Estimating from Fig. 10.8b, the ratio is about 1:4, leading to the factor of 4 between the wavelength and the radius at which the maximum frequency asymmetry is obtained.

10.4 Curvature-Induced Nonreciprocities in Frequency Linewidth and Decay Length

An asymmetric spin-wave dispersion generally refers to the broken symmetry of the spin-wave propagation. From a technological perspective the nonreciprocal propagation is a desirable feature and could provide the condition for the unidirectional propagation of spin-wave packages in order to avoid the formation of standing spin waves. This is fundamental for enhancing the input/output efficiency of the transport of information encoded in the phase, frequency, and amplitude of spin waves in magnonic devices. In this section we will show that the asymmetric dispersion relation of spin waves in nanotubes is generally accompanied by asymmetries in lifetimes and decay lengths, which are further contributing to the performance enhancement of communications and logic devices based on spin waves [8, 9]. To do so, we derive analytically the wave vector–dependent dynamic susceptibility tensor or Polder tensor [33], from which the frequency linewidth and decay length can be calculated. We show that the asymmetry of these parameters originates from the dynamic dipole–dipole interactions, similar to the dispersion relation derived before, and is directly related to the mean curvature of the nanotube through the $\mathcal{K}_n(k_z)$ function from Eq. 10.7.

To analytically calculate the dynamic susceptibility and the frequency linewidth in the presence of damping, the LLG equation of motion has to be linearized again as for the dispersion relation calculation, with the only difference that an RF external field $\vec{h}_{\text{rf}}$ is also taken into consideration:

$$\dot{\vec{\Omega}} = -\gamma_0\, \vec{\Omega} \times \left(\vec{H}_{\text{eff}} + \vec{h}_{\text{rf}}\right) - \alpha_G \left(\dot{\vec{\Omega}} \times \vec{\Omega}\right). \tag{10.11}$$

The dynamic Eq. 10.3 for the excitation $\vec{m}$ extended with the damping term can be written in the Fourier space in the following form:

$$\begin{pmatrix} \mathcal{R}_n(k_z) \\ \mathcal{Z}_n(k_z) \end{pmatrix} = \aleph_{k_z}^n \begin{pmatrix} \langle h_{\text{rf}}\rangle_\rho^{n,k_z} \\ \langle h_{\text{rf}}\rangle_z^{n,k_z} \end{pmatrix} \tag{10.12}$$

where $\aleph_{k_z}^n$ is the dynamic susceptibility (Polder tensor[33]), which is expressed as

$$\aleph_{k_z}^n = \begin{pmatrix} \chi_{\rho\rho}^{n,k_z} & \chi_{\rho z}^{n,k_z} \\ -\chi_{\rho z}^{n,k_z} & \chi_{zz}^{n,k_z} \end{pmatrix} = \begin{pmatrix} \frac{i\mathbb{B}}{\mathbb{A}^2+\mathbb{BC}} & \frac{i\mathbb{A}}{\mathbb{A}^2+\mathbb{BC}} \\ -\frac{i\mathbb{A}}{\mathbb{A}^2+\mathbb{BC}} & \frac{i\mathbb{C}}{\mathbb{A}^2+\mathbb{BC}}, \end{pmatrix} \tag{10.13}$$

where $\mathbb{A} \equiv \omega - \gamma M_s \mathcal{A}_n(k_z)$, $\mathbb{B} \equiv -\alpha_G\omega + i\omega_{M_s}\mathcal{B}_n(k_z)$, $\mathbb{C} = -\alpha_G\omega + i\omega_{M_s}\mathcal{C}_n(k_z)$ with $\mathcal{A}_n(k_z)$, $\mathcal{B}_n(k_z)$, and $\mathcal{C}_n(k_z)$ being the stiffness fields given as:

$$\begin{aligned} \mathcal{A}_n(k_z) &= \mathcal{K}_n(k_z), \\ \mathcal{B}_n(k_z) &= l_{\text{ex}}^2 k_z^2 + (n^2-1)h_u + h_0 + \mathcal{L}_n(k_z), \\ \mathcal{C}_n(k_z) &= l_{\text{ex}}^2 k_z^2 + n^2 h_u + h_0 + \mathcal{J}_n(k_z), \end{aligned} \tag{10.14}$$

with the hypergeometrical functions defined by Eqs. 10.7.

The spin-wave absorption can be expressed in a Lorentzian form by using the Polder tensor defined in Eq. 10.13 as $\mathcal{S}_{\rho\rho}^{n,k_z}(\omega) = \text{Im}(\mathcal{X}_{\rho\rho}^{n,k_z})$

$$\mathcal{S}_{\rho\rho}^{n,k_z}(\omega) = \frac{\mathcal{W}_{k_z}^n(\omega, \alpha_{\text{G}})}{\left(\omega^2 - \omega_n^2(k_z)\right)^2 + \left(\frac{1}{2}\Delta_{k_z}^n(\omega)\right)^2} \tag{10.15}$$

where the eigenfrequency $\omega_n(k_z)$ and the terms $\Delta_{k_z}^n(\omega)$ and $\mathcal{W}_{k_z}^n(\omega, \alpha_{\text{G}})$ are given as

$$\omega_n(k_z) = \omega_{M_s}\left(\mathcal{K}_n(k_z) + \sqrt{\mathcal{C}_n(k_z)\,\mathcal{B}_n(k_z)}\right), \tag{10.16}$$

$$\Delta_{k_z}^{n}(\omega) = 2\omega_{M_s}\alpha_G\omega\left(\mathcal{C}_n(k_z) + \mathcal{B}_n(k_z)\right)\left(\frac{\omega + \omega_n(k_z)}{\omega - \bar{\omega}_n(k_z)}\right) \tag{10.17}$$

and

$$\mathcal{W}_{k_z}^{n} = -\frac{(\omega + \omega_n(k_z))}{(\omega - \bar{\omega}_n(k_z))}\left(\alpha_G\omega_{M_s}\omega\left(\omega^2 - \omega_n^2(k_z)\right) + \frac{\omega_{M_s}^2}{2}\Delta_{k_z}^{n}\mathcal{B}_n(k_z)\right), \tag{10.18}$$

with $\bar{\omega}_n(k_z) = \omega_{M_s}\left(\mathcal{K}_n(k_z) - \sqrt{\mathcal{C}_n(k_z)\mathcal{B}_n(k_z)}\right)$. The term $\mathcal{K}_n(k_z)$ vanishes for planar thin films, and Eq. 10.16 simplifies to the dispersion relation well known from the literature [30, 31]. The analytical expression for the group velocity can be obtained from the k_z-derivative of the eigenfrequency, $v_g(n, k_z) = \partial\omega_n(k_z)/\partial k_z$, and the frequency linewidth Γ can be obtained from the absorption Eq. 10.15. Note that the Lorentzian susceptibility should be inversely proportional to the linewidth as a quadratic term, that is, $\mathcal{S}_{\rho\rho}^{n,k_z}(\omega) \propto \left((\omega - \omega_n(k_z))^2 - (\Gamma/2)^2\right)^{-1}$. Using Eqs. 10.15 and 10.17, (when $\Delta_{k_z}^{n}(\omega)/(\omega + \omega_n(k_z))$ for the eigenfrequencies $\omega = \omega_n(k_z)$), it is straightforward to deduce the equation of the frequency linewidth:

$$\Gamma(n, k_z) = \omega_{M_s}\alpha_G\left(\mathcal{C}_n(k_z) + \mathcal{B}_n(k_z)\right)\left(1 + \frac{\mathcal{K}_n(k_z)}{\sqrt{\mathcal{B}_n(k_z)\mathcal{C}_n(k_z)}}\right) \tag{10.19}$$

One can realize that the obtained equation is very similar to that deduced for thin films [20, 32] with the addition of the last term. If the nanotube radius goes to infinity, $\mathcal{K}_n(k_z) = 0$ and the above expression reduces to that of the frequency linewidth of 2D thin films. In this case, the linewidth directly depends on the stiffness fields and the Gilbert damping parameter as well known from the literature [2, 12, 15, 20, 25, 32].

Equation 10.19 has the same mathematical formulations as the linewidth derived for planar 2D heavy metal/magnetic metal sandwiches with interfacial DMI (see, for example, Eq. 8 in Ref. [12]). However, the asymmetries in Γ are introduced by the sign dependence of $\mathcal{K}_n(k_z)$ on the propagation direction, that is, $\mathcal{K}_n(k_z) \neq -\mathcal{K}_n(k_z)$ broken mirror symmetry. As already discussed this term arises from the dynamic magnetic charges of the dipole–dipole interaction.

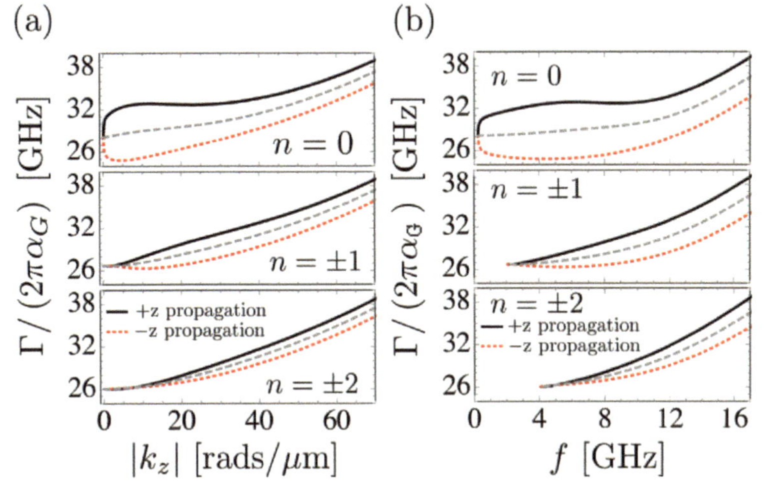

Figure 10.9 Frequency linewidth as a function of (a) wave vector k_z and (b) eigenfrequency f for a permalloy nanotube with 80 nm outer and 70 nm inner radii and a circular field of $\mu_0 H_0 = 6$ mT. Curves are presented for the azimuthal wavenumbers $n = 0, \pm 1, \pm 2$. The black continuous and red dotted lines correspond to the frequency linewidth of counterpropagating spin waves calculated for the infinitely long permalloy nanotube. The gray dashed lines correspond to the case of an extended 2D thin film with 10 nm thickness and with an in-plane magnetization oriented parallel to the magnetic field $\mu_0 H_0 = 6$ mT. Reprinted (figure) with permission from Ref. [46]. Copyright (2018) by the American Physical Society.

The results of the analytical model on the linewidth are first presented assuming an infinitely long permalloy nanotube with outer (inner) radius $R = 80$ nm ($r = 70$ nm). The equilibrium state, a vortex state, is stabilized by an Oersted field of $\mu_0 H_0 = 6$ mT. The frequency linewidth Γ scaled by the Gilbert damping α_G as a function of the wave vector k_z and of the eigenfrequency is shown in Figs. 10.9a and 10.9b, respectively. Clearly, the linewidth is asymmetric for modes with the same wave vector or the same eigenfrequency but opposite propagation directions. As a comparison we show the linewidth obtained for a planar permalloy thin film with a magnetic configuration similar to that in an unrolled view of the nanotube. Therefore, an extended 2D thin film is assumed with an in-plane applied magnetic field with the same value as the Oersted field applied for the nanotube, namely $\mu_0 H_0 =$ 6 mT. In Fig. 10.9 the frequency linewidth for the planar permalloy thin film is plotted with gray dashed lines, and the black solid and red dotted lines represent the linewidth for the counterpropagating spin waves in nanotubes. Note that the range of Γ values predicted by our analytical calculations is consistent with the values measured experimentally for permalloy [21, 40]. Indeed, assuming typical values for the Gilbert damping parameter α_G between 0.001 and 0.01, one can observe the frequency linewidth ranging between 200 and 300 MHz for the given range of eigenfrequencies and wave vectors.

To quantify the asymmetry in the linewidth the following expression can be defined $\Delta\Gamma(n, k_z) \equiv |\Gamma(n, k_z) - \Gamma(n, -k_z)|$, which has the form

$$\Delta\Gamma(n, k_z) = 2\omega_{M_s}\alpha_G|\mathcal{C}_n(k_z) + \mathcal{B}_n(k_z)|\frac{|\mathcal{K}_n(k_z)|}{\sqrt{\mathcal{B}_n(k_z)\mathcal{C}_n(k_z)}} \tag{10.20}$$

Note that for the FMR mode, $k_z = 0$, the frequency linewidth asymmetry is zero $\Delta\Gamma(n, k_z) = 0$ as a consequence of the odd parity of $\mathcal{K}_n(k_z)$.

The linewidth asymmetry $\Delta\Gamma$, defined as the difference in the linewidth between counterpropagating spin waves with the same wave vector and eigenfrequency, are shown in Figs. 10.10a and 10.10b, respectively. For all modes, $\Delta\Gamma$ shows a maximum $\Delta\Gamma^{(\mathrm{Max})}$ at a particular frequency ($f = f_\Gamma$) and wave vector ($k_z = k_\Gamma$). The evolution of the quantities $\Delta\Gamma^{(\mathrm{Max})}$, f_Γ, and $\lambda_\Gamma = 2\pi/k_\Gamma$ as a function of the nanotube radius is summarized in Fig. 10.10c. The linear relation between λ_Γ and R indicates that the asymmetry is directly proportional to the mean curvature ($\sim 1/R$) of the nanotube. A similar linear dependence was shown between the maximum asymmetry in the dispersion relation and the wavelength λ_{SW} of the spin wave modes [28, 30]. The changes in the linewidth asymmetries between counterpropagating spin waves can be estimated by comparing the values of $\Delta\Gamma^{\mathrm{Max}}$ in Fig. 10.10c with the values for Γ in Fig. 10.9. Indeed, changes approximately greater than 10% and up to 20% for a wide range of nanotube radii (between 30 nm and 260 nm) are observed, which

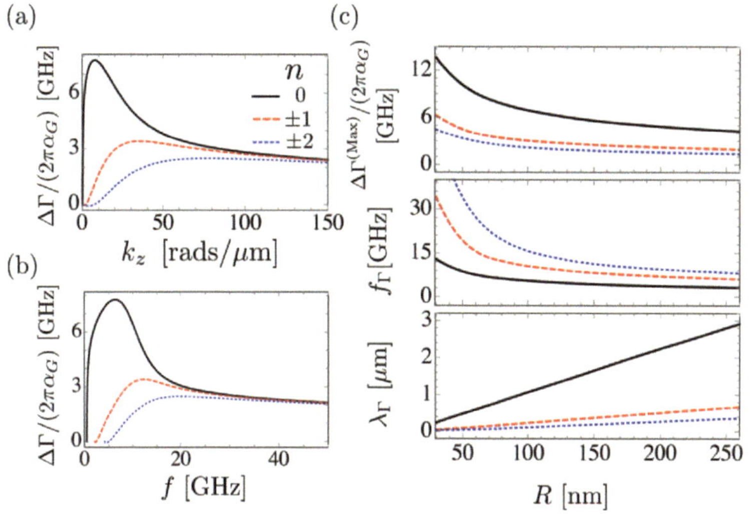

Figure 10.10 Frequency linewidth asymmetries (defined by Eq. 10.20) at the fundamental $n = 0$ (black lines) and at the first two higher-order azimuthal modes $n = \pm 1$ (red dashed line) and $n = \pm 2$ (blue dotted line). Evolution of the frequency linewidth asymmetry as a function of (a) the wave vector k_z and (b) eigenfrequency of spin waves. (c) Radial dependence of the maximum linewidth asymmetry $\Delta\Gamma^{(\mathrm{Max})}$ (upper curve) of the wavelength λ_Γ (middle curve) and of the eigenfrequency f_Γ (bottom figure) of spin waves at which the asymmetry reaches its maximum. All cases presented here are calculated for the nanotube thickness $d = 10$ nm. An infinitely large nanotube with outer radius $R = 80$ nm and applied circular field $\mu_0 H_0 = 6$ mT are assumed in (a) and (b). An applied circular field slightly larger than the critical field ($H_0 = 1.01 H_c$) [28] is assumed for calculating the curves in (c). Reprinted (figure) with permission from Ref. [46]. Copyright (2018) by the American Physical Society.

is comparable with those extracted from experiments on heavy metal/magnetic metal sandwiches wherein linewidth asymmetry results from interfacial DMI [12].

For completeness, we summarized in Fig. 10.11 the evolution of the linewidth asymmetry as a density plot in a function of the radius and the propagation vector. Basically, it shows the same tendency as the asymmetry of the dispersion relation.

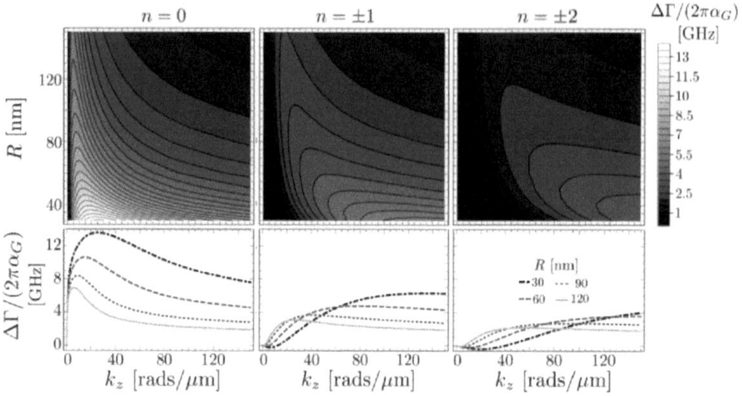

Figure 10.11 Density plot of the linewidth asymmetry $\Delta\Gamma$ as a function of the nanotube radius R and wave vector k_z for the zero-order $n = 0$, first-order $n = \pm 1$, and second-order $n = \pm 2$ azimuthal modes. The linewidth asymmetries for four different radii $R = (30, 60, 90, 120)$ nm are shown. The nanotube is considered to be made of permalloy, infinitely long with thickness $d = 10$ nm.

Using the linewidth and the group velocity, the spin-wave decay length, $l_D(n, k_z) \equiv v_g(n, k_z)/\Gamma(n, k_z)$ and its asymmetry $\Delta l_D(n, k_z) = l_D(n, k_z) - l_D(n, -k_z)$ can also be calculated analytically and are shown in Fig. 10.12. According to the results changes ranging between 10% and 22% in the decay length in the range of frequencies between 2 GHz and 10 GHz can be observed. The decay length is shown for three different nanotube radii and the changes are estimated by $\Delta l_D(n, k_z)/l_D(n, k_z)$. For the zero-order mode $n = 0$ for a nanotube with $R = 80$ nm radius, a 20% difference is in the decay length of the counterpropagating spin waves for frequencies between 4 GHz and 6 GHz. Such difference is also present for nanotubes with large radii as $R = 300$ nm at a frequency range of 2–3 GHz. For higher-order modes ($n \neq 0$) the percentage difference decreases to approximately 13% due to the reduced curvature-induced asymmetry of group velocity v_g. As a consequence, a nanotube with a fixed diameter can be operated to show the largest asymmetries in the dispersion, linewidth, and decay length in a well-defined frequency and wave vector range.

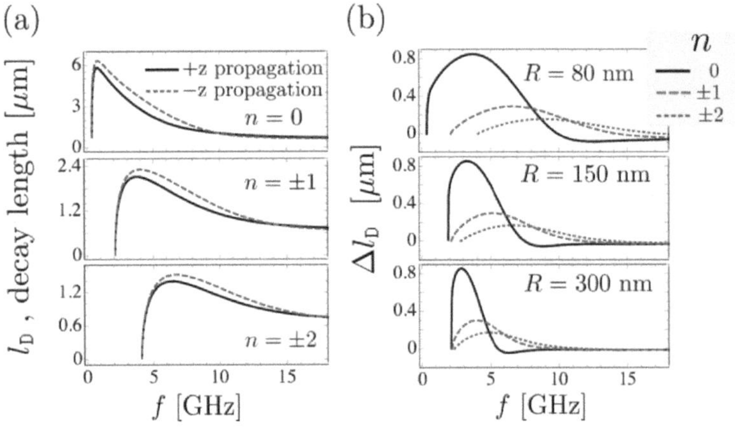

Figure 10.12 Decay length of counterpropagating spin waves (a) and decay length asymmetries (b) as a function of spin-wave eigenfrequency and for wavenumber $n = 0, \pm1, \pm2$. An infinitely large nanotube with outer radius $R = 80$ nm and inner radius $r = 70$ nm is assumed in (a). An applied circular magnetic field $\mu_0 H_0 = 6$ mT is assumed in (a) and (b). Note that for all radii in (b) the circular field magnitude is larger than the critical field necessary for stabilizing the circular magnetization.[28] A Gilbert damping parameter $\alpha_G = 0.001$ for the permalloy nanotube was considered. Reprinted (figure) with permission from Ref. [46]. Copyright (2018) by the American Physical Society.

10.5 Asymptotic Analytical Expressions

We believe it is important to show that for specific cases the well-known dispersion relations from the literature are recovered. The following cases will be discussed: spin-wave propagation along the azimuthal direction only or $k_z = 0$ the thin-film limit or $k_z \gg 1/R$, and finally the spin-wave transport with finite but small wave vectors referred to as $k_z \neq 0$.

10.5.1 For $k_z = 0$

It can be easily shown that for $k_z = 0$ the terms in Eqs. 10.7 will be as follows: $\mathcal{K}_n(k_z) = \mathcal{L}_n(k_z) = 0$ at $k_z = 0$, whereas $\mathcal{J}_n(k_z = 0)$ can be written analytically as

$$\mathcal{J}_n(0) = \begin{cases} 1 & ;\ n = 0 \\ \frac{1}{2}\left(1 + \frac{r^2}{b^2}\right) & ;\ |n| = 1 \\ \frac{1}{n^2 - 1}\left(-1 + |n|\frac{\pi}{S}\left(R^2 + r^2 - \frac{r^{|n|+1}}{R^{|n|-1}}\right)\right) & ;\ |n| > 1 \end{cases} \tag{10.21}$$

Replacing these into the spin-wave dispersion (see Eq. 10.5) at $k_z = 0$ the following analytical form is obtained:

$$\omega_n(0) = \gamma_0\mu_0\Big\{\left(H_0 + (n^2 - 1)H_u\right) \times \left(H_0 + n^2 H_u + M_s\mathcal{J}_n(0)\right)\Big\}^{1/2}. \tag{10.22}$$

This expression is analogous to the backward-volume (BV) spin-wave dispersion of a thin film with in-plane magnetization parallel to the applied field H_0. The azimuthal wave vector k_φ in our curvilinear frame of reference is the equivalent to the wave vector in the BV wave vector for thin films, which is quantized as $k_\varphi = n/\rho$ (with $r \leq \rho \leq R$), given the periodic azimuthal boundary conditions, being $n = 0, \pm 1, \pm 2, \ldots$ the azimuthal quantum number. The effective azimuthal wave vector is defined as

$$\bar{k}_\varphi \equiv (2\pi/S)\int_r^R k_\varphi\, \rho d\rho = n/\bar{\rho}, \tag{10.23}$$

instead of k_φ, to remove the radial dependence on the azimuthal wave vector. For thin magnetic layers this can be done without losing generality. The $\bar{\rho} = (R + r)/2$ is the nanotube mean radius. The dipolar term $M_s\mathcal{J}_n(0)$ in the spin-wave dispersion 10.22 is the same as the dipolar field $M_s F(kd)$ in the BV configuration for thin films with thickness d, wave vector k, and $F(x) = (1 - e^{-x})/x$ [2, 15, 25]. At very large azimuthal wave vectors $\bar{k}_\varphi \gg 1/\bar{\rho}$ (at the limit $|n| \gg 1$), the expression in 10.22 reduces to

$$\omega_{n\gg 1}(0) \approx \gamma_0\mu_0\left\{\left(H_0 + n^2 H_u\right) \times \left(H_0 + n^2 H_u + \left(\frac{R^2 + r^2}{R^2 - r^2}\right)\frac{M_s}{n}\right)\right\}^{1/2}, \tag{10.24}$$

wherein for a large radius $R \gg d$, the mean curvature of the nanotube can be reduced up to the point where the dispersion relation shown in Eq. 10.24 converges asymptotically to the exchange-dominated BV dispersion relation in thin films:

$$\omega_{n\gg1}(0) \approx \gamma_0\mu_0\left\{ \left(H_0 + M_s l_{\text{ex}}^2 \bar{k}_\varphi^2\right) \times \left(H_0 + M_s l_{\text{ex}}^2 \bar{k}_\varphi^2 + \frac{M_s}{\bar{k}_\varphi d}\right) \right\}^{1/2}, \tag{10.25}$$

If we also remove the azimuthal propagation, at $n = 0$ one should obtain the well-known Kittel formula for FMR. Indeed, the spin-wave dispersion relation show in Eq. 10.22 for magnetization parallel to the applied field but perpendicularly to H_u arising from the exchange due to the circular magnetization profile has the FMR form [23]

$$\omega_{\text{FMR}} \equiv \omega_0(k_z = 0) = \gamma_0\mu_0\sqrt{(H_0 - H_u)(H_0 + M_s)}. \tag{10.26}$$

Here, the role of the nanotube radius can be clearly identified: The term b in Eq. 10.8 increases with increasing the transversal nanotube size, and consequently, the critical field H_u decreases. At large enough nanotube radius R, the situation $H_u \ll H_0$ occurs, giving place to the FMR expression

$$\omega_{\text{FMR}} \approx \gamma_0\mu_0\sqrt{H_0(H_0 + M_s)}, \tag{10.27}$$

which is the well-known FMR Kittel formula [23] of a thin film with a homogeneously in-plane magnetization parallel to the applied magnetic field H_0. Accordingly, the FMR response of nanotubes in a vortex magnetization state is equivalent to the FMR response of a planar thin film whenever the nanotube radius is large enough, which is the limit where the chiral effects proportional to the mean curvature of the nanotube are negligible.

10.5.2 For $k_z \gg 1/R$

The hyperfunctions $\mathcal{J}_n$ and $\mathcal{K}_n$ in Eqs. 10.7 can be neglected for very large wave vectors $k_z \gg 1/R$, whereas $\mathcal{L}_n(k_z \gg 1/R) \approx 1$. Accordingly, the dispersion relation can be approximated as

$$\begin{aligned}\omega_n(k_z) \approx \gamma_0\mu_0\Big\{ &\left(M_s l_{\text{ex}}^2 k_z^2 + (n^2 - 1)H_u + H_0 + M_s\right) \\ &\times \left(M_s l_{\text{ex}}^2 k_z^2 + n^2 H_u + H_0\right) \Big\}^{1/2},\end{aligned} \tag{10.28}$$

which for $n = 0$ is identical to the exchange-dominated dispersion relation [11] of a planar thin film in DE configuration with in-plane magnetization oriented perpendicularly to the in-plane easy axis. Similar to the case of FMR ($k_z = 0$), the exchange field H_u (see Eq. 10.8) can be negligible at large enough nanotube radii, and the dispersion relation becomes identical to the one of a thin film in DE configuration with in-plane magnetization:

$$\omega_0(k_z) \approx \gamma_0\mu_0\Big\{\left(M_s l_{ex}^2 k_z^2 + H_0 + M_s\right) \times \left(M_s l_{ex}^2 k_z^2 + H_0\right)\Big\}^{1/2}. \qquad (10.29)$$

10.5.3 For $k_z \neq 0$ ($k_z \ll 1/R$)

In the range of finite wave vectors the dispersion relation has no simple analytical formula but can be calculated by a numerical integrations of the quantities $\mathcal{J}_n$, $\mathcal{K}_n$ and $\mathcal{L}_n$ from Eq. 10.7. These expressions consist of improper integrals that, at first sight, might comprehend large computational consuming time to be performed. Nevertheless, it can be shown that the improper integrals can be computed by well-defined proper integrals, hence optimizing the computational time. The full derivation is shown in Ref. [31].

The important point is that for small wave vector $k_z \ll \pi/R$ the integrals $\mathcal{J}_0$, $\mathcal{K}_0$ and $\mathcal{L}_0$ can be approximated by the following simple functions:

$$\mathcal{J}_0 \approx 1 - u\, k_z^2, \qquad (10.30)$$

$$\mathcal{K}_0 \approx v\, k_z, \qquad (10.31)$$

$$\mathcal{L}_0 \approx \mathrm{w}\, k_z^2\,, \qquad (10.32)$$

where the proportionality quantities u, v, and w depend on the outer (inner) nanotube radius R (r), namely on the mean curvature of the nanotube, and have to be calculated numerically. The linear dependence of $\mathcal{K}_0$ on the wave vector k_z at $k_z \ll \pi/R$ with the proportionality constant depending on the mean curvature leads to the asymmetric dispersion relation similar to that observed for thin films with interfacial DMI.

Taking into account Eqs. 10.30–10.32 the dispersion relation (from Eq. 10.5) in the limit of very small wave vectors $k_z \ll \pi/R$

will have the following form:

$$\frac{\omega_0(k_z)}{\gamma_0\mu_0} \approx M_s\nu k_z + \sqrt{\left\{M_s\left(l_{ex}^2 + w\right)k_z^2 - H_u + H_0\right\}\left\{M_s\left(l_{ex}^2 - u\right)k_z^2 + H_0 + M_s\right\}}. \tag{10.33}$$

After expanding up to the second order in k_z,

$$\omega \approx \omega_{FMR} + \mu_0\gamma_0 M_s\nu k_z + \frac{(\mu_0\gamma_0)^2 M_s}{2\omega_{FMR}}\left\{(l_{ex}^2 + w)(H_0 + M_s) + (l_{ex}^2 - u)(H_0 - H_u)\right\}k_z^2, \tag{10.34}$$

where the FMR frequency response ω_{FMR} is given in Eq. 10.26 and the exchange field H_u in Eq. 10.8. The dispersion relation for small k_z obtained in Eq. 10.34 resembles the DE dispersion relation of a planar thin film [2, 22, 25]. The main difference is due to the tubular structure of our system. In planar thin films without intrinsic nonreciprocities, in DE configuration the terms linear in the modulus of the wave vector k_z appear as a result of the dipolar dominated spin waves, whereas orders higher than k_z^2 are usually neglected. In Eq. 10.34, the term linear in k_z depends on its sign as a consequence of the cylindrical topology, hence playing the role of the curvature-induced nonreciprocity.

References

1. Arias, R., Chu, P. and Mills, D. L. (2005). Dipole exchange spin waves and microwave response of ferromagnetic spheres, *Phys. Rev. B*, **71**, 224410, doi:10.1103/PhysRevB.71.224410, URL https://link.aps.org/doi/10.1103/PhysRevB.71.224410.
2. Arias, R. and Mills, D. L. (1999). Extrinsic contributions to the ferromagnetic resonance response of ultrathin films, *Phys. Rev. B*, **60**, 7395–7409, doi:10.1103/PhysRevB.60.7395, URL https://link.aps.org/doi/10.1103/PhysRevB.60.7395.
3. Arias, R. and Mills, D. L. (2001). Theory of spin excitations and the microwave response of cylindrical ferromagnetic nanowires, *Phys. Rev. B*, **63**, 134439, doi:10.1103/PhysRevB.63.134439, URL https://link.aps.org/doi/10.1103/PhysRevB.63.134439.

4. Arias, R. and Mills, D. L. (2003). Theory of collective spin waves and microwave response of ferromagnetic nanowire arrays, *Phys. Rev. B*, **67**, 094423, doi:10.1103/PhysRevB.67.094423, URL https://link.aps.org/doi/10.1103/PhysRevB.67.094423.
5. Arias, R. and Mills, D. L. (2004). Magnetostatic modes in ferromagnetic nanowires, *Phys. Rev. B*, **70**, 094414, doi:10.1103/PhysRevB.70. 094414, URL https://link.aps.org/doi/10.1103/PhysRevB.70.094414.
6. Berger, M. (1992). Riemannian manifolds from curvature to toplogy, in *Chern: A Great Geometer of the 20th Century* (International Press, Boston), pp. 184–238.
7. Brown, W. F. (1963). *Micromagnetics* (Interscience).
8. Chumak, A., Vasyuchka, V., Serga, A. and Hillebrands, B. (2015). Magnon spintronics, *Nat. Phys.*, **11**(6), 453–461.
9. Chumak, A. V., Serga, A. A. and Hillebrands, B. (2014). Magnon transistor for all-magnon data processing, *Nat. Commun.*, **5**.
10. Cortés-Ortuño, D. and Landeros, P. (2013). Influence of the Dzyaloshinskii–Moriya interaction on the spin-wave spectra of thin films, *J. Phys.: Condens. Matter*, **25**(15), 156001, doi:10.1088/0953-8984/25/15/156001, URL https://doi.org/10.1088%2F0953-8984%2F25% 2F15%2F156001.
11. Damon, R. W. and Eshbach, J. R. (1961). Magnetostatic modes of a ferromagnet slab, *J. Phys. Chem. Solids*, **19**(3), 308–320.
12. Di, K., Zhang, V. L., Lim, H. S., Ng, S. C., Kuok, M. H., Yu, J., Yoon, J., Qiu, X. and Yang, H. (2015). Direct observation of the Dzyaloshinskii–Moriya interaction in a pt/co/ni film, *Phys. Rev. Lett.*, **114**(4), 047201.
13. Dzyaloshinsky, I. (1958). A thermodynamic theory of weak ferromagnetism of antiferromagnetics, *J. Phys. Chem. Solids*, **4**(4), 241–255.
14. Gaididei, Y., Kravchuk, V. P. and Sheka, D. D. (2014). Curvature effects in thin magnetic shells, *Phys. Rev. Lett.*, **112**, 257203, doi:10. 1103/PhysRevLett.112.257203, URL http://link.aps.org/doi/10.1103/PhysRevLett.112.257203.
15. Gallardo, R., Banholzer, A., Wagner, K., Korner, M., Lenz, K., Farle, M., Lindner, J., Fassbender, J. and Landeros, P. (2014). Splitting of spin-wave modes in thin films with arrays of periodic perturbations: theory and experiment, *New J. Phys.*, **16**(2), 023015.
16. Golovchan, A. V., Kruglyak, V. V., Tkachenko, V. S. and Kuchko, A. N. (2018). Magnonic band spectrum of spin waves in an elliptical helix, *R. Soc. Open Sci.*, **5**(1), 172285, doi:10.1098/rsos.172285, URL https://royalsocietypublishing.org/doi/abs/10.1098/rsos.172285.

17. Hertel, R. (2013). Curvature-induced magnetochirality, *SPIN*, **3**(3), 1340009, URL http://www.worldscientific.com/doi/abs/10.1142/S2010324713400092.
18. Hurben, M. and Patton, C. (1995). Theory of magnetostatic waves for inplane magnetized isotropic films, *J. Magn. Magn. Mater.*, **139**(3), 263–291.
19. Hurben, M. and Patton, C. (1996). Theory of magnetostatic waves for in-plane magnetized anisotropic films, *J. Magn. Magn. Mater.*, **163**, 39–69.
20. Hurben, M. and Patton, C. (1998). Theory of two magnon scattering microwave relaxation and ferromagnetic resonance linewidth in magnetic thin films, *J. Appl. Phys.*, **83**(8), 4344–4365.
21. Kalarickal, S. S., Krivosik, P., Wu, M., Patton, C. E., Schneider, M. L., Kabos, P., Silva, T. J. and Nibarger, J. P. (2006). Ferromagnetic resonance linewidth in metallic thin films: comparison of measurement methods, *J. Appl. Phys.*, **99**(9), 093909.
22. Kalinikos, B. and Slavin, A. (1986). Theory of dipole-exchange spin wave spectrum for ferromagnetic films with mixed exchange boundary conditions, *J. Phys. C: Solid State Phys.*, **19**(35), 7013.
23. Kittel, C. (1986). *Introduction to Solid State Physics*, 6th edn. (John Wiley & Sons, Inc., New York).
24. Kravchuk, V. P., Sheka, D. D., Kákay, A., Volkov, O. M., Rößler, U. K., van den Brink, J., Makarov, D. and Gaididei, Y. (2018). Multiplet of skyrmion states on a curvilinear defect: Reconfigurable skyrmion lattices, *Phys. Rev. Lett.*, **120**(6), 067201, doi:10.1103/PhysRevLett.120.067201, URL https://link.aps.org/doi/10.1103/PhysRevLett.120.067201.
25. Landeros, P., Arias, R. E. and Mills, D. (2008). Two magnon scattering in ultrathin ferromagnets: the case where the magnetization is out of plane, *Phys. Rev. B*, **77**(21), 214405.
26. McClarty, P. A., Kruger, F., Guidi, T., Parker, S. F., Refson, K., Parker, A. W., Prabhakaran, D. and Coldea, R. (2017). Topological triplon modes and bound states in a Shastry–Sutherland magnet, *Nat. Phys.*, **13**(8), 736–741.
27. Moriya, T. (1960). Anisotropic superexchange interaction and weak ferromagnetism, *Phys. Rev.*, **120**(1), 91.
28. Otálora, J. A., Cortés-Ortuño, D., Görlitz, D., Nielsch, K. and Landeros, P. (2015). Oersted field assisted magnetization reversal in cylindrical core-shell nanostructures, *J. Appl. Phys.*, **117**(17), 173914, doi:10.1063/1.4919746, URL https://doi.org/10.1063/1.4919746.

29. Otálora, J. A., López-López, J. A., Landeros, P., Vargas, P. and Núñez, Á. S. (2013). Breaking of chiral symmetry in vortex domain wall propagation in ferromagnetic nanotubes, *J. Magn. Magn. Mater.*, **341**, 86–92, URL http://www.sciencedirect.com/science/article/pii/S0304885313002333.
30. Otálora, J. A., Yan, M., Schultheiss, H., Hertel, R. and Kákay, A. (2016). Curvature-induced asymmetric spin-wave dispersion, *Phys. Rev. Lett.*, **117**(22), 227203.
31. Otálora, J. A., Yan, M., Schultheiss, H., Hertel, R. and Kákay, A. (2017). Asymmetric spin-wave dispersion in ferromagnetic nanotubes induced by surface curvature, *Phys. Rev. B*, **95**, 184415, doi:10.1103/PhysRevB.95.184415, URL https://link.aps.org/doi/10.1103/PhysRevB.95.184415.
32. Patton, C. E. (1968). Linewidth and relaxation processes for the main resonances in the spin-wave spectra of NiFe alloy films, *J. Appl. Phys.*, **39**(7), 3060–3068, doi:10.1063/1.1656733, URL https://doi.org/10.1063/1.1656733.
33. Polder, D. (1949). VIII. On the theory of ferromagnetic resonance, *London Edinburgh Dublin Philos. Mag. J. Sci.*, **40**(300), 99–115.
34. Pylypovskyi, O. V., Kravchuk, V. P., Sheka, D. D., Makarov, D., Schmidt, O. G. and Gaididei, Y. (2015). Coupling of chiralities in spin and physical spaces: the Mobius ring as a case study, *Phys. Rev. Lett.*, **114**, 197204, doi:10.1103/PhysRevLett.114.197204, URL http://link.aps.org/doi/10.1103/PhysRevLett.114.197204.
35. Rychly, J., Tkachenko, V. S., K los, J. W., Kuchko, A. and Krawczyk, M. (2018). Spin wave modes in a cylindrical nanowire in crossover dipolar-exchange regime, *J. Phys. D: Appl. Phys.*, **52**(7), 075003, doi:10.1088/1361-6463/aaf2fc, URL https://doi.org/10.1088%2F1361-6463%2Faaf2fc.
36. Sheka, D. D., Kravchuk, V. P., Yershov, K. V. and Gaididei, Y. (2015). Torsion-induced effects in magnetic nanowires, *Phys. Rev. B*, **92**, 054417, doi:10.1103/PhysRevB.92.054417, URL https://link.aps.org/doi/10.1103/PhysRevB.92.054417.
37. Shindou, R., Matsumoto, R., Murakami, S. and Ohe, J.-I. (2013a). Topological chiral magnonic edge mode in a magnonic crystal, *Phys. Rev. B*, **87**(17), 174427.
38. Shindou, R., Ohe, J.-I., Matsumoto, R., Murakami, S. and Saitoh, E. (2013b). Chiral spin-wave edge modes in dipolar magnetic thin films, *Phys. Rev. B*, **87**(17), 174402.

39. Stano, M. and Fruchart, O. (2018). *Magnetic Nanowires and Nanotubes*, Vol. 27 (Elsevier).
40. Twisselmann, D. and McMichael, R. D. (2003). Intrinsic damping and intentional ferromagnetic resonance broadening in thin permalloy films, *J. Appl. Phys.*, **93**(10), 6903–6905.
41. Vázquez, M. (2016). *Magnetic Nano- and Microwires*, 1st edn. (Elsevier).
42. Yan, M., Andreas, C., Kákay, A., Garćia-Sánchez, F. and Hertel, R. (2011). Fast domain wall dynamics in magnetic nanotubes: suppression of walker breakdown and cherenkov-like spin wave emission, *Appl. Phys. Lett.*, **99**(12), 122505.
43. Yan, M., Andreas, C., Kákay, A., Garćia-Sánchez, F. and Hertel, R. (2012). Chiral symmetry breaking and pair-creation mediated walker breakdown in magnetic nanotubes, *Appl. Phys. Lett.*, **100**(25), 252401.
44. Zakeri, K., Zhang, Y., Prokop, J., Chuang, T.-H., Sakr, N., Tang, W. X. and Kirschner, J. (2010). Asymmetric spin-wave dispersion on Fe(110): direct evidence of the Dzyaloshinskii–Moriya interaction, *Phys. Rev. Lett.*, **104**(13), 137203, URL http://link.aps.org/doi/10.1103/PhysRevLett.104.137203.
45. Zimmermann, M., Meier, T. N. G., Dirnberger, F., Kákay, A., Decker, M., Wintz, S., Finizio, S., Josten, E., Raabe, J., Kronseder, M., Bougeard, D., Lindner, J. and Back, C. H. (2018). Origin and manipulation of stable vortex ground states in permalloy nanotubes, *Nano Lett.*, **18**(5), 2828–2834, doi:10.1021/acs.nanolett.7b05222, URL https://doi.org/10.1021/acs.nanolett.7b05222, pMID: 29620910.
46. Otálora, J. A., Kákay, A., Lindner, J., Schultheiss, H., Thomas, A., Fassbender, J. and Nielsch, K. (2018). Frequency linewidth and decay length of spin waves in curved magnetic membranes, *Phys. Rev. B*, **98**, 014403.

Chapter 11

Strong Coupling in Cavity Magnonics

Angelo Leo, Silvia Rizzato, Anna Grazia Monteduro, and Giuseppe Maruccio

Department of Mathematics and Physics, University of Salento, Via per Arnesano, 73100, Lecce, Italy

CNR NANOTEC - Istituto di Nanotecnologia, Via per Arnesano, 73100 Lecce, Italy

giuseppe.maruccio@unisalento.it

11.1 Introduction

For decades, the prospect of quantum technologies attracted large research efforts for their expected revolutionary performances in computation [1], communications [2, 3], as well as high-precision sensing [4]. More recently, *circuit and cavity quantum electrodynamics* (cQED) [5] were established as a new branch of physics and an ideal platform for the study of the coherent interaction between radiation and matter and the manipulation of individual quantum systems. Notably, a strong coupling regime was achieved in the resulting hybrid system characterized by a coupling force between the two subsystems greater than their respective energy losses. Among the most promising candidates, the hybrid systems consisting of spin ensembles and microwave (MW) photons

Three-Dimensional Magnonics: Layered, Micro- and Nanostructures
Edited by Gianluca Gubbiotti

ISBN 978-981-4800-73-0 (Hardcover), 978-0-429-29915-5 (eBook)
www.jennystanford.com

permitted achieving high coherence times [6], useful for information storage.

In this respect, after focusing on dilute spin ensembles and paramagnetic materials [7, 8], the attention moved toward the study of the interaction between photon and ferro-/ferrimagnetic (FM) crystalline materials [9, 10], as they reach regimes of strong interaction in less restrictive environments, such as room rather than cryogenic temperatures. This happens since FM crystals have a net spin density 3 orders of magnitude higher than paramagnetic ones and coupling strength scale with it. On the photon counterpart, the use of high-quality factor (Q-factor) resonators allows efficient confinement of the electromagnetic field.

In *cavity magnonics*, yttrium iron garnet (YIG; $Y_3Fe_5O_{12}$) is an ideal material system because of its low dielectric loss, narrow resonance line width of 2.3×10^{-6} T in the MW region and a good saturated magnetization value of 0.176 T, related to the Fe^{+3} ions in the $6S_{5/2}$ ground state and to the net spin density of 4.22×10^{27} m^{-3}. The crystalline form of this magnetic insulator, with an exchange constant α of 3×10^{-12} cm^2 [11] and a Curie temperature of 550 K, has an easy $<111>$ and a hard $<110>$ magnetization direction due to magnetocrystalline anisotropy [12]. Furthermore, YIG crystals exhibit intriguing weak interaction with optical photons.

The objective of this chapter is to give an overview of the emerging field of cavity magnonics, analyzing the ideas, key results, and perspectives. In Section 11.2, the basics concepts about magnon–photon coupling are provided. Then the fingerprints of a magnon–photon strongly coupled system are presented in Section 11.3, before summarizing the various coupling regimes in Section 11.4. Subsequently, in Sections 11.5, 11.6, and 11.7, other key results and experiments are reviewed concerning respectively coupling to high-order magnetostatic modes (MSMs), in the case of multiple YIG spheres, and to transmon qubits. Finally, recent frontiers concerning cavity magnomechanics and optomagnonics are discussed in Section 11.8. Due to the rapid growth in this field, this review cannot be considered as exhaustive but as a support for researchers approaching the field.

11.2 Theory of Magnon–Photon Coupling

Assuming that interactions occur only among nearest neighbors, the Hamiltonian of a magnetic lattice can be expressed as the sum of a first term accounting for FM exchange interaction between neighbor spins and a second that refers to the Zeeman effect [12]:

$$\widehat{\mathcal{H}} = -2\frac{J}{\hbar^2}\sum_{j,\delta}\hat{S}_j \cdot \hat{S}_{j+\delta} - \frac{g\mu_B H_0}{\hbar}\sum_j \hat{S}_{jz} \tag{11.1}$$

where J is the exchange constant, $\hat{S}_j$ and $\hat{S}_{j+\delta}$ are the spin operators acting on spins in sites j and $j+\delta$, respectively, with δ representing the vector that refers to the nearest neighbor of j, while $\hat{S}_{jz}$ is the operator referring to the spin component z along the magnetostatic field H_0. Moreover, g is the Landé g-factor, μ_B is the Bohr magneton, and $\hbar$ is the reduced Planck constant.

Following the Holstein–Primakoff method [13] to model the spin flip on the jth site with respect to the Z nearest neighbors, one can introduce the operators $\hat{m}_{\mathbf{k}}^+$ and $\hat{m}_{\mathbf{k}}$ that, respectively, create and annihilate magnons with wave vector $\mathbf{k}$. Then, by naming the probability to have magnons with wave vector $\mathbf{k}$ as $\gamma_{\mathbf{k}} = \frac{1}{Z}\sum_\delta e^{i\mathbf{k}\cdot\delta}$ and by introducing the magnons number $\hat{n}_{\mathbf{k}} = \hat{m}_{\mathbf{k}}^+\hat{m}_{\mathbf{k}}$ with wave vector $\mathbf{k}$ and energy at frequency $\omega_{\mathbf{k}}$ as

$$\hbar\omega_{\mathbf{k}} = 4J\,sZ\,(1-\gamma_{\mathbf{k}}) + g\mu_B H_0, \tag{11.2}$$

the Hamiltonian becomes

$$\widehat{\mathcal{H}}_0 = \sum\nolimits_{\mathbf{k}} \hbar\omega_{\mathbf{k}}\left(\hat{m}_{\mathbf{k}}^+\hat{m}_{\mathbf{k}} + \frac{1}{2}\right) = \sum\nolimits_{\mathbf{k}} \hbar\omega_{\mathbf{k}}\left(\hat{n}_{\mathbf{k}} + \frac{1}{2}\right). \tag{11.3}$$

Therefore, as in the cases of phonons and photons, the energetic spectrum of spin waves is quantized and the energy of a mode that involves $n_{\mathbf{k}}$ single excitations with frequency $\omega_{\mathbf{k}}$ is

$$\varepsilon_{\mathbf{k}} = (n_{\mathbf{k}} + 1/2)\,\hbar\omega_{\mathbf{k}} \tag{11.4}$$

For an infinite cubic lattice with a lattice parameter a_0, the dispersion relation in the long wavelength limit is given by

$$\varepsilon_{\mathbf{k}} = \hbar\omega_{\mathbf{k}} \approx g\mu_B H_0 + 2sJ\,a_0^2k^2 \tag{11.5}$$

In magnetostatic limit occurring when $\mathbf{k}_0 \ll \mathbf{k} \ll \pi/a_0$, where $\mathbf{k}_0^2 = \omega^2\mu_0\varepsilon$ with ε the permittivity and μ_0 permeability of free space and a_0 is the spacing between spins, the coupling between the spins is dominated by the dipolar fields from the magnetic moments rather than the exchange interaction and the resulting MSMs are also known as dipolar spin waves [14].

To describe the magnon–photon coupling, the Jaynes–Cummings model developed for the interaction among an optical cavity's modes and atomic systems in quantum photonics [15] can be adopted. The contribution of the confined electromagnetic field to the total Hamiltonian can be treated like a quantum harmonic oscillator, whose Hamiltonian in second quantization is

$$\widehat{\mathcal{H}}_{\mathrm{a}} = \hbar\omega_{\mathrm{a}}\left(\hat{a}^{\dagger}\hat{a} + \frac{1}{2}\right) \tag{11.6}$$

where $\omega_{\mathrm{a}} = \sqrt{\mathbf{k}/m}$, while $\hat{a}^{\dagger}$ and $\hat{a}$ are annihilation and creation photonic operators, respectively.

Since MSMs are linear systems in a similar way to cavity modes, they contribute to the total Hamiltonian as

$$\widehat{\mathcal{H}}_{\mathrm{m}} = \hbar\omega_{\mathrm{m}}\left(\hat{m}^{\dagger}\hat{m} + \frac{1}{2}\right), \tag{11.7}$$

with collective spin oscillation frequency ω_{m} [16]; here for simplicity we omitted the label $\mathbf{k}$ and indicated $\omega_{\mathbf{k}}$ with ω_{m}. The first fundamental (uniform) MSM is known as Kittel mode [17] and is experimentally observed as ferromagnetic resonance (FMR).

The dynamics of the hybrid system are governed by an interaction Hamiltonian of the form

$$\widehat{\mathcal{H}}_{\mathrm{int}} \cong \hbar g_{\mathrm{m}}\left(\hat{a}^{+}\hat{m} + \hat{m}^{+}\hat{a}\right), \tag{11.8}$$

in the approximation where nonresonant terms can be neglected and just resonant interaction among subsystems to form a two-level system can be considered [9]. Thus, the total Hamiltonian assumes the form

$$\widehat{\mathcal{H}} = \widehat{\mathcal{H}}_{\mathrm{a}} + \widehat{\mathcal{H}}_{\mathrm{m}} + \widehat{\mathcal{H}}_{\mathrm{int}} \tag{11.9}$$

$$\frac{\widehat{\mathcal{H}}}{\hbar} = \omega_{\mathrm{a}}\hat{a}^{\dagger}\hat{a} + \omega_{\mathrm{m}}\hat{m}^{\dagger}\hat{m} + g_{\mathrm{m}}\left(\hat{a}^{\dagger}\hat{m} + \hat{m}^{\dagger}\hat{a}\right), \tag{11.10}$$

where $\hat{a}^{\dagger}$ ($\hat{a}$) is the creation (annihilation) operator for the MW photon at frequency ω_{a}, and the boson operator $\hat{m}^{\dagger}$ ($\hat{m}$) refers to standing spin-wave excitations in the Holstein–Primakoff approximation.

The term $g_{\rm m}$ describes coupling among the two subsystems, namely the electromagnetic field and the magnonic system (Fig. 11.1a) and is related to the strength of interaction g_0 of coupling of a single spin with the confined oscillating magnetic field

$$\frac{g_0}{2\pi} = \frac{\left(\eta\Gamma\sqrt{\frac{(\mu_0\hbar\omega_{\rm a})}{V_{\rm a}}}\right)}{4\pi} \tag{11.11}$$

by the relation

$$g_{\rm m} = g_0\sqrt{N_s} \quad [18], \tag{11.12}$$

where N_s refers to the number of net spins s in the sample, which is one of the reasons to prefer FM material rather than diluted spin systems. In Eq. 11.11, the quantity $\Gamma = g\mu_{\rm B}/\hbar$ is the gyromagnetic ratio of the spin ensemble, where g is the Landé g-factor, $\mu_{\rm B}$ is the Bohr magneton, $\hbar$ is the reduced Planck constant, and μ_0 is the vacuum permeability. $V_{\rm a}$ is the volume interested by the electromagnetic field, in particular by the magnetic component of the oscillating field, ω_a is the angular frequency of the oscillating field, $\eta = \int_{V_m} \frac{\vec{h}\cdot\vec{M}}{|\vec{h}_{\rm max}||\vec{M}_{\rm max}|} dV \leq 1$ is the spatial overlap between states of the two subsystems, with $\vec{h}$ referring to the magnetic field of the oscillating mode, $\vec{M}$ the magnetization vector of the magnetic material, while $|\vec{h}_{\rm max}|$ and $|\vec{M}_{\rm max}|$ are the maxima inside the interested volume. Following Ref. [9], if the ferromagnet is placed on the magnetic antinode of the cavity and the acting MW magnetic field can be considered isotropic, then inducing spin-wave formation inside the whole magnetic sample volume $V_{\rm m}$, the total coupling strength is $g_{\rm m} \propto \sqrt{\omega_{\rm a} V_{\rm m}/V_{\rm a}}$. To maximize the coupling strength between the two subsystems, the magnetic sample is typically placed at the location of maximum radiofrequency (RF) magnetic field intensity (typically with the local MW field perpendicular to the static magnetic field, although this last setting depends by the aim to probe the excitation of particular spin waves), and in this respect, finite element simulations facilitate the design of the experimental layout (Fig. 11.1c).

Beyond hybridization strength, another relevant parameter for describing the system is the damping in each subsystem: the photonic system is characterized by a photon relaxation rate $\kappa_{\rm a}$, which is proportional to energy loss for each field oscillation, while

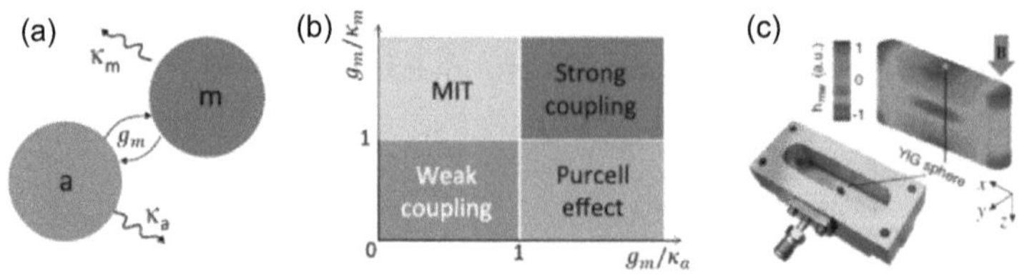

Figure 11.1 (a) Schematic of magnon–photon coupling: g_m is the coupling strength, and κ_a, κ_m are respectively dissipation rates of microwave cavity and magnon modes [9]. (b) Magnon–photon interaction regimes as a function of ratio between coupling strength g_m and dissipation rates, respectively of the cavity (κ_a) and magnons (κ_m). (c) MW cavity loaded by a YIG sphere and simulation of oscillating magnetic field (h_{mw}) distribution within the cavity for the TE_{101} mode. Reprinted (figure) with permission from Ref. [9]. Copyright (2014) by the American Physical Society.

the magnetic material is subject to a damping of the magnetization κ_m (Fig. 11.1a). Thus, key aspects for achieving a strong coupling regime are a high net spin density in low-magnetic-damping materials (like YIG) and an MW resonator with high quality factor (greater than 10^3). As a further figure of the coupling regime, the cooperativity of the two levels system is defined as $C_i = g_m^2/\kappa_m\kappa_a$.

11.3 Fingerprints of a Magnon–Photon Coupled System

Strong coupling among cavity photons and magnons was observed at cryogenic temperatures with an experimental setup similar to that sketched in Fig. 11.2a,b [19], where a magnetically saturated single-crystal YIG sphere is placed in the magnetic antinode of a 3D cavity resonator and a vector network analyzer (VNA) is connected for recording the MW response, while a magnet provides the external bias field. For measurements in quantum regimes, mK temperatures are important as well as the use of cryogenic components in order also to reduce the noise associated with thermal fluctuations. Notably, as compared to opto- or electromechanical systems, cavity magnonics presents the high advantage of tunability because of the dependence of magnon mode frequency on the magnetic field.

A characteristic fingerprint of the formation of a hybrid system composed by strongly coupled photons and magnons is the observation of an avoided crossing (Rabi splitting as a consequence of level repulsion) in the MW spectrum (Fig. 11.2c), which can be described by means of an input-output formalism [19–23] where the transmission $T(\omega)$ is expressed as

$$T(\omega) = \frac{\kappa_C}{i(\omega - \omega_{\rm a}) - \frac{1}{2}(2\kappa_{\rm c} + \kappa_{\rm int}) + \sum_j \frac{|g_{{\rm m},j}|^2}{-\frac{1}{2}\kappa_{{\rm m},j} + i(\omega - \omega_j)}}. \quad (11.13)$$

Here $\kappa_{\rm a}/2\pi = (2\kappa_c + \kappa_{int})/2\pi = \omega_a/2\pi\, Q_L$ is the photonic damping for each mode resonating at $\omega_a/2\pi$, which takes into account the damping κ_c through the single connectors and the internal damping κ_{int} associated with the mere cavity resonator; i is the imaginary unit number; and the index j identifies a specific MSM, having frequency linewidth $\kappa_{{\rm m},j}$ and interaction (coupling) strength $\frac{g_{{\rm m},j}}{2\pi} = \frac{g_{0,j}\sqrt{N_s}}{2\pi}$ between the two (magnonic and photonic) states. Experimentally, $\frac{g_{{\rm m},j}}{2\pi}$ can be evaluated as the frequency mismatch among two peaks corresponding to the hybrid magnon–photon modes (Fig. 11.2d), which split at energies $\hbar\omega_{\rm a} \pm g_{\rm m}$ at resonance, when $\omega_{\rm a}/2\pi = \omega_{\rm m}/2\pi$. Notably, Fig. 11.2e shows the discussed dependence of the coupling strength on the number of spins.

Further evidence of interaction between the two (photonic and magnonic) resonators is provided by the presence of oscillations in time-resolved measurements near strong coupling conditions (Fig. 11.3a), namely the energy stored inside cavity decays in an exponential manner (Fig. 11.3b), but with periodic oscillations associated with a coherent exchange of energy among photon and magnon modes. Indeed, if $|\Psi(t)\rangle = a(t)|a\rangle + m(t)|m\rangle$ is the state of the hybrid system, where $|a\rangle$ and $|m\rangle$ are the photonic and magnonic states with their time-dependent coefficients $a(t)$ and $m(t)$, the dynamics of the two levels subsystem on resonance can be described by the coupled equations [9]

$$\begin{pmatrix} \dot{a}(t) \\ \dot{m}(t) \end{pmatrix} = \begin{pmatrix} -\kappa_{\rm a} & -ig_{\rm m} \\ -ig_{\rm m} & -\kappa_{\rm m} \end{pmatrix} \begin{pmatrix} a(t) \\ m(t) \end{pmatrix} = {\rm M} \begin{pmatrix} a(t) \\ m(t) \end{pmatrix} \quad (11.14)$$

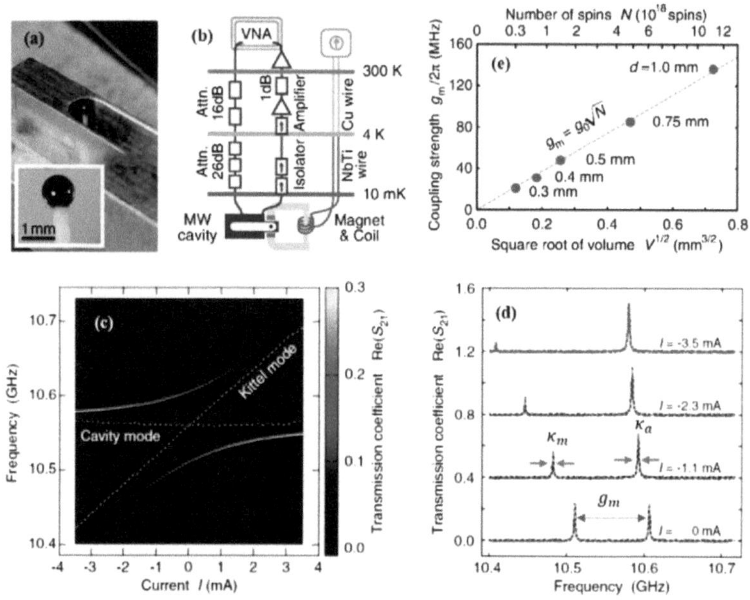

Figure 11.2 (a) Experimental setup and (b) apparatus for MW transmission measurement at cryogenic temperature. An external bias magnetic field is applied by a magnet and a current-controlled coil employed for tuning the bias. The signal traveling through the input coaxial line is attenuated, while the probe line is isolated and amplified along the various stages. (c) Amplitude of the transmitted signal [Re(S_{21})] through the cavity as a function of the probe MW frequency and the static bias magnetic field reported as the coil current. The horizontal dashed line shows the cavity resonant frequency, while the diagonal dashed line corresponds to the sample FMR frequency (Kittel uniform precession mode). (d) Cross sections of transmitted signal as a function of frequency for different values of external bias magnetic field. (e) Coupling strength among Kittel mode and microwave cavity mode as a function of the square root of the YIG sample volume. Reprinted (figure) with permission from Ref. [19]. Copyright (2014) by the American Physical Society.

having solution of the type

$$\begin{pmatrix} a(t) \\ m(t) \end{pmatrix} = e^{\mathrm{M}t} \begin{pmatrix} a(0) \\ m(0) \end{pmatrix}. \tag{11.15}$$

For $\kappa_\mathrm{a} \approx \kappa_\mathrm{m}$, cavity photons evolution follows the trend $a(t) = e^{-\kappa_\mathrm{a} t} \cos g_\mathrm{m} t$. Thus we can write the power of the cavity field as

$$P\,(\mathrm{dBm}) \propto \log\left[\cos^2(g_\mathrm{m} t) \cdot \exp(-2\kappa_\mathrm{a} t)\right] \tag{11.16}$$

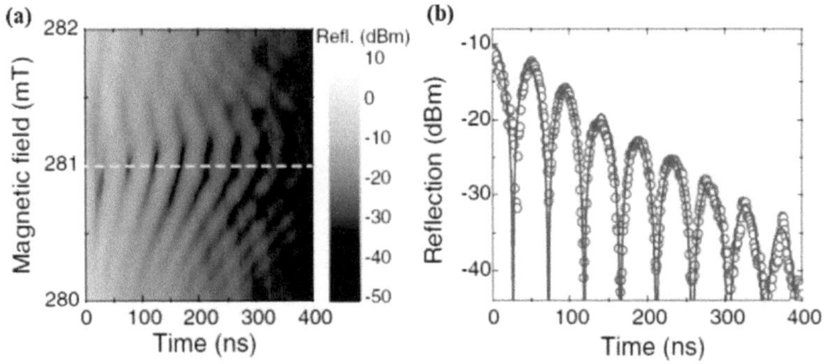

Figure 11.3 (a) Evolution of the cavity response after a pulsed signal modulated at cavity's eigenfrequency as a function of bias magnetic field. (b) Measured Rabi-like oscillation signal at bias magnetic field such that YIG FMR occurs. Circles: experimental data. Solid line: theoretical curve obtained by using parameters from the normal splitting spectrum. Reprinted (figure) with permission from Ref. [9]. Copyright (2014) by the American Physical Society.

which describes Rabi oscillations [9] of signal relaxation, as shown in Fig. 11.3b.

In real systems, and especially for systems in which spin waves with different characteristics are excited, cavity decay does not coincide with magnonic relaxations, leading to a more complex dynamic.

11.4 Magnon–Photon Coupling Regimes

So far, the strong coupling regime was discussed. However, depending on the interaction between magnons and MW photons, different coupling regimes can be distinguished in *cavity magnonics*, as summarized in Fig. 11.1b [9], namely the above-discussed *strong coupling*, the *magnetically induced transparency* (MIT) (Fig. 11.4a–c), the *Purcell effect* (Fig. 11.4d–f), and *weak coupling*. The distinctive factor is the relation between the magnonic–photonic damping and the coupling strength between the subsystems. If decoherence prevails, a weak coupling regime is exhibited with $\kappa_m, \kappa_a \gg g_m$, while if $\kappa_m, \kappa_a < g_m$, the cavity–magnon–polariton system is stable,

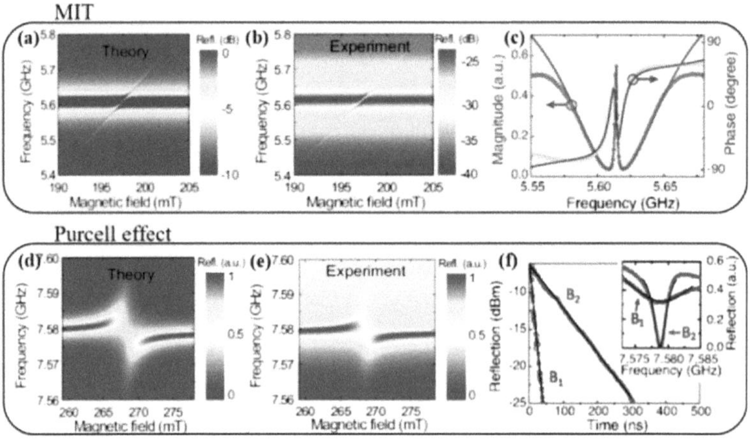

Figure 11.4 (a and b) Simulated and experimental response in the MIT regime depending on an external magnetostatic field and cavity frequency [9]. (c) Cross sections at a zero detuning magnetic field in the MIT regime. (d and e) Simulated and experimental response in the case of the Purcell effect [9]. (f) Cavity relaxation at two different bias magnetic fields B_1 (with coupling to magnon) and B_2 (without coupling). Inset: Cavity reflection spectra at B_1 and B_2. Reprinted (figure) with permission from Ref. [9]. Copyright (2014) by the American Physical Society.

presenting the characteristic features of strong coupling and Rabi splitting. The other two regimes, namely MIT and the Purcell effect, are characterized by a bad confinement of MW photons or a fast relaxation of spin waves, respectively. Both these regimes were reported by Zhang and coworkers [9]. In particular, the MIT regime ($\kappa_m < g_m < \kappa_a$; Fig. 11.4a–c) was achieved by introducing an MW absorber (Eccosorb LS-30/SS3) in order to increase MW dissipation in the cavity up to becoming the dominant process. For entering the Purcell regime ($\kappa_a < g_m < \kappa_m$; Fig. 11.4d–f), instead, lossy (e.g., rare-earth-doped) YIG spheres or iron filing glued to the YIG sphere surface can be employed in order to have lossy magnons that increase the decay of the coupled MW cavity photons; as a result, after pulsed excitation an accelerated exponential decrease is manifested instead of a Rabi oscillation.

In addition to these four regimes, an ultrastrong coupling regime when $g_m > 10\kappa_m$, $10\kappa_a$ was also described by Zhang and coworkers

[9] and Goryachev et al. [24] as a complex magnonic spectrum and obtained by increasing the size of the YIG sphere up to the case of a cavity with modal volume comparable with the dimension of the (bulk) ferromagnet.

More recently, Harder et al. extended the frame beyond the above-discussed coherent magnon–photon coupling, reporting a novel dissipative coupling regime associated with a so-called *cavity Lenz effect* due to a back-action damping by the induced RF cavity current on the magnetization dynamics when magnons are excited [25]. Specifically, the system was described by a combination of the resistor-inductor-capacitor circuit and Landau–Lifthitz–Gilbert equations for the cavity current I and dynamic magnetization M:

$$\begin{pmatrix} \omega^2 - \omega_c^2 + i2\kappa_a\omega_c\omega & i\omega^2 K_F \\ -i\omega_m(K_A - K_L) & \omega - \omega_r + i\kappa_m\omega \end{pmatrix} \begin{pmatrix} I \\ M \end{pmatrix} = 0. \qquad (11.17)$$

In Eq. 11.17 $\omega_m = \Gamma M_S$ is the angular frequency of the sample magnetization, which is proportional to the gyromagnetic ratio Γ of the material; ω_r and ω_c are the magnon mode and the cavity mode, respectively; while ω refers to the oscillation frequency of the magnetic RF field. The three terms K_F, K_A, and K_L are associated, respectively, with Faraday and Ampère law contributions (already observed before) and the cavity Lenz effect. The last two have opposite signs and the authors changed the position of a YIG sphere in a cylindrical waveguide moving from the TE_{11} mode magnetic antinode (position A) to the node (position B), as shown in Fig. 11.5a, in order to tune the Ampère term up to its suppression. As a function of the sphere angular position, a transition from level repulsion ($K_A - K_L > 0$) to level attraction ($K_A - K_L < 0$) (Fig. 11.5b) was observed in the transmission spectra, causing a systematic change in the coupling strength $g_m/2\pi$ (Fig. 11.5c) and a rearrangement (inversion) of the two branches in the anticrossing features, as shown in Fig. 11.5b. A modified Hamiltonian was also proposed, which includes a coupling phase Φ to describe the competing coherent ($\Phi = 0$) and dissipative couplings ($\Phi \neq 0$):

$$\widehat{\mathcal{H}} = \hbar\omega_a\hat{a}^\dagger\hat{a} + \hbar\omega_m\hat{m}^\dagger\hat{m} + \hbar g_m\left(\hat{a}^\dagger\hat{m} + e^{j\Phi}\hat{m}^\dagger\hat{a}\right). \qquad (11.18)$$

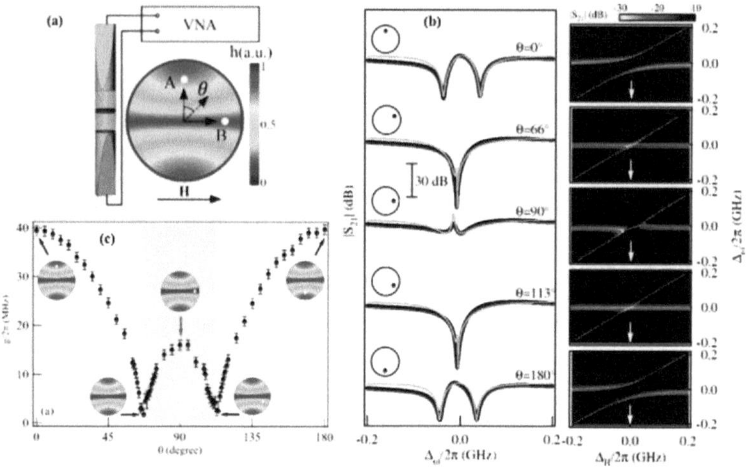

Figure 11.5 (a) Experimental setup used for evaluation of cavity Lenz effect. A cylindrical waveguide is connected to a VNA, which collects the response of the resonator loaded by a YIG sphere placed in different positions with respect to an azimuthal plane. Different colors of the azimuthal plane refer to the relative intensity of RF magnetic field h. (b) Phenomenon of level attraction in the 2D spectra as a function of loading sample placement and corresponding 2D maps. (c) Net coupling strength $g_m/2\pi$ measured at different YIG positions with respect to the RF magnetic antinodes. Reprinted (figure) with permission from Ref. [25]. Copyright (2018) by the American Physical Society.

11.5 Coupling to High-Order Magnetostatic Modes

A large part of the literature focused on hybridization among the cavity and the fundamental (uniform/Kittel) mode of YIG spheroids. However, higher-order non-uniform MSMs [26] were also investigated in other publications coupling a magnetically saturated single-crystal bulk YIG sphere to photons in 3D cavities [27–29] as well as in a coaxial cable and in a coplanar waveguide resonator (Fig. 11.6) [20, 30].

The MSMs in FM spheres were described by White [31] and Fletcher and Bell [32], who obtained a characteristic equation in associated Legendre functions P_n^m:

$$n+1+\xi_0\frac{P_n^{m'}(f,H_0)}{P_n^m(f,H_0)}\pm m\frac{\Gamma f M_S}{\Gamma^2 H_{\text{int}}-f^2}=0, \tag{11.19}$$

where $H_{\text{int}} = H_0 - M_S/3$ and $P_n^{m'}(f,H_0) = \frac{dP_n^m(f,H_0)}{d\xi_0}$, with $\xi_0 = (1+\frac{H_{\text{int}}}{M_S}-\frac{f^2}{\Gamma^2 M_S H_{\text{int}}})^{1/2}$; the index $n \in$ N labels localization of MSMs on the surface with respect to the external magnetostatic field $\vec{H}_0$, while m ($|m| \leq n$) refers to the angular momentum. Another index (r, not shown in this treatment) refers to the degeneracy of sustained MSMs. The MSMs relative to indexes n and m are labeled as (nm) and are grouped in families as a function of the value $n - |m|$. Generally, the resonant frequencies of MSMs are dispersive, except the mode families $n - |m| = 0$ and 1. In these cases, Eq. 11.19 assumes the simplified forms

$$\frac{f}{\Gamma M_S}-\frac{H_{0,mm}}{M_S}+\frac{1}{3}=\frac{m}{2m+1} \quad (n=m), \tag{11.20}$$

$$\frac{f}{\Gamma M_S}-\frac{H_{0,m(m+1)}}{M_S}+\frac{1}{3}=\frac{m}{2m+3} \quad (n=m+1), \tag{11.21}$$

where f refers to the magnon characteristic frequency. The fundamental FMR frequency $\omega_{\text{FMR}}/2\pi$ (Kittel mode) corresponding to the uniform precession mode in the sphere [12] is obtained by fixing $n = m = 1$, obtaining

$$f_{\text{FMR}}=\frac{\omega_{\text{FMR}}}{2\pi}=\Gamma H_0 \tag{11.22}$$

At this frequency, the magnetization relative to a lattice site precesses in a coherent way with the magnetization of the other sites. In general, when the FM sphere is immersed in a confined magnetic field oscillating at frequency $\omega_a/2\pi$ and a strong coupling regime is reached at H_0, the resonance frequencies of the two subsystems must match, and by imposing this condition it is possible to determine the indexes m and n of the MSMs associated with each anticrossing. To facilitate the analysis of the standing spin-wave patterns, the introduction of a novel functional variable to plot the data was recently proposed [33].

Experimentally, strong coupling to MSMs was observed by coupling a magnetically saturated single-crystal bulk YIG sphere

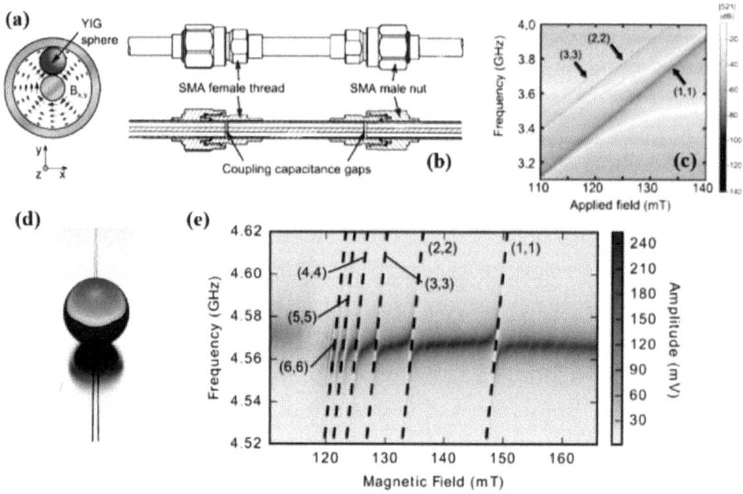

Figure 11.6 (a) YIG sphere in a non-uniform magnetic field inside a coaxial cavity [20]. (b) Schematic and longitudinal cross section of the coaxial cavity. (c) Cavity spectrum showing normal modes splitting related to strong coupling between cavity and MSM modes. (d) Schematic of a YIG sphere (diameter 250 μm) on a coplanar waveguide resonator (center conductor width 10 μm). (e) Transmitted signal amplitude S_{21} as a function of magnetic field and frequency (input power −80 dBm). Anticrossings between cavity modes and high-order magnetostatic modes are visible. Reprinted from Ref. [20], with the permission of AIP Publishing.

to both 3D cavities and 2D circuits in the presence of a bias magnetostatic field (Fig. 11.6). For example, the non-uniform magnetic field profile in the dielectric gap of coaxial cavity and the inhomogeneous magnetic field at the center of a coplanar waveguide resonator allowed the cavities to couple to uniform and non-uniform magnon modes of a YIG sphere [20, 30]. In particular, the employment of 2D cavities allowed to establish the coupling strength dependence on the distance among the sphere at FMR and a stripline resonator [34], as well as to investigate localized magnon states in YIG using superconducting planar waveguides [35] or split-ring resonators (SRRs) [36].

MSM families' profiles were also reported in MW 3D cavities at cryogenic temperatures [28, 29]. Approaches proper of cavity optomechanics, such as a dynamic detuning [37] or a high-amplitude off-resonance stationary signal [38–40], were also employed for further investigation/insight. For systems with a material having a large net spin density, such as YIG, when an additional drive field is applied, the Hamiltonian becomes

$$\widehat{\mathcal{H}} = \hbar\omega_a \hat{a}^\dagger \hat{a} + \hbar\omega_m \hat{m}^\dagger \hat{m} + \hbar g_m \left(\hat{a}^\dagger \hat{m} + \hat{m}^\dagger \hat{a}\right) + \hbar K \hat{m}^\dagger \hat{m} \hat{m}^\dagger \hat{m} + \hbar\Omega_d (\hat{m}^\dagger e^{-i\omega_d t} + \hat{m} e^{i\omega_d t}) \quad (11.23)$$

where K is the Kerr coefficient characterizing the strength of the nonlinear magnon effect, while Ω_d denotes the coupling strength of the drive field with the magnon mode. In Ref. [29], the authors demonstrated a frequency shift of the central cavity mode and also more considerable shifts of magnon modes by the application of an additional drive signal at a frequency ω_d (very off the resonance with the cavity mode) beside the probe cavity one ω_a, due to the magnon Kerr effect. In particular, Fig. 11.7c shows the transmitted signal of the cavity (excited and probed at its resonant frequency $\sim$10 GHz) changes as a function of the drive field frequency. Whenever this frequency is resonant with the magnon modes (Kittel or high-order MSMs), the transmitted signal presents a dip. In addition, by increasing the driving power, these dips became more intense and simultaneously undergo a right shift (Fig. 11.7d). This means that the magnon modes have a blueshift increasing the drive power. The analytical relation $[\Delta_m^2 + (\frac{\kappa_m}{2})^2]\Delta_m - cP = 0$ between the magnon frequency shift Δ_m and the drive power P fits very well the experimental data for the fundamental mode, but there is no agreement with results of higher MSMs, due to their inhomogeneous magnetization.

The magnon–photon coupling was also investigated at cryogenic temperature in an MW cavity with a double-post structure (Fig. 11.7e) [28]. The two posts can be considered as two different MW resonators, each one with a specific inductance and capacitance. Depending on the relative direction of the current through the post, two different modes can be excited: a bright mode (labeled ↑↓) and a dark mode (↑↑). In the first case, the currents flowing opposite through the posts generate a parallel magnetic field H at the center of the cavity and almost antiparallel elsewhere. As a result, the total magnetic field is enhanced between the posts and cancelled outside. On the contrary, in the dark mode, the currents through the

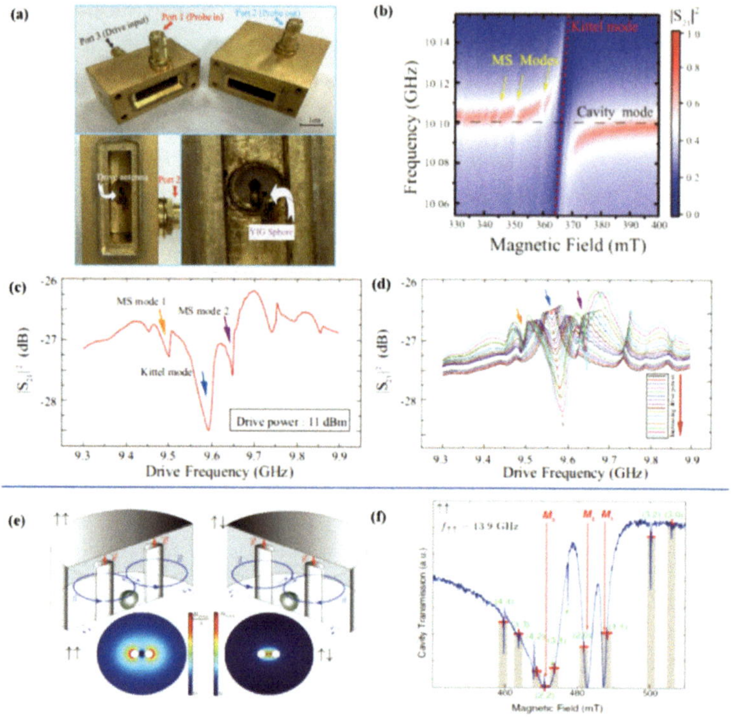

Figure 11.7 (a) 3D cavity with three ports used in Ref. [29]. (b) Transmission spectrum as a function of the bias magnetic field and frequency. The larger anticrossing is due to strong coupling between the FMR mode and the cavity mode, while smaller ones are due to the coupling of MSM modes with the cavity mode. (c) Transmission spectrum of the cavity measured as a function of the drive-field frequency. The blue arrow indicates the response of the Kittel mode, whereas the orange and purple arrows indicate two further MS modes. (d) Transmission spectrum of the cavity measured as a function of the drive frequency by successively increasing the driving power. (a–d) Reprinted (figure) with permission from Ref. [29]. Copyright (2016) by the American Physical Society. (e) Top: Cross-sectional 3D cavities with two resonance posts. Cavity ↑↑ is in the "dark" mode configuration, and cavity ↑↓ in the "bright" mode. Bottom: Top view of the magnetic field distribution computed in the equatorial plane of the YIG sphere. (f) Magnon modes observed near the dark cavity mode. (e and f) Reprinted (figure) with permission from Ref. [28]. Copyright (2014) by the American Physical Society.

posts are concordant, and in turn the vectors of the magnetic field from the two posts are antiparallel between the posts, cancelling the field. Elsewhere, the field is enhanced since the field vectors are almost parallel. In the first configuration, the focusing of the magnetic field among the posts, where the YIG sphere is placed, allows the achievement of an ultrastrong coupling between the bright resonance and the fundamental magnon mode. In the other configuration, a three-mode coupling is obtained between the dark mode and the magnon-mode doublet M_3. The coupling with the first-order mode is not visible since dynamic magnetization is perfectly uniform over the volume of the sphere, while the MW magnetic field of the dark mode is perfectly antisymmetric with respect to the middle of the distance between the cavity posts.

11.6 Coupling with Multiple YIG Spheres

Initially introduced to overcome the short coherence time of memory schemes based on dilute spin ensembles, the YIG magnon–MW photon strong coupling has even shown further advantages for what concerns storage and retrieval processes. In 2015, the system complexity and functionality and lifetime were increased by Zhang et al., who loaded multiple identical YIG spheres in the same MW cavity, demonstrating the realization of a magnon gradient memory (MGM) [41] (Fig. 11.8a) in a similar way to the atomic frequency comb memories [42].

Specifically, multiple magnon modes were coherently coupled to the MW resonance mode of the cavity, leading to the observation of collective effects, namely magnon dark (subradiant) and bright (superradiant) modes (Fig. 11.8a), respectively, corresponding to the hybridization of magnon modes that precess in and out of phase and exhibiting a quenched and enhanced interaction with the cavity mode. For this reason, the (decoupled) dark mode presents advantages for information storage due to its long lifetime, while the bright mode can be employed for retrieval of information.

In this frame, one can generalize the Hamiltonian of the linear coupling between the FMR uniform mode for each individual jth sphere and the cavity resonance photons to the case of N spheres, obtaining

$$\widehat{\mathcal{H}} = \hbar\omega_a \hat{a}^\dagger \hat{a} + \hbar \sum_{j=1}^{N} \left(\omega_{m,j} \hat{m}_j^\dagger \hat{m}_j + g_{m,j} \hat{a}^\dagger \hat{m}_j + g_{m,j}^* \hat{a} \hat{m}_j^\dagger \right), \quad (11.24)$$

and an evenly distributed spectrum can be achieved for $\omega_{m,j}$ by biasing each sphere at different magnetic fields $H_j = H_0 + [j - (N + 1)/2]\Delta H$, where H_0 is a common external magnetic field, while the magnetic field difference ΔH among adjacent spheres can be obtained experimentally using small coils (Fig. 11.8a). The collective magnon modes can be then described as one temporal bright mode

$$\hat{B} = \frac{1}{\sqrt{N}} \sum_{j=1}^{N} \hat{m}_j, \quad (11.25)$$

and $N - 1$ temporal dark modes

$$\hat{D}_n = \frac{1}{\sqrt{N}} \sum_{j=1}^{N} \hat{m}_j e^{2\pi i (j - \frac{N-1}{2}) \cdot n/N} \quad (11.26)$$

which remain detectable in contrast with the stationary dark mode since they are periodically converted back to the bright mode (here n is the dark mode index).

Experimental evidence is reported in Fig. 11.8b and Fig. 11.8c, where the presence of a number of anticrossings in the spectra that are related to the number of involved spheres (respectively, two and eight) is visible. In the case of two spheres, at a fixed external bias field H_0, as a function of the magnetic field difference among the spheres, three resonances were observed near the resonance condition but only two in the case of a negligible detuning field ΔH when both modes are simultaneously on resonance (Fig. 11.8b). Figure 11.8d–h, instead, shows 2D maps at a fixed detuning field while sweeping the external bias field H_0 near resonance conditions, showing two avoided crossings as a result of strong coupling among the two magnon modes and the MW cavity mode except for the case of the negligible detuning field when the resonance associated with the dark magnon mode disappears since it decouples from the MW cavity mode.

Figure 11.8c, instead, is concerned with the case of eight spheres, resulting in nine equally spaced hybrid modes in the spectrum at H_0 = 2687 Oe (and eight avoided crossing in the 2D maps in the inset), while the dashed curve corresponds to H_0 = 0 Oe. The operation of the MGM was then investigated using a 15 ns excitation pulse at the cavity resonance frequency 7.522 GHz under H_0 = 2687 Oe and retrieving the stored pulse after 100 ns, showing that a significant improvement in the performance can be achieved by optimizing the external coupling condition by impedance matching. The authors also investigated the dependence of delay time and retrieval efficiency on the frequency gradient and demonstrated also multipulse storage in the MGM. The advantages of using MGMs are related to the possibility to store information without considering the framework in which it occurs, besides the fact that the employed material intrinsically does not suffer from inhomogeneous broadening issues as in the dilute spin ensembles. The authors demonstrated a 100 ns storage time limited by magnon lifetime and emphasized that the linearity of the interaction opens the way to scaling down to quantum regime at mK temperatures. An early work in this direction was carried out by coherently coupling two YIG spheres in a coaxial cavity [43].

11.7 Coupling to Transmons

An important step for the establishment of quantum magnonic transducers was performed in 2017 by Lachance-Quirion et al., who demonstrated the capability of encoding qubits in magnon states by achieving coherent coupling between a magnon Kittel mode and a transmon qubit [44]. In their approach, while the magnon Kittel mode is placed at the magnetic field antinode (coupled via its magnetic dipole and tuned by the direct current bias magnetic field H_0), the transmon superconducting qubit is strongly coupled to the cavity electric field and resonates at a fixed frequency of about 8 GHz. By acting on the coil current controlling the direct current bias magnetic field, magnon and qubits components can be brought in a strong coupling regime, as demonstrated by the observed avoided crossing. The qubit–magnon interaction is mediated by the virtual photon excitation in an MW cavity that acts like a bus, mainly through the TE_{102} cavity mode (coupler mode). The hybrid qubit–

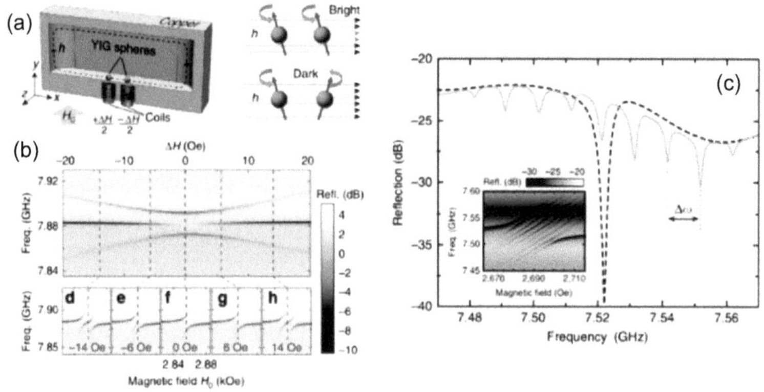

Figure 11.8 (a) Left: Cavity loaded by two YIG spheres whose magnetization is tuned by two small coils. Right: Illustration of the magnon bright and dark modes, which are coupled to and isolated from the magnetic component of the field confined in the cavity. (b) Top: Reflection spectrum as a function of the detuning between the external fields ΔH applied on each sphere. Bottom: Reflection spectra as a function of bias magnetic field H_0 at a fixed magnetic gradient. (c) Spectrum of the cavity with 8 spheres at a bias field of 2687 Oe (solid line) and 0 Oe (dashed line). Inset: Reflection (Refl.) 2D map of the MGM as a function of external magnetic field and cavity working frequency [41].

magnon system can be described by the Jaynes–Cummings-type Hamiltonian

$$\widehat{\mathcal{H}}_{\mathrm{qm}} = \hat{\sigma}^{+}\hat{m} + g^{*}_{\mathrm{qm}}\hat{\sigma}^{-}\hat{m}^{\dagger}, \tag{11.27a}$$

where $g_{\mathrm{qm}} = g_{\mathrm{q}}g_{\mathrm{m}}/\Delta$ is the effective qubit–magnon coupling strength, g_{q} and g_{m}, respectively, being the coupling strength of the qubit and the magnon with the cavity mode and Δ the detuning corresponding to the difference between the bare frequencies of the qubit and the cavity mode. $\hat{\sigma}$ and $\hat{m}$ are the qubit excitation and magnon operators, respectively.

In a successive study, the same group investigated operation in the off resonant, dispersive regime (detuning $\omega_{\mathrm{q}}^{\mathrm{bare}} - \omega_{\mathrm{m}}^{\mathrm{bare}}$ much larger than g_{qm}) in order to quench the exchange of excitation quanta among the two subsystems through virtual photons and exploited a different (TE_{103}) mode (probe mode) to read the qubit avoiding measurement-induced dephasing. In this case, the dispersive part of the system Hamiltonian reads

$$\widehat{\mathcal{H}}_{\mathrm{q-m}} = \hbar\chi_{\mathrm{q-m}}\hat{\sigma}_z\hat{m}^\dagger\hat{m}, \tag{11.27b}$$

where $\hat{\sigma}_z = |e\rangle\langle e| - |g\rangle\langle g|$, with $|g\ (e)\rangle$ being the ground (excited) state of the transmon qubit and $\chi_{\mathrm{q-m}}$ the qubit–magnon dispersive shift. In this regime, the dispersive interaction makes the qubit and magnon frequencies dependent on the state of the other subsystem [45], revealing new characteristic features of many-system coupling, and the superconducting qubit can be employed to probe and control magnon states. Experimentally, spectroscopic measurements of the qubit were carried out while exciting the Kittel mode at frequency ω_{mw}, detuned by $\Delta_{\mathrm{mw}} = \omega_{\mathrm{m}}^{g} - \omega_{\mathrm{mw}}$ from the dressed magnon frequency ω_{m}^{g}. In this case the two-photon parametric transition that confirms coupling between systems is observed at the mean frequency among magnetization precession of FM bulk and transition frequency of qubits. The adopted approach shows an analogy with experiments in optoelectromechanics [46] and provides a versatile tool for quantum control of the single-magnon excitation.

11.8 Cavity Magnomechanics and Optomagnonics

Following analogies with optomechanical systems, cavity magnomechanics was proposed as a new field where magnetostrictive (radiation pressure-like) forces are exploited for achieving coherent phonon–magnon interactions in addition to photon–magnon coupling via magnetic dipole interaction. As a result, tripartite entanglement among magnons, cavity photons, and phonons was demonstrated [10, 47], which, in the computation field, can provide an alternative information transduction platform. In this respect, YIG continues to present optimal characteristics with very low mechanical damping (loss), enabling a long lifetime for acoustic phonons as additional information carriers, a basic prerequisite for achieving coherent magnon–phonon coupling. However, a strong parametric drive scheme is required to compensate for the large frequency mismatch among the GHz magnon and MHz phonon modes [47].

The Hamiltonian of the system reads as

$$\widehat{\mathcal{H}} = \hbar g_{\mathrm{mb}} \hat{m}^{\dagger} \hat{m} (\hat{b}^{\dagger} + \hat{b}), \tag{11.28}$$

where $\hat{b}^{\dagger}$ ($\hat{b}$) refers to the boson creation (annihilation) operator of the phonon mode and g_{mb} is the single magnon–phonon coupling strength (orders of magnitude lower than the magnon–photon one). In their seminal work, Zhang et al. focused on the $S_{1,2,2}$ spheroidal phonon mode exhibiting the largest magnetostrictive coupling strength (related to the overlap among the uniform magnon mode and the phonon mode) and employed a sphere with diameter 250 μm as a trade-off among the larger magnetostrictive coupling strength and the concomitant lower responsivity of the higher phonon frequencies resulting from decreasing the sphere diameter [47]. Specifically, in their study, the phonon frequency was $\omega_{\mathrm{b}}/2\pi =$ 11.42 MHz, while the coupling strength $g_{\mathrm{mb}}/2\pi$ was $\leq$9.9 mHz.

Spectra are acquired by sweeping the probe frequency ω_{s}, plotting the reflection signal as a function of the two-photon (probe-drive) detuning Δ_{sd} at a fixed magnetic field and drive frequency ω_{d}, which is red (blue)-detuned for the lower (upper) hybrid mode. In both cases, the coherent magnomechanical interaction is manifested in the spectra as sharp phonon-induced resonances at $\Delta_{\mathrm{sd}} = \omega_{\mathrm{b}}/2\pi =$ 11.42 MHz, which in the zooms exhibit a Fano-like shape with opposite symmetry for the red- and blue-detuned drives. The authors demonstrated a magnetically controllable coherent magnomechanical interaction. Moreover, they investigated also the dependence on the drive power and reported evidence of magnomechanically induced transparency (MMIT), magnomechanically induced absorption (MMIA), and magnomechanical parametric amplification (MMPA), phonon lasing, and resonant photon–magnon–phonon coupling [10, 47].

The aforementioned work illustrates how an MW resonator (photon) and an FM bulk crystal (magnon) can be combined to form a hybrid system in which to store securely the information and transfer it to a third subsystem (phonon). Similar to cavity magnomechanics, the new field of optomagnonics was also recently proposed, where information can be shared with optical photons for implementing long-distance quantum communication and quantum-noise-limited MW amplifiers by means of a coherent exchange between MW signals and optical photons (exploiting YIC transparency to light above 600 nm). Nonlinearity is a typical requirement for frequency conversion of an electromagnetic field, and various strategies were reported ([48] and references therein). Nonlinearity in the dispersive regions of an optically transparent crystal was employed for realizing electro-optic modulators, for example, using ferroelectric crystals. Magneto-optic modulators were also implemented, exploiting the sharp spin resonance in paramagnets, such as erbium-doped crystals, for achieving enhanced nonlinearity. Parametrically coupled nanomechanical resonators provided, so far, maximal conversion efficiency from MW to light.

In this respect, in the frame of optomagnonics, bidirectional conversion between MWs and light via FM magnons was recently reported [48, 49] if, for example, an itinerant MW field is coupled to the hybrid system through an MW cavity, while a traveling optical signal field addresses the Kittel mode [48] via the Faraday [50] or the Kerr effect [51]. A major difficulty in achieving efficient coherent conversion lies in the large frequency difference between magnons and optical photons. At such frequencies, the magnetic dipole interactions with light are 3 orders lower than the electric dipole–optical waves interaction. However, in analogy to optomechanics, a three-particle interaction can be used involving two photons and a magnon (instead of a phonon) resonant with the two photons' energy difference. Notably, compared to phonons in optomechanics, magnons in optomagnonics can be more easily decoupled from the environment and magnetically manipulated [48, 49]. A scheme for optical–MW signal bidirectional conversion via FMR was proposed by Hisatomi and Usami [48] and is shown in Fig. 11.9. Here, the hybridized system composed by the cavity and the FM material is coupled with both MW and optical modes. In particular, the input and output itinerant modes (labeled, respectively, with $\hat{a}_\mathrm{i}$ and $\hat{a}_\mathrm{o}$) are coupled to the cavity mode at a rate $2\kappa_\mathrm{c}$, while a traveling optical signal addresses the hybrid mode via the Faraday effect or the spin-Raman effect [52]. It follows that the polarization of the light beam (expressible in terms of creation and annihilation operators $\hat{l}_\mathrm{i}^\dagger$ and $\hat{l}_\mathrm{i}$) focused on the sample is affected by the magnetization oscillation of

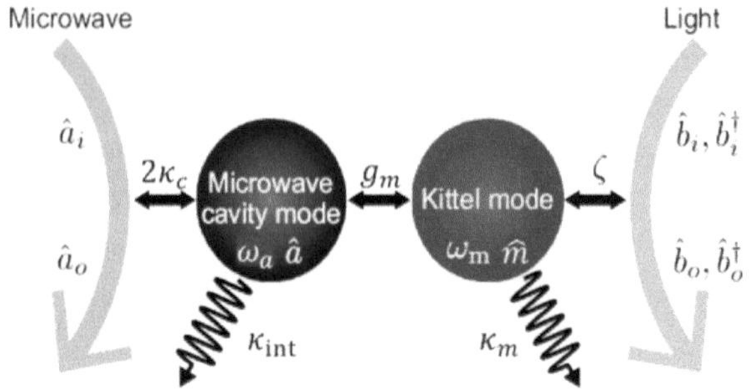

Figure 11.9 Schematic representation of the bidirectional conversion between MW and optical photons by strong coupling of a cavity mode with Kittel mode. Reprinted (figure) with permission from Ref. [48] Copyright (2016) by the American Physical Society.

the material itself, inducing a variation in the optical mode described by operators $\hat{l}_0^\dagger$ and $\hat{l}_0$ related to the transverse polarization of the optical wave. In this picture, the light–magnon coupling rate ζ depends by the dimensions of sample and by the power and angular frequency of the optical field.

The model described above expects that the Hamiltonian of interaction between the cavity MW mode, the Kittel mode in the FM sample, and the optical field contains an additional term associated with the interaction between the magnonic mode and the traveling optical photon mode at angular frequency Ω_0. This Hamiltonian of magnon–optical photon coupling $\mathcal{H}_{\mathrm{ml}}$ at moment t is calculated over all the angular frequencies Ω of the optical signal:

$$\mathcal{H}_{\mathrm{ml}}(t) = -i\hbar\sqrt{\zeta} \int_{-\infty}^{+\infty} \frac{d\Omega}{2\pi} (\hat{m} + \hat{m}^\dagger) \left[\hat{l}_{\mathrm{i}}(\Omega) e^{i\Omega_0 t} - \hat{l}_i^\dagger(\Omega) e^{-i\Omega_0 t}\right]. \tag{11.29}$$

Beyond MSMs, a YIG sphere can also support photonic whispering gallery modes (WGMs), and MW-to-light conversion can be achieved through spin–orbit interaction [53, 54]. In this respect, recent developments of cavity optomagnonics were carried out by impinging an infrared 1.5 μm laser with Ω_0 frequency on a YIG sphere strongly coupled in a 3D resonator using the experimental setup shown in Fig. 11.10a, where light through an optical fiber

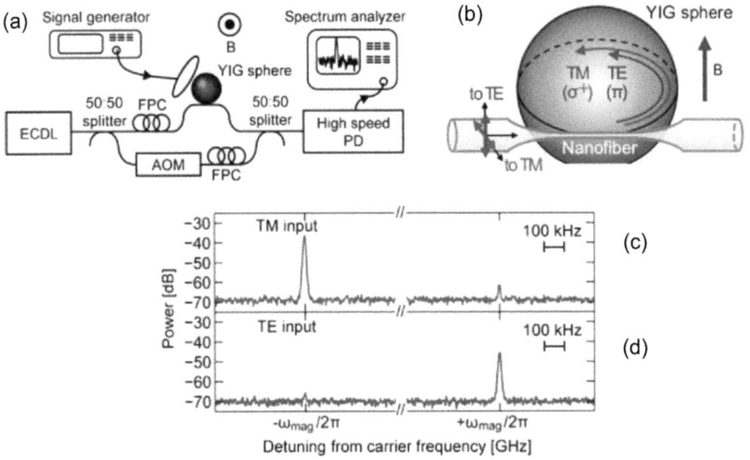

Figure 11.10 (a) Setup for observing sideband signals generated by the FMR magnon-induced Brillouin light scattering of light. Laser light is split into two paths, and the photons in one of the paths act as local oscillators (LOs), whose frequency is shifted by an acousto-optic modulator (AOM). For the heterodyne measurement, the LO photons are mixed with the photons in the other path, which are coupled to the WGM resonator via the nanofiber. MWs from a signal generator resonantly excite the magnons, and the heterodyne signal is sent to a spectrum analyzer. (b) Correspondence of input polarizations, polarizations in the WGM resonator, and a schematic-level diagram of Brillouin scattering. (c and d) Observed sideband-signal power spectra for the TM-mode and TE-mode inputs. Reprinted (figure) with permission from Ref. [54] Copyright (2016) by the American Physical Society.

interacts with the WGMs of a magnetically saturated YIG sphere (shown in Fig. 11.10b). The spatially uniform spin precession of YIG spheres is the basis of the pronounced nonreciprocity and asymmetry in the sideband signals generated by the FMR magnon-induced Brillouin light scattering of light (Fig. 11.10c), which is reported in this work [54]. The thermal bistability of magnons excited in the same configuration was demonstrated [55]. All the aforementioned properties arise from the geometric birefringence of the WGMs photons in YIG at FMR under a polarized optical signal, driving the MW photon–FM magnon–optical photon system in a nonreciprocal triple-resonance condition. Haigh et al.

employed prism coupling to the magnetic sphere for precise mode identification [56].

11.9 Conclusions

In conclusion, cavity magnonics emerged as a very attractive field for applications in quantum technologies, demonstrating the ability to integrate different systems and excitations together and the possibility to store information in the magnetic subsystem (as well as other subsystems) and transfer it to strongly coupled cavity modes as well as phonons, optical photons, and transmon qubits. In this respect, YIG was the material of choice for several studies due to its exceptional properties. Recently, investigation of other configurations (such as with 2D resonators) and materials is expanding basic knowledge, and future research is expected to provide further steps toward the implementation of robust quantum protocols.

References

1. Nielsen, M. A. and Chuang, I. (2002). *Quantum Computation and Quantum Information* (AAPT).
2. Wehner, S., Elkouss, D. and Hanson, R. (2018). Quantum internet: a vision for the road ahead, *Science*, **362**(6412), eaam9288.
3. Kimble, H. J. (2008). The quantum internet, *Nature*, **453**(7198), 1023.
4. Degen, C. L., Reinhard, F. and Cappellaro, P. (2017). Quantum sensing, *Rev. Mod. Phys.*, **89**(3), 035002.
5. Raimond, J.-M., Brune, M. and Haroche, S. (2001). Manipulating quantum entanglement with atoms and photons in a cavity, *Rev. Mod. Phys.*, **73**(3), 565.
6. Kubo, Y., Ong, F., Bertet, P., Vion, D., Jacques, V., Zheng, D., Dréau, A., Roch, J.-F., Auffèves, A. and Jelezko, F. (2010). Strong coupling of a spin ensemble to a superconducting resonator, *Phys. Rev. Lett.*, **105**(14), 140502.
7. Bushev, P., Feofanov, A., Rotzinger, H., Protopopov, I., Cole, J., Wilson, C., Fischer, G., Lukashenko, A. and Ustinov, A. (2011). Ultralow-power

spectroscopy of a rare-earth spin ensemble using a superconducting resonator, *Phys. Rev. B*, **84**(6), 060501.

8. Saito, S., Zhu, X., Amsüss, R., Matsuzaki, Y., Kakuyanagi, K., Shimo-Oka, T., Mizuochi, N., Nemoto, K., Munro, W. J. and Semba, K. (2013). Towards realizing a quantum memory for a superconducting qubit: storage and retrieval of quantum states, *Phys. Rev. Lett.*, **111**(10), 107008.
9. Zhang, X., Zou, C.-L., Jiang, L. and Tang, H. X. (2014). Strongly coupled magnons and cavity microwave photons, *Phys. Rev. Lett.*, **113**(15), 156401.
10. Zhang, X., Zou, C.-L., Jiang, L. and Tang, H. X. (2016). Cavity magnomechanics, *Sci. Adv.*, **2**(3), e1501286.
11. Serga, A., Chumak, A. and Hillebrands, B. (2010). YIG magnonics, *J. Phys. D: Appl. Phys.*, **43**(26), 264002.
12. Stancil, D. D. and Prabhakar, A. (2009). *Spin Waves* (Springer).
13. Holstein, T. and Primakoff, H. (1940). Field dependence of the intrinsic domain magnetization of a ferromagnet, *Phys. Rev.*, **58**(12), 1098.
14. Stancil, D. and Prabhakar, A. (2009). *Spin Waves: Theory and Applications* (Springer).
15. Shore, B. W. and Knight, P. L. (1993). The jaynes-cummings model, *J. Mod. Opt.*, **40**(7), 1195–1238.
16. Zhang, X., Liu, T., Flatté, M. E. and Tang, H. X. (2014). Electric-field coupling to spin waves in a centrosymmetric ferrite, *Phys. Rev. Lett.*, **113**(3), 037202.
17. Kittel, C. (1948). On the theory of ferromagnetic resonance absorption, *Phys. Rev.*, **73**(2), 155.
18. Agarwal, G. (1984). Vacuum-field Rabi splittings in microwave absorption by Rydberg atoms in a cavity, *Phys. Rev. Lett.*, **53**(18), 1732.
19. Tabuchi, Y., Ishino, S., Ishikawa, T., Yamazaki, R., Usami, K. and Nakamura, Y. (2014). Hybridizing ferromagnetic magnons and microwave photons in the quantum limit, *Phys. Rev. Lett.*, **113**(8), 083603.
20. Lambert, N., Haigh, J. and Ferguson, A. (2015). Identification of spin wave modes in yttrium iron garnet strongly coupled to a co-axial cavity, *J. Appl. Phys.*, **117**(5), 053910.
21. Walls, D. F. and Milburn, G. J. (2007). *Quantum Optics* (Springer Science & Business Media).
22. Clerk, A. A., Devoret, M. H., Girvin, S. M., Marquardt, F. and Schoelkopf, R. J. (2010). Introduction to quantum noise, measurement, and amplification, *Rev. Mod. Phys.*, **82**(2), 1155.

23. Harder, M., Bai, L., Match, C., Sirker, J. and Hu, C. (2016). Study of the cavity-magnon-polariton transmission line shape, *Sci. China Phys. Mech. Astron.*, **59**(11), 117511.
24. Bourhill, J., Kostylev, N., Goryachev, M., Creedon, D. and Tobar, M. (2016). Ultrahigh cooperativity interactions between magnons and resonant photons in a YIG sphere, *Phys. Rev. B*, **93**(14), 144420.
25. Harder, M., Yang, Y., Yao, B., Yu, C., Rao, J., Gui, Y., Stamps, R. and Hu, C.-M. (2018). Level attraction due to dissipative magnon-photon coupling, *Phys. Rev. Lett.*, **121**(13), 137203.
26. Walker, L. (1957). Magnetostatic modes in ferromagnetic resonance, *Phys. Rev.*, **105**(2), 390.
27. Zhang, D., Wang, X.-M., Li, T.-F., Luo, X.-Q., Wu, W., Nori, F. and You, J. Q. (2015). Cavity quantum electrodynamics with ferromagnetic magnons in a small yttrium-iron-garnet sphere, *npj Quantum Inf.*, **1**, 15014.
28. Goryachev, M., Farr, W. G., Creedon, D. L., Fan, Y., Kostylev, M. and Tobar, E. (2014). High-cooperativity cavity QED with magnons at microwave frequencies, *Phys. Rev. Appl*, **2**(5), 054002.
29. Wang, Y.-P., Zhang, G.-Q., Zhang, D., Luo, X.-Q., Xiong, W., Wang, S.-P., Li, T.-F., Hu, C.-M. and You, J. (2016). Magnon Kerr effect in a strongly coupled cavity-magnon system, *Phys. Rev. B*, **94**(22), 224410.
30. Morris, R. G. E., van Loo, A. F., Kosen, S. and Karenowska, A. D. (2017). Strong coupling of magnons in a YIG sphere to photons in a planar superconducting resonator in the quantum limit, *Sci. Rep.*, **7**(1), 11511.
31. White, R. L. (1960). Use of magnetostatic modes as a research tool, *J. Appl. Phys.*, **31**(5), S86–S94.
32. Fletcher, P. and Bell, R. (1959). Ferrimagnetic resonance modes in spheres, *J. Appl. Phys.*, **30**(5), 687–698.
33. Leo, A., Monteduro, A. G., Rizzato, S., Martina, L. and Maruccio, G. Identification and time-resolved study of YIG spin wave modes in a MW cavity in strong coupling regime, ArXiv, submit/2660257.
34. Rao, J., Kaur, S., Fan, X., Xue, D., Yao, B., Gui, Y. and Hu, C.-M. (2017). Characterization of the non-resonant radiation damping in coupled cavity photon magnon system, *Appl. Phys. Lett.*, **110**(26), 262404.
35. Morris, R., van Loo, A., Kosen, S. and Karenowska, A. (2017). Strong coupling of magnons in a YIG sphere to photons in a planar superconducting resonator in the quantum limit, *Sci. Rep.*, **7**.
36. Kaur, S., Yao, B., Rao, J., Gui, Y. and Hu, C.-M. (2016). Voltage control of cavity magnon polariton, *Appl. Phys. Lett.*, **109**(3), 032404.

37. Haigh, J., Lambert, N., Doherty, A. and Ferguson, A. (2015). Dispersive readout of ferromagnetic resonance for strongly coupled magnons and microwave photons, *Phys. Rev. B*, **91**(10), 104410.
38. Kirchmair, G., Vlastakis, B., Leghtas, Z., Nigg, S. E., Paik, H., Ginossar, E., Mirrahimi, M., Frunzio, L., Girvin, S. M. and Schoelkopf, R. J. (2013). Observation of quantum state collapse and revival due to the single-photon Kerr effect, *Nature*, **495**(7440), 205.
39. Vlastakis, B., Kirchmair, G., Leghtas, Z., Nigg, S. E., Frunzio, L., Girvin, S. M., Mirrahimi, M., Devoret, M. H. and Schoelkopf, R. J. (2013). Deterministically encoding quantum information using 100-photon Schrödinger cat states, *Science*, **342**(6158), 607–610.
40. Yuan, M., Singh, V., Blanter, Y. M. and Steele, G. A. (2015). Large cooperativity and microkelvin cooling with a three-dimensional optomechanical cavity, *Nat. Commun.*, **6**, 8491.
41. Zhang, X., Zou, C.-L., Zhu, N., Marquardt, F., Jiang, L. and Tang, H. X. (2015). Magnon dark modes and gradient memory, *Nat. Commun.*, **6**.
42. Usmani, I., Afzelius, M., De Riedmatten, H. and Gisin, N. (2010). Mapping multiple photonic qubits into and out of one solid-state atomic ensemble, *Nat. Commun.*, **1**, 12.
43. Lambert, N. J., Haigh, J., Langenfeld, S., Doherty, A. and Ferguson, A. (2016). Cavity-mediated coherent coupling of magnetic moments, *Phys. Rev. A*, **93**(2), 021803.
44. Lachance-Quirion, D., Tabuchi, Y., Ishino, S., Noguchi, A., Ishikawa, T., Yamazaki, R. and Nakamura, Y. (2017). Resolving quanta of collective spin excitations in a millimeter-sized ferromagnet, *Sci. Adv.*, **3**(7), e1603150.
45. Tabuchi, Y., Ishino, S., Noguchi, A., Ishikawa, T., Yamazaki, R., Usami, K. and Nakamura, Y. (2015). Coherent coupling between a ferromagnetic magnon and a superconducting qubit, *Science*, **349**(6246), 405–408.
46. Aspelmeyer, M., Kippenberg, T. J. and Marquard, F. (2014). Cavity optomechanics, *Rev. Mod. Phys.*, **86**(4), 1391–1452.
47. Li, J., Zhu, S.-Y. and Agarwal, G. (2018). Magnon-photon-phonon entanglement in cavity magnomechanics, *Phys. Rev. Lett.*, **121**(20), 203601.
48. Hisatomi, R., Osada, A., Tabuchi, Y., Ishikawa, T., Noguchi, A., Yamazaki, R., Usami, K. and Nakamura, Y. (2016). Bidirectional conversion between microwave and light via ferromagnetic magnons, *Phys. Rev. B*, **93**(17), 174427.

49. Liu, T., Zhang, X., Tang, H. X. and Flatté, M. E. (2016). Optomagnonics in magnetic solids, *Phys. Rev. B*, **94**(6), 060405.
50. Pershan, P. (1967). Magneto-optical effects, *J. Appl. Phys.*, **38**(3), 1482–1490.
51. Wettling, W. (1976). Magneto-optics of ferrites, *J. Magn. Magn. Mater.*, **3**(1–2), 147–160.
52. Shen, Y. and Bloembergen, N. (1966). Interaction between light waves and spin waves, *Phys. Rev.*, **143**(2), 372.
53. Moriya, T. (1968). Theory of absorption and scattering of light by magnetic crystals, *J. Appl. Phys.*, **39**(2), 1042–1049.
54. Osada, A., Hisatomi, R., Noguchi, A., Tabuchi, Y., Yamazaki, R., Usami, K., Sadgrove, M., Yalla, R., Nomura, M. and Nakamura, Y. (2016). Cavity optomagnonics with spin-orbit coupled photons, *Phys. Rev. Lett.*, **116**(22), 223601.
55. Chai, C.-Z., Hu, X.-X., Zou, C.-L., Guo, G.-C. and Dong, C.-H. (2019). Thermal bistability of magnon in yttrium iron garnet microspheres, *Appl. Phys. Lett.*, **114**(2), 021101.
56. Haigh, J., Nunnenkamp, A., Ramsay, A. and Ferguson, A. (2016). Triple-resonant Brillouin light scattering in magneto-optical cavities, *Phys. Rev. Lett.*, **117**(13), 133602.

Index